YOUR COMPUTER IS ON FIRE

YOUR COMPUTER IS ON FIRE

EDITED BY THOMAS S. MULLANEY, BENJAMIN PETERS,
MAR HICKS, AND KAVITA PHILIP

THE MIT PRESS   CAMBRIDGE, MASSACHUSETTS   LONDON, ENGLAND

This book was set in ITC Stone and Avenir by New Best-set Typesetters Ltd. Printed and bound in the United States of America.

Library of Congress Cataloging-in-Publication Data

Names: Mullaney, Thomas S. (Thomas Shawn), editor. | Peters, Benjamin, 1980– editor. | Hicks, Mar, editor. | Philip, Kavita, 1964– editor.
Title: Your computer is on fire / edited by Thomas S. Mullaney, Benjamin Peters, Mar Hicks, Kavita Philip.
Description: Cambridge, Massachusetts : The MIT Press, [2020] | Includes bibliographical references and index.
Identifiers: LCCN 2019059152 | ISBN 9780262539739 (paperback)
Subjects: LCSH: Computers—Social aspects. | Information technology—Social aspects.
Classification: LCC QA76.9.C66 Y67 2020 | DDC 303.48/34—dc23
LC record available at https://lccn.loc.gov/2019059152

10 9 8 7 6 5 4 3

CONTENTS

INTRODUCTIONS

YOUR COMPUTER IS ON FIRE

Thomas S. Mullaney

Humanists and social scientists who work on computing and new media are subject to daily reminders about how little most technologists reflect upon our work or take our scholarship under advisement. We are gadflies, it would seem, always dwelling upon the negative sides of technology rather than its unquestionable positives. We are technophobes who, either because of our failure to understand the technologies we critique, or perhaps out of envy for the power of those who build them, are akin to millenarian street preachers waving signs that read "The End Is Nigh." Our critiques are made in bad faith, moreover, as we are so often spotted using in our classrooms, our essays, and our books the very systems, platforms, and devices that we lament.

The profound self-confidence and self-assurance of these same technologists are on daily display as well. Their "bias toward action" is considered refreshing; their predilection to "move fast and break things" is celebrated. Even when systems fail in spectacular fashion, a seemingly endless supply of assuring words are at the ready to defuse or deflect.

Technology firms selling bleeding-edge surveillance systems to authoritarian regimes at home and abroad?

Bad actors.

Latest-generation web cameras incapable of recognizing the faces of African-American users (yet functioning flawlessly with Caucasian users)?[1]

Bad training data.

Rampant and historically unprecedented growth of corporate surveillance?

Terms of usage.

No matter the problem, it seems, a chorus of techno-utopian voices is always at the ready to offer up "solutions" that, remarkably enough, typically involve the same strategies (and personnel) as those that helped give rise to the crisis in the first place. We can always code our way out, we are assured. We can make, bootstrap, and science the shit out of this.

A largely unreflective rush to automating immense sectors of the labor economy out of existence?

Creative destruction.

Widely documented gender, class, and ethnoracial inequalities across the IT labor force?

A pipeline problem.

Camera "blink detection" technology repeatedly flagging users of Asian descent with the prompt "Did Someone Blink?"[2]

A bug, not a feature.

Google's image-recognition algorithm tagging the faces of African-American individuals with the word "gorillas"?[3]

They fixed it in the next release.

From the standpoint of technologists, the appeal of this argument is not hard to understand. If one believes in it, after all, then one also believes that technologists should continue to enjoy a monopoly, not only over the first-order creation of an ever-increasing number of computationally grounded political, social, cultural, and economic frameworks, but also over the second-order repair and replacement of said systems when they (repeatedly) fall short or break down. Who would want to relinquish such a plum position?

The contributors to this volume—scholars who approach computing and new media from a variety of critical perspectives derived from humanistic, social scientific, and STEM disciplines—have come together to issue a manifesto that can be summarized as follows:

Your computer is on fire.

Humankind can no longer afford to be lulled into complacency by narratives of techno-utopianism or technoneutrality, or by self-assured and oversimplified evasion. Every time we hear the call of a lullaby—soothing words such as "human error," "virtual reality," "the Cloud," or others meant to coax us back to sleep, leaving the "adults" to continue driving—our response should be a warning siren, alarming us and those around us into a state of alertness and vigilance. Every established or emerging norm needs to be interrogated—whether the taken-for-granted whiteness of humanoid robots, the ostensibly "accentless" normative speech of virtual assistants, the near invisibility of the human labor that makes so many of the ostensibly "automated" systems possible, the hegemonic position enjoyed by the English language and the Latin alphabet within modern information-processing systems, the widespread deployment of algorithmic policing, the erosion of publicly governed infrastructures at the hands of private (and ultimately ephemeral) mobile platforms, the increasing flirtation with (if not implementation of) autonomous weapons systems capable of selecting and engaging targets independently, and the list goes on. The long-standing dismissal or evasion of humanistic and social scientific critiques of computing and new media is over. It has to be over, because to allow it to continue is simply too dangerous.

NOTHING IS VIRTUAL

When we speak of "fire" in this volume, we do so in three interconnected ways. Our first usage is literal. Despite widespread tropes that portray computing and new media as immaterial and disembodied—whether through an emphasis on "virtual" reality, "telepresence," "the Cloud," "streaming," the "postindustrial" economy, or otherwise—computing and new media are nothing if not entirely physical, material, and organic. They are physical machines, propelled by fire both material and metabolic. When they run, they run hot; and when they work hard, they run hotter. Data centers alone account for more than 2 percent of global energy use, energy consumption predicted to grow with the expansion of the Internet of Things.[4] (Google emitted over 50 kilograms of CO_2 in the time it took for you to read this sentence.) Through the studies of platforms and infrastructure, bitcoin mining, programming languages, underground cable networks, and much more, this volume drives home what is often termed the "materiality of the digital"—that is, the physicality of computational and new media technologies that are too often described in ethereal terms.

Computing and new media depend upon flesh-and-bone metabolism. Our "virtual worlds" are made possible by battalions of human beings (as well as nonhuman organisms): cable layers, miners, e-waste recyclers, content moderators, call-center operators, data-entry technicians, and repair technicians, many of whom come from marginalized class, racial, and gendered positions. Computing and new media run on a vast metabolic conflagration. In certain cases, the work of fabricating so-called virtual experiences exposes laborers to a daily regimen of toxic by-products of electronics manufacturing and disposal and, in other cases, to forms of post-traumatic stress disorder (PTSD) that grow out of long working days spent concentrating on realms of the digital world that the rest of us rely upon them to keep out of view: still-frame and live-action portrayals of extreme violence, child pornography, and vicious or hateful speech.

With all of this in mind, we wrote this book to remind ourselves and our readers: Every single thing that "happens online," "virtually," and "autonomously" happens offline first—and often involves human beings whose labor is kept deliberately invisible. Everything is IRL. Nothing is virtual.

THIS IS AN EMERGENCY

Fire also signals a state of crisis. Whether in the context of criminal justice practice, accelerating ecological crisis, access to credit and capital, governance, or elsewhere, computation and new media increasingly have life-or-death consequences. They are becoming ever more woven into the fabric of our social, political, and economic lives, and as this happens the conscious and unconscious values of system builders become encoded into the very algorithms that undergird said systems. Inequality, marginalization, and biases are transposed from the social and political world, where they can (at least in theory) be struggled over in the light of day, and rendered more durable, invisible, unavailable to human audit, and thus largely unassailable within the realm of computation. In this process, inequalities of gender, race, class, religion, and body type find their way into robotics, automated decision-making systems, virtual assistants, code academy curricula, search algorithms, and much more. Expanding beyond the first theme above, then, not only is it essential to remind ourselves that "nothing is virtual" but more broadly that there are dangerous consequences whenever one describes computational and new media forms using sanitized, deodorized, and neutralized vocabularies that exempt them from critical analysis.

WHERE WILL THE FIRE SPREAD?

Fire also propagates—in which direction and with what velocity is nearly impossible to forecast with certainty. It can spread steadily and predictably, moving out along fronts across contiguous territory. But it can also leap across space, riding atop a single airborne ember, and rapidly surround us.

We invoke this dimension of fire to emphasize the profound difficulties we face when deciding where we as scholars and activists should concentrate our attention and deploy our resources most effectively. Where are the frontiers of computing and new media, whether in terms of emerging and future forms of hegemony or, by contrast, novel forms of subversion and liberatory possibility? Many have already begun to realize that such frontiers may not be located where we might assume. If the vanguard of artificial intelligence research once resided in the defeat of Russian Garry Kasparov by chess-playing Deep Blue, more recently it lies in the 2016 defeat of Korean Lee Sedol at the "hands" of Google's Go/*Weiqi*/*Baduk*-playing system AlphaGo.[5] In the same year, China took top prize as the global leader in supercomputing for the seventh time in a row, with its Sunway TaihuLight clocking a theoretical peak performance of 125.4 petaflops, or 1,254 trillion floating point calculations per second.[6] Meanwhile, banking systems long reliant upon state-governed infrastructures are rapidly being displaced on the African continent by the cellphone-based money transfer system M-Pesa—the largest mobile-money business in the world.[7] And bringing us full circle, many of the cellphones upon which this multibillion-dollar economy runs still use T9 text-input technology—a technology that, although invented in North America, found its first active user base in Korea via the Korean Hangul alphabet. Where should the fires be fought? Where should fires be kindled (perhaps even lit)?

THE TIME FOR EQUIVOCATION IS OVER

Each essay in this book is framed as a direct and uncompromising declaration of fact. When you peruse the table of contents, you will see immediately that the essays do not argue that the Cloud "can be thought of" as a factory or that sexism "can be considered" a feature rather than a bug in the history of computing. Instead, the essayists draw upon their formidable research experience and deep knowledge of the technical subject matter at hand to cut directly to the heart of the matter. The Cloud *is* a factory. Your AI *is* a human. Sexism *is* a feature, not a bug.[8]

Our choice of phrasing is deliberate. Collectively, these essays are meant to serve as a wake-up call to both our colleagues and students in STEM fields, articulating these points in ways that, by being unequivocal, do not permit equivocation from our readers. Debate, certainly. Disagreement, quite likely. But evasion or facile techno-optimism? *No.*

These declarations of fact are also meant as a wake-up call to humanists and social scientists. By insisting on empirically rich, nuanced, yet *unapologetically direct and bold arguments* from all of our contributors, this volume represents a recognition, as it were, of a kernel of truth in the widespread dismissal of our work by those who are closest to the centers of production in the computing and new media economy. Critics like us, we must admit, have pulled our punches far too often. In spite of years of research and intellectual labor, a deep and hard-earned command of the technical systems under examination, as well as rich evidentiary and theoretical bases, we too often stop short of where we could take our arguments and conclusions. Too often, we hedge our bets and use formulations like "can be thought of as," or "is analogous to," or "is comparable to," and so forth—formulations that leave readers wondering: Does this author really mean what they say, or are they merely proposing a "way of seeing"? We are in a position to make urgent and reasonable demands of ourselves, our elected officials, and our most decidedly *unelected* industry leaders whose actions (and inactions) in large part define the lives we lead. The time for equivocation is over.

It is also important to stress that the goal of this volume is *not* to leave readers with the impression of a world, or a future, that is foreclosed and dark. Each contributor strikes a balance between certain sober realities about computing and new media, while always acknowledging their ubiquity, often their indispensability, and the sometimes awe-inspiring nature of technologies that have or are quickly becoming part of our everyday lives. Likewise, while the contributors grapple with questions of inequality—inequalities of gender, race, language, and more—we also remain vigilant about not inadvertently reproducing such inequalities by means of stereotype threat. While we feel it is essential to take an uncompromising look at ongoing dilemmas of gender and racial inequality, for example, the goal is not to dissuade readers—whether women, persons of color, first-generation students, or others—from pursuing their passions within the context of one or another STEM field. This volume is not, in other words, crafted as a call of despair but as a call to arms. We very much hope that the students who read this book will go on to take up positions in STEM fields, and then to agitate therein on behalf of the issues we raise.

ACKNOWLEDGMENTS

I wish to thank my coeditors, all participants in the two Stanford conferences, and one conference at the Computer History Museum, that helped give rise to this volume; our editor Katie Helke, and her colleagues at the MIT Press; and Joseph Rulon Stuart, for his capable work on the index.

NOTES

1. "HP Computers Are Racist," YouTube (December 10, 2009), https://www.youtube.com/watch?v=t4DT3tQqgRM.

2. https://www.flickr.com/photos/jozjozjoz/3529106844.

3. Kate Crawford, "Artificial Intelligence's White Guy," *New York Times* (June 25, 2016), https://www.nytimes.com/2016/06/26/opinion/sunday/artificial-intelligences-white-guy-problem.htm.

4. Fred Pearce, "Energy Hogs: Can World's Huge Data Centers Be Made More Efficient?" *Yale Environment 360* (April 3, 2018), https://e360.yale.edu/features/energy-hogs-can-huge-data-centers-be-made-more-efficient.

5. Cade Metz, "Google's AI Wins Fifth and Final Game against Go Genius Lee Sedol," *Wired* (March 15, 2016), https://www.wired.com/2016/03/googles-ai-wins-fifth-final-game-go-genius-lee-sedol/.

6. Thomas S. Mullaney, "The Origins of Chinese Supercomputing, and an American Delegation's Mao-Era Visit," *Foreign Affairs* (August 4, 2016), https://www.foreignaffairs.com/articles/china/2016-08-04/origins-chinese-supercomputing; Thomas S. Mullaney, "'Security Is Only as Good as Your Fastest Computer': China Now Dominates Supercomputing. That Matters for U.S. National Security," *Foreign Policy* (July 21, 2016), http://foreignpolicy.com/2016/07/21/china-taihulight-sunway-encryption-security-is-only-as-good-as-your-fastest-computer/.

7. See Paul N. Edwards, "Platforms Are Infrastructures on Fire," ch. 15 in this volume.

8. The essays in this volume address a wide variety of subject matter and case studies, ranging from human-in-the-loop content moderation to Bitcoin mining, from mobile banking platforms to game design, and from undersea cable networks to keyboard interfaces, among many other examples. At the same time, the volume does not aspire to be a "conspectus" on computing and new media. There are noteworthy omissions, such as high-frequency trading, 5G and the question of cellphone radiation, disability studies, biohacking, election hacking, and a great many other subjects. Our hope is that this volume, rather than trying to exhaust all possible avenues of discussion, provides a clear enough road map to help inspire the student and reader to grapple with subject matter we have not addressed. In this sense, the volume is well positioned for use in a classroom and for providing a well-structured yet open space for students to take some of the issues explained here into new areas of consideration. Moreover, the essays are written in a crisp, propulsive tone, and the word count is kept intentionally low, to enable instructors to assign them in clusters to their students in both STEM fields and the humanities. In this way, our goal is to help instructors help their students think comparatively and expansively.

WHEN DID THE FIRE START?

Mar Hicks

My institution is a tech-and-engineering-focused school, and I teach a course there called Disasters. It tends to attract engineering students looking for a humanities elective, despite—or because of—the disturbing subject matter. When I designed the course, I knew that it would be a major draw because, even if a person might not wish for bad things to happen, when bad things *do* happen, it's often difficult to look away.

I never imagined that I would be teaching the course in the middle of a global pandemic.

As we shifted our class online in March 2020 in order to safeguard our health, our lives, and the lives of those around us, and to try to slow a deadly and highly transmissible virus, I found myself thinking about other types of destructive virality, and about October 2019, when I watched Mark Zuckerberg testify before Congress. During his questioning by Congresswoman Alexandria Ocasio-Cortez about misinformation on his platform, Zuckerberg tried to give a series of vague nonanswers but occasionally tripped up in ways that showed that he truly did not understand the scope and scale of the problems he had helped unleash on society.

The questions asked of him in Congress mostly took for granted that, as the lead architect of Facebook, Zuckerberg could also be the driving force behind solving its problems, if only he could be convinced to take responsibility. Ocasio-Cortez's questioning, however, showed that he actually didn't even have enough knowledge of politics or the world around him to really understand the problems of misinformation on Facebook, never mind the will to fix them.

Recalling his answers, I thought about how online technologies, now so crucial to prevent the spread of deadly misinformation, had been functionally asleep at the wheel for years, in large part because their leaders, and even many of their workers, had indulged in the fiction that the technology that shapes our lives can somehow be neutral or apolitical even though it has clear and massive impact on our social relations.

This exchange in front of Congress, with a progressive Congresswoman grilling a former tech ingenue who is now one of the world's most powerful men, made clear what students in my disasters course already knew: that we are in the midst of witnessing a disaster right now sharpened by so many of the problems with "Big Tech" that we have seen growing for at least a decade. What is transpiring with the giants of Silicon Valley and their impact on our national and international infrastructure deserves a full unit in my disasters course, because the call for change in high tech, and the need for solutions, has never been more urgent.

You might think that taking a course on disasters in the middle of a disaster would be too much to bear. But from what I heard from my students, completing the course in these circumstances was grim but useful. Instead of just being a litany of horrible events, my disasters course shows students how disasters—awful as they are—can help create the kind of political, industrial, and regulatory change that is needed for a stable and functioning democracy. Students learn how the horror of a disaster can often be the final push that catalyzes much needed changes made by legislators, politicians, and citizens. Even though the problems that cause disasters are never unforeseen, it sometimes takes a disaster, replete with loss of life and other kinds of destruction, to convince the populace and government to respond—to actually *force* change on corporations and industries that have not been subject to enough oversight. But the hundred thousand US dead at the time of this writing in May 2020 (who are disproportionately Black, Latinx, and Indigenous Peoples), and the many more people in the streets protesting police brutality against Black citizens, are an urgent reminder of just how much destruction and loss of life are often excused—especially if those dying are Black, disabled, poor, or elderly.

Right now we are seeing the complete upending of ethical, privacy, political, and economic norms by powerful Silicon Valley corporations who are almost singlehandedly deciding what counts as misinformation, what counts as hate speech, and how much privacy you are allowed to have. In some cases, single individuals are making these decisions: as in the case of Mark Zuckerberg's decision to leave a president's incitement to violence up on Facebook during civil unrest (while Twitter, for the first time, attached a warning and filter).[1]

We've seen inexorable industry overreach into the conduct of our lives and institutions before—from the robber barons of the Gilded Age to the irresponsible destruction of lives and the environment at the hands of the auto industry and chemical corporations in the mid-twentieth century. We've even seen telecommunications and computing companies that have gotten too big for their, and our, own good before. The breakup of AT&T and the antitrust proceedings against IBM and Microsoft in the 1980s and 1990s, respectively, significantly shaped our current high-tech landscape, paving the way both for the mega-corporations that replaced them as well as for those corporations' similar, eventual falls from grace.

But somehow, each time a disaster like this happens, or a corporation gets too big, it seems like a new problem. That is because although the patterns of neglect, greed, and harm are the same, the devil is in the details of each technology—whether it's cars, oil, or computer platforms. The technology involved with each new disaster changes just enough to defamiliarize the situation and to allow people to ignore striking similarities with the past. This creates enough time for new corporations to gain runaway power and form new infrastructures that we all have to live with and use, no matter how imperfect or problematic they may be. Over the past twenty years, we have seen that very process play out with the internet and the digital economy, until now, when we have reached a crisis point.

WE DIDN'T START THE FIRE

I had a favorite history teacher who used to say that there was nothing new under the sun—but that didn't mean that you knew about it all yet. This is perhaps the most important thing to remember about the utility of history: there is always more to learn about the past that will help us shape the future. The second most important thing is perhaps realizing that failure stories—not just feel-good narratives that make us feel like things are bound to always get better in the long run—are critically important if we, as citizens and workers, are going to play a productive role in the struggle for a better future.

So when *did* the fire start? Well, if the fire refers to our complicated relationship with technologies designed to make our lives better but that often backfire in unforeseen ways and create new problems, that fire has been burning for centuries. The disasters course I mentioned above begins with the 1854 London cholera outbreak. This was the first cholera episode deemed bad enough—particularly because it happened in a better-off neighborhood previously thought immune to the disease—to jolt London into creating sanitary sewers and massive underground infrastructures that would once and for all be able to ensure the city's water supply was not perpetually poisoned by its own production of waste. London was literally eating—well, drinking—its own shit. While this episode may seem a long way off, it should be humbling to recall that less than two hundred years ago instituting technologies to keep sewage away from drinking water was considered first optional, and then revolutionary. As the United States stumbles in our scientific and social understandings of COVID-19 and the best way to end the pandemic, we might look back with more understanding and sympathy for these confused, cholera-ridden Londoners, drinking the wastewater that was killing them because, for most ordinary citizens at the time, there simply was no other choice.

Most of the technological advancements we've seen over the course of the nineteenth and twentieth centuries, from skyscraper cities to the ubiquity of the World Wide Web, involve some amount of literal or figurative shit-eating by average citizens, and also long periods of time during which obvious, systemic problems are mostly ignored by those at the top echelons of industry and government. Throughout history, we see technologies often deployed at scale for real-life beta testing, and the ensuing problems this inevitably presents. Since those problems disproportionately harm those with the least power in society, there is usually a long lag between the problems being noticed or cared about by people in charge and becoming seen as important enough or disturbing enough to warrant solving. One dirty secret is that often these problems are foreseen, but those in charge don't really care about them too much, seeing them as unrelated to the "real" problem that they are hoping to solve or the product they are hoping to make.

Another thing about disasters is that they don't simply happen by accident but rather are constructed: or, one could even say, designed. Mark Zuckerberg's first attempt at constructing a social media platform was a "hot or not" rip-off site that objectified his women classmates, using pictures stolen off the internal Harvard facebook servers. Zuckerberg called his site FaceMash, and the copy at the top of the site read: "Were we let into Harvard for our looks? No. Will we be judged on them? Yes." In doing this, Zuckerberg found a way to amplify, speed up, and scale up that process of voyeurism and objectification long possible with the original, paper, college facebooks. He made this element central to his design for his first social platform. Given this history, it may come as less of a surprise that his later platform designs had a kind of built-in voyeurism and a tendency to objectify and profit off of users, and that doing this without their full, knowing consent was baked into the nature and business model of the platform.[2]

This is not unusual—there are almost always red flags and warning signs before a disaster, if one cares to look. But often the people who sound the alarm, or notice these red flags, are the one who are least likely to have status, least likely to be listened to, and least likely to be believed. Their narratives form a constant counterhistory, but one that is usually only unearthed and taken seriously after the worst has come to pass.[3] Instead of coming as a total surprise, most disasters are more akin to a wake-up call: they have a way of showing which problems have been actively hidden for too long and which voices of warning have been ignored.

This is the process we have been witnessing since the first internet bubble of the late 1990s: a programmer or set of programmers hacking together an imperfect prototype, then unleashing it on the world with the help of venture capitalists whose only objective is maximum profit, and seeing where the chips fall. "Move fast and break things" ran the motto of these self-styled disruptors, because when you are rich enough, and privileged enough, it might seem like all the breaking you're doing doesn't have negative consequences. This is especially the case when the people you're hurting don't have a seat at the table and aren't the sort of people you would tend to include on your programming team, in your boardrooms, or in your shareholders meetings.

An early AI researcher, Joseph Weizenbaum (the creator of the ELIZA chatbot; see fig. I.1), once said: "I think the computer has from the beginning been a fundamentally conservative force. It has made possible the saving of institutions pretty much as they were, which otherwise might have had to be changed. Superficially, it looks as if [things have] been revolutionized by the computer. But only very superficially." Weizenbaum was a child refugee of the Holocaust who became an MIT computer science professor, and he participated in programming many of the systems he came to critique, noting that at the time he was doing the programming it was hard work and he was having "too much fun" in "attacking those hard problems" to think much about broader effects. Although he had narrowly escaped the mechanized death of Nazi Germany, which relied on centralized information technologies to help round up and kill over 6 million Jewish people, Roma, homosexuals, disabled people, and others judged inferior by the Nazis, he went on to build many similar, centralizing information technologies for US industry and government. "It never occurred to me at the time that I was cooperating in a technological venture which had certain social side effects which I might come to regret," he later said.[4]

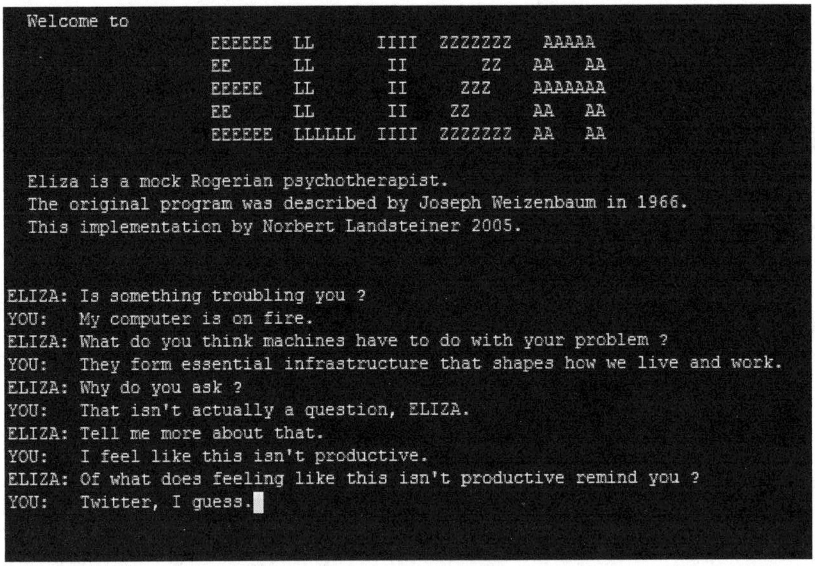

```
Welcome to
              EEEEEE  LL     IIII  ZZZZZZZ  AAAAA
              EE      LL      II        ZZ  AA   AA
              EEEEE   LL      II       ZZZ  AAAAAAA
              EE      LL      II      ZZ    AA   AA
              EEEEEE  LLLLLL  IIII  ZZZZZZZ  AA   AA

   Eliza is a mock Rogerian psychotherapist.
   The original program was described by Joseph Weizenbaum in 1966.
   This implementation by Norbert Landsteiner 2005.


ELIZA: Is something troubling you ?
YOU:   My computer is on fire.
ELIZA: What do you think machines have to do with your problem ?
YOU:   They form essential infrastructure that shapes how we live and work.
ELIZA: Why do you ask ?
YOU:   That isn't actually a question, ELIZA.
ELIZA: Tell me more about that.
YOU:   I feel like this isn't productive.
ELIZA: Of what does feeling like this isn't productive remind you ?
YOU:   Twitter, I guess.█
```

Figure I.1   Screen capture of author's conversation with ELIZA while writing this introduction.

Weizenbaum might be forgiven for this lack of foresight: for instance, as he was helping banks computerize, he failed to see the dangers of the increasing centralization of the banking industry that these computerized "solutions" would foster. When Weizenbaum experimented with AI, creating his first chatbot, ELIZA, he began to understand the dangers in what he was doing because he saw how people interacted with his creation. This provoked a lifelong crisis of conscience and a new consciousness of his responsibilities as a programmer. But decades later, most programmers creating websites that monetize people's private information, while abrogating their privacy and data mining their entire identities, did not have a similar crisis of conscience. Zuckerberg once said mockingly to a friend who asked why people were uploading all of this private information, "I don't know why. They 'trust me.' Dumb fucks."[5] Decisions about which technologies should exist cannot simply be left to programmers.

With the benefit of hindsight and disciplines other than computer science we can clearly see how the internet economy has long generated wealth through taking advantage of existing inequalities, existing infrastructure, and by using venture capital funding to reinvent the wheel, only to crush people who aren't in the driver's seat. Some of the earliest corporations to do this appeared to be offering novel solutions—and in some ways were. Yahoo and other early email and online news hubs helped reinvent and scale mail and news delivery mechanisms. Google found a new way to serve and profit from advertising while indexing the web. Friendster and Myspace, and later Facebook, Twitter, and Instagram, reinvented social networks that had earlier relied on physical or telephone interactions. Most seemed beneficial, or at least relatively harmless, at the start. Or did they?

If we look back with a critical eye, we recall that Yahoo, through its auctions site, early on ran afoul of hate speech laws in other countries by hosting, and defending, the sale of Nazi paraphernalia. Yahoo did this on ahistorical grounds, putting their corporation's business needs over the rights of citizens likely to be harmed by hate speech, or killed by genocidal political ideals. Google, especially after the advent of street view and Gmail, provoked major concerns about the centralization of so many citizens' information in the hands of a private corporation, as well as concerns about how it was casually undoing expectations that many people had about the privacy of their personal information. Google's efforts to index and make visible the entirety of the web were presented as neutral and helpful, but at the same time the company's goal was to monetize all of that information, and our searches through it, for ad sales. Google's now-discarded original motto "Don't be evil" perhaps seems, in retrospect, to hint at a deeper understanding on the part of its founders that what the company was doing could go badly off the rails without proper oversight. Lastly, Facebook's roots in sexism and ongoing privacy violations are at this point well known. We understand now that it had a business model that seemingly had harm built into it at the ground level, even if a now-older Zuckerberg has tried to rewrite history and sanitize the story of Facebook's origins and his corporate strategy.[6]

In some ways the stories above are unsurprising, and fit a pattern: the entire history of electronic computing is, as is the case with many technologies, intertwined with efforts at domination. Early computers were used in the service of warfare, and were also operated by a workforce of women who were silenced and submerged in the historical record for decades. Although the Colossus code-breaking computers helped the UK and Allies win World War II, they also showed the enormous might of computers and their alignment with militarism and political power.[7] While men at the top made claims about the wonderful new future that computers could bring during the Cold War, women—and in the US, particularly Black women—were unacknowledged grist for the mill of computing, grinding away at some of the hardest problems that needed to be solved. In the words of Margot Shetterly, Black women were helping to put a white man on the moon while Black people often still couldn't even safely drive to the next state.[8] This continued through the Cold War and the Space Race, when, as Meredith Broussard has discussed, the budding field of artificial intelligence began to promise, with little evidence and no proof of concept, that computers could transform every aspect of our society *for the better*.[9]

As Broussard points out, this kind of "blue sky" thinking about computers' supposedly fantastic ability to foster social progress was encouraged with massive government funding and grants that simply took it on faith that computing experts were somehow ideally poised to tackle societal problems, not just technical ones. That these computing technologies were heavily funded by the government, and thought to be instrumental to successful warfare, was no coincidence. These assumptions helped quickly ramp up fields like AI while building in virtually no accountability.

This feedback loop between industry, government, and academic computer science progressively sought to heighten our dependence on computers without any proof that those with technical skills could solve, or even understand, social, political, economic, or other problems. Indeed, often there was little evidence they could even deliver on the *technical* solutions that they promised. The visions of general AI outlined by Alan Turing, Marvin Minsky, and others in the twentieth century still have barely materialized, Broussard points out, and where they have, they have come with devastating technical flaws too often excused as being "bugs" rather than fundamental system design failures.

In addition to a lack of accountability, power imbalances continued to be a bug—or, if you prefer, a feature—in the drive to computerize everything. Even as swords turned to plowshares in the twentieth century and electronic computers moved out of the realm of warfare and squarely into business and government administration, computing corporations showed repeatedly how computers tended to privilege those with the most money and power already, and provided centralized solutions that could easily lead to devastating loss of life or of people's civil rights. Technologies provided by IBM, the bellwether of the tech market in the mid-twentieth century, played a role in the Holocaust, then later supported the white supremacist apartheid government of South Africa.[10] Bill Gates engaged in such out-of-proportion attempts to torpedo competitors and get a stranglehold on the information network of the early World Wide Web in the 1990s and early 2000s that multiple governments opened antitrust proceedings against Microsoft. And throughout these episodes, the managerial workforces at these corporations remained relatively homogeneous, with their leadership made up almost exclusively of white upper- and middle-class men.

IT WAS ALWAYS BURNING

As surveillance scholar Simone Browne has shown, the interrelationships between power, surveillance, and white supremacy were not technological accidents; they are a historical process that has formed a fundamental part of technological design in US history.[11] The largest, most respectable, and most middle-of-the road computing companies not only practiced racist and sexist hiring and promotion for most of their histories but also, at a fundamental level, had an alignment with those in power globally which helped exacerbate existing social and political harms in their own nations and others.[12] These were not accidents or bugs in the system, but examples of business as usual. As Halcyon Lawrence describes later in this volume, technologies have long built on existing power relationships—particularly ones that extend and normalize empire, colonialism, and the cultural control that comes with imperial domination. If early electronic computers were powerful political tools, ones that determined the course of a world war, is it any wonder that ever more advanced electronic computers have been helpful for consolidating and wielding power?

Disasters build, fulminate, and eventually explode into mainstream consciousness because problems that many people take for granted as necessary for a system to work start to get out of hand. Disasters happen because the people at the top who claim that the system works—that their way of doing things is the best way—tend to proceed down paths that are destructive for many other people in society, even if many people with less power point out the coming danger. Disasters have a way of showing how lopsided our views of technological progress are, highlighting the problems that were always there but never got fixed, because the incentives to leave the system half-broken were too great.

If the history of disasters shows us nothing else, it is that the old adage about trusting the fox to guard the henhouse will always end up the same way. Increasingly, as corporations have been able to place themselves in the role of arbiter of their own products and value, it has meant that democratic input into the process of deciding which technologies are safe, useful, or worthwhile has been short-circuited. Even with established technologies with good safety records, like commercial airliners, this can create a disaster surprisingly quickly. Before the automated MCAS system forced two Boeing 737 Max airplanes to drop out of the sky in quick succession in 2019, killing hundreds of people, Boeing engineers had argued that the system was unsafe. But they were overruled by management. Boeing was able to force the dangerous new feature through, undetected, because the government agency meant to regulate them was instead letting the corporation largely call the shots. After the first crash, Boeing's CEO continued to insist the system was safe, blaming the crash on pilot error even though Boeing had removed the relevant parts of the manual that might have allowed the pilots to recover from the malfunction. Had it not been for whistleblowers and dedicated investigative journalists, Boeing may have gotten away with this, and more.

The larger lesson, however, is that regulatory agencies that should have prevented disasters like the 737 Max tragedy had been stripped of their ability to do so, and whole sectors that require regulation, like web technologies and online communication platforms, effectively have no external oversight bodies. In the US since 2016, instead of more regulatory safeguards put into place by a democratically elected government, we have seen runaway centralization and the destruction of regulatory and safety agencies under an increasingly authoritarian federal government. During a global pandemic this has meant that even healthcare supplies and vital public information have been withheld from citizens and state governments. Scientists and epidemiologists at the CDC and NIH have been increasingly forced out or muzzled, and many of the teams dedicated to pandemic response had been dissolved prior to the COVID-19 crisis, leaving a gaping hole of expertise where our disaster response should have been.

As the president screamed on Twitter, taking his misinformation about state leaders, the progress of virus research, and elections directly to the public, suddenly an unregulated tech company with no competence or background in history, politics, journalism, rhetoric, the psychology of online interaction, or fact checking was thrust into the position of trying to unbuild the minsinformation machine that has been long in the making. Many people thought sites like Twitter and Facebook were harmless, until they helped give a platform to one of the most dangerous presidents in US history, and helped authoritarianism and genocide advance in multiple countries.

We are witnessing a period in which it is becoming ever more urgent to recognize that technological progress without social accountability is not real progress—and that in fact it is destructive to the democratic institutions and norms we have long held up as ideals. As Sarah T. Roberts shows in her chapter in this book, the fiction that platforms that are our main arbiters of information are also somehow neutral has effectively destroyed the public commons. As Safiya Noble has shown in *Algorithms of Oppression*, trusting an advertising corporation to be a neutral purveyor of information when their profits depend on manipulating that information fundamentally misunderstands our capitalist marketplace as well as the value and nature of unbiased information. This insight extends to every platform that makes a profit through telling people what they are expected to *want* to hear, or want to click on, rather than the often inconvenient, unprofitable, or disturbing truth.[13]

As scholar David Golumbia points out, in the US in the twenty-first century, there is widespread belief that governments should not have access to privacy-invading technologies like facial recognition, yet many ignore the fact that multibillion-dollar corporations are already deploying these technologies with no democratic oversight or citizen input.[14] Because corporations are not elected, they cannot be voted out, and yet they have become pseudogovernmental by virtue of their wealth, power, and the reach of their technological systems. Their leaders insist that they, and they alone, know what is best for us—from what information we should see to how much privacy we should retain. Increasingly, these companies have placed themselves in the role of determining how we move about in the world, literally and figuratively, and their power to define our reality increasingly extends to the power to decide elections in the US and other nations, taking away our most fundamental rights as citizens to self-determination. These corporations tacitly assert that our future should be decided in the end by what is most profitable and efficient for them, rather than being left to the messy process of democracy unfiltered by the technocratic class. They tell us to trust them and repeatedly assure us that the tech industry will police themselves and fix their own mistakes.

Unfortunately, as a historian, I can tell you this never works. If the crisis we are currently living through has not already made it apparent, we can see it clearly in the recent technological history of our closest historical cousin. The UK context provides a clear cautionary tale—as I discuss later in this volume, and at further length in my book, *Programmed Inequality*. That country, after inventing and successfully deploying the first electronic programmable computers to change the outcome of World War II, managed to destroy its own trained technical labor force, and with it its computing industry through sexist labor practices. We now see this clearly taking shape in high tech in the US, where sexism and racism have not only held back major advances (by starving these industries of talent and frustrating the career outcomes of people who were born with the "wrong" gender and/or skin color), but have also played an increasing role in our public discourse and our political process. That those at the top often cannot understand the social, economic, and political harms they are perpetrating is unfortunately not a new story in the history of computing.

Even when it loses money, a broken system that consolidates more power will not be discarded. Civil rights, workers' rights, and care for consumers are never as convenient or profitable as oppression and exploitation, because oppression is about power as much as it is about profit. It is hard enough to dislodge and correct such broken systems when we do have a say—a legally protected vote—in how they should function. It is exponentially harder when the very basis on which the system operates means we do not. After getting pushback from their own employees on the company's ethical failures, post-"don't be evil" Google management appointed a problematic "AI ethics board" without giving employees a voice. Employee pushback got the board disbanded, but the company now has no AI ethics board, nor is it required to have one.[15] Labor unrest and consumer complaints can only pressure corporations into better behavior when governments enforce laws and regulations meant to protect our lives instead of the companies' bottom lines.

In 1911, workers throwing themselves out of the upstairs windows of a burning factory in New York City sparked a worldwide labor movement. These workers, mostly young immigrant women, jumped from the top floors of the Triangle Shirtwaist Factory because the building was on fire and they had been locked in by their employers. In this case, the horror of seeing more than one hundred crumpled, charred bodies laid out on the city streets was able to shock the world, and particularly the wealthiest and most powerful in New York society, into believing how bad things had gotten, winning them over to the workers' side. But in the courts, the factory owners essentially got away with mass murder. Only continued labor struggles and the power of workers at the ballot box eventually changed the laws that governed safe working conditions and saved many future workers' lives. Holding corporations accountable has never been easy, and shocking citizens in a wealthy nation into action requires concerted, organized efforts. A disaster alone is never enough.

WE DIDN'T LIGHT IT, BUT WE TRIED TO FIGHT IT

And herein lies the best hope we have for extinguishing the fire currently engulfing us: we need to take advantage of this moment of disaster to understand how connected our systems are, and to leverage grassroots action and worker organization to change the ways we work, live, and govern ourselves. For decades, the corporations that built the digital economy have tried to make us think that we could consume without end, speak without consequence, and outsource our responsibilities for ethical systems to someone at a higher pay grade. We now see how wrong that was.

For decades, computing companies have tried to convince all white-collar workers that they were management, or aligned with management, and so did not need to argue with those at the top, or need unions to help press for change. As recent events in the tech industry have shown, nothing could be further from the truth. From Google to Kickstarter, tech workers have begun to see that, if they don't have a real voice in deciding the direction of the company, they don't have any control over the harms created by the products they make. As individuals, their voices can be easily ignored, as the Boeing example shows.

As a group, however, tech workers—and citizens—have power. As the cases of Project Dragonfly and Project Maven at Google show, worker pushback can shut down projects and save lives. As the Google Walkout showed, until more people speak up it is still more acceptable to pay a sexual harasser millions of dollars to leave than it is to pay women an equal wage or give them equal opportunities to stay. Unionization efforts at Kickstarter and other corporations provide a blueprint for the next steps we need to take back our democracy and to make it possible for people to speak in favor of what is right in a broader sense.[16] And as we saw from the rejection of the transphobic and xenophobic Google ethics board due to intense pressure from employee organization and protest, workers can begin to call the shots about what actually makes good and ethical technology if they work together and fight.[17] But they can only do this within a framework of a stable, elected democratic government that has, at its core, a commitment to protecting citizens and workers instead of seeing them as expendable.

As we enter this new phase of the digital era, it is important to remember the historical lesson that computers, and technologies more generally, have always been about control and power. Once a pillar of democratic society, the US press has increasingly been weakened by attempts by platforms like Twitter and Facebook to step into the role of news media. As social media platforms cut the financial support out from under professional journalists, the stakes have never been higher. From Amazon worker strikes, to the Uber driver airport blockade in opposition to Trump's "Muslim ban," to campaigns led by Black women online like #yourslipisshowing to help root out online propaganda, the power of people acting not as individuals but as members of organized social groups with clear goals is critical, and a free press to report on those actions is equally vital.[18]

The path ahead is difficult but clear: in order to undo our current multilevel disaster we have to support workers, vote for regulation, and protest (or support those protesting) widespread harms like racist violence. The strength of the ballot box is not enough when a country's informational infrastructure is in ruins. Supporting older, more stable technologies that enhance our society, like the postal system, traditional news media, and citizen-funded public health is as important as rejecting newer technologies that threaten to disrupt and divide.

And if you work in tech, whether you write code for Uber or drive for it, whether you design network-attached storage devices or military drones, take the initiative to start organizing for the future even as you try to build it: because when you cannot make a difference as an individual you can still make a difference in a group. Talk to your coworkers and share your salary information to help all of you get equal pay. Schedule a meeting with a union organizer. Volunteer and join coalitions in your community. A fire can be extinguished, but it can't be extinguished one cup of water at a time.

Going forward, we have the same task in front of us that people trying to recover from a disaster always have: gather support from the bottom to force change at the top. We must pool our resources to resist, refuse, and push back at the highest levels of corporate and government power if we are going to have a chance to be heard. But on an individual basis, we must also make difficult ethical choices: If your job does not allow you to sleep at night, find a new one. Don't spend your life as a conscientious cog in a terribly broken system. If you can't leave, do your best to interrupt the harms your work is creating. Even if you cannot avert a train wreck, simply slowing down the runaway train can save lives: if you can do nothing else, go slow. Don't think "if I don't do it, they'll just get someone else to do it." Remember instead that the power, and the trap, of neoliberal thinking is that it divides and conquers and makes us feel that there is no way out of the current system, when there is.

With effort, and a willingness to question and upend the systems we currently take for granted, we can change the infrastructures we've built, and we can put out the fires for our future selves and the next generation. The bad news is that these problems will be hard to solve, and those in control have enormous power that they will not easily give up. The good news is that we know we can do it, because it's been done many times before, with many different industries and government administrations that have become overbearing and excessively powerful to the point of corrupting democracy. This same pattern has played out numerous times: history, after all, doesn't repeat itself, but it rhymes. Our power only exists in its exercise: we have to use it in order to make it real.

To shape the future, look to the past. For all its horrors, history also contains hope. By understanding what has come before, we gain the knowledge we need to go forward.

NOTES

1. Davey Alba, Kate Conger, and Raymond Zhong, "Twitter Adds Warnings to Trump and White House Tweets," *New York Times* (May 29, 2020), https://www.nytimes.com/2020/05/29/technology/trump-twitter-minneapolis-george-floyd.html.

2. Kate Losse, "The Male Gazed," *Model View Culture* (January 2014), https://modelviewculture.com/pieces/the-male-gazed.

3. See, for instance, the way that Black women understood and warned about the dynamics of online trolling and abuse far sooner than white women, yet got far less credit for sounding the alarm. Rachelle Hampton, "The Black Feminists Who Saw the Alt-Right Threat Coming," *Slate* (April 23, 2019), https://slate.com/technology/2019/04/black-feminists-alt-right-twitter-gamergate.html.

4. Diana ben-Aaron, "Interview: Weizenbaum Examines Computers and Society," *The Tech* (April 9, 1985), 2.

5. Laura Raphael, "Mark Zuckerberg Called People Who Handed Over Their Data 'Dumb F****,'" *Esquire* (March 20, 2018), https://www.esquire.com/uk/latest-news/a19490586/mark-zuckerberg-called-people-who-handed-over-their-data-dumb-f/.

6. In October 2019, Zuckerberg began to claim that he started Facebook as a way for people to share news and opinions about the US invasion of Iraq. For more on Facebook's early days from one of its first employees, see Kate Losse, *The Boy Kings* (New York: Free Press, 2005).

7. Mar Hicks, *Programmed Inequality: How Britain Discarded Women Technologists and Lost Its Edge in Computing* (Cambridge, MA: MIT Press, 2017).

8. Margot Lee Shetterly, *Hidden Figures: The American Dream and the Untold Story of the Black Women Mathematicians Who Helped Win the Space Race* (New York: William Morrow, 2016).

9. Meredith Broussard, *Artificial Unintelligence: How Computers Misunderstand the World* (Cambridge, MA: MIT Press, 2018).

10. Edwin Black, *IBM and the Holocaust* (New York: Crown Books, 2001), and Clyde W. Ford, *Think Black: A Memoir* (New York: Amistad, 2019).

11. Simone Browne, *Dark Matters: On the Surveillance of Blackness* (Durham, NC: Duke, 2015). See also Ruha Benjamin, *Race after Technology* (New York: Polity, 2019).

12. Clyde W. Ford, *Think Black: A Memoir* (New York: Amistad, 2019).

13. Safiya Noble, *Algorithms of Oppression* (New York: NYU Press, 2018).

14. David Golumbia, "Do You Oppose Bad Technology, or Democracy?," *Medium* (April 24, 2019), https://medium.com/@davidgolumbia/do-you-oppose-bad-technology-or-democracy-c8bab5e53b32.

15. Nick Statt, "Google Dissolves AI Ethics Board Just One Week after Forming It: Not a Great Sign," *The Verge* (April 4, 2019), https://www.theverge.com/2019/4/4/18296113/google-ai-ethics-board-ends-controversy-kay-coles-james-heritage-foundation.

16. April Glaser, "Kickstarter's Year of Turmoil," *Slate* (September 12, 2019), https://slate.com/technology/2019/09/kickstarter-turmoil-union-drive-historic-tech-industry.html.

17. Googlers Against Transphobia, "Googlers Against Transphobia and Hate" *Medium.com* (April 1, 2019), https://medium.com/@against.transphobia/googlers-against-transphobia-and-hate-b1b0a5dbf76.

18. Eli Blumenthal, "The Scene at JFK as Taxi Drivers Strike Following Trump's Immigration Ban," USA Today (January 28, 2017), https://www.usatoday.com/story/news/2017/01/28/taxi -drivers-strike-jfk-airport-following-trumps-immigration-ban/97198818/; Josh Dzieza, "'Beat the Machine': Amazon Warehouse Workers Strike to Protest Inhumane Conditions," *The Verge* (July 16, 2019), https://www.theverge.com/2019/7/16/20696154/amazon-prime-day-2019-strike -warehouse-workers-inhumane-conditions-the-rate-productivity; Rachelle Hampton, "The Black Feminists Who Saw the Alt-Right Threat Coming," *Slate* (April 23, 2019), https://slate.com/ technology/2019/04/black-feminists-alt-right-twitter-gamergate.html.

I

NOTHING IS VIRTUAL

1

THE CLOUD IS A FACTORY

Nathan Ensmenger

In the future histories to be written about the digital revolution of the late twentieth and early twenty-first centuries, there will inevitably appear a chapter on Amazon .com. One of the rare dot-com-era startups that survived beyond its infancy, Amazon leveraged its early success in its intended market space (book sales) into broader dominance in electronic retail more generally. Amazon is not only the largest of the top tech firms in both revenue and market value, but it competes successfully with traditional retail giants like Walmart.[1] On any given day, its founder and CEO Jeff Bezos stands as the richest man in America (on the other days he is second only to Bill Gates, who will no doubt also demand a chapter of his own in our imagined future history). The carefully cultivated story of both Bezos and the firm he created perfectly captures the dominant narrative of success in the digital economy (with the sole exception that Bezos actually managed to complete his Ivy League degree). If you were to ask the average American how daily life has changed for them in the internet era, they would almost certainly reference their experience with Amazon. And seemingly every day Amazon is expanding into new arenas, from entertainment to home automation to artificial intelligence.

And yet, despite Amazon's undisputed centrality in the contemporary digital economy, a close look at its core business model reveals it to be surprisingly conventional. At least a century prior to the invention of e-commerce, mail-order catalog companies like the Sears, Roebuck Company had accustomed American consumers to purchasing goods sight unseen from vendors with whom they communicated solely via information technology. Like Amazon, Sears neither manufactured goods nor owned inventory but functioned solely as information intermediary (and, as we will see shortly, a logistics and transportation company). What both companies provided was a layer of network infrastructure that links consumers to producers via a single unified interface. In the case of Sears, this interface was a paper catalog, for Amazon a website, but the basic services provided are identical. By organizing, consolidating, and filtering information, both the catalog and website served to simplify otherwise complicated and time-consuming informational activities and establish and maintain networks of trust across geographically dispersed networks of strangers. Concealed behind these seemingly simple user interfaces was a complex infrastructure of information-processing and communications systems, from display and advertising technologies to payment-processing systems to user support and service. And here again, it was arguably Sears a century earlier who was the most original and innovative; the systems that Amazon uses are perhaps more automated but are conceptually very similar (and, as in the case of the postal network, essentially unchanged). It is true that in the early twentieth century Sears handled "only" millions of commercial transactions annually, whereas today Amazon processes billions, but that is simply a difference in scale, not in kind.

But although both Sears and Amazon saw themselves essentially as information organizations, the messy reality of retail, even information-technology-mediated retail, is that eventually the goods need to be delivered. Although their sophisticated information systems could provide a competitive advantage when processing transactions, the costs associated with the management of information paled next to the costs of handling, storing, and transporting physical materials, and so both mail-order and e-commerce firms often find themselves reluctantly expanding along the distribution chain. For Sears, this meant coordination (and occasionally partnership) with railroad companies and national postal networks and the construction of massive warehouses and distribution centers. For Amazon, this meant the coordination with (or, increasingly, ownership of) trucking companies and shipping fleets, partnership with national postal networks, and the construction of massive warehouses and distribution centers. Within a decade of their establishment, both firms had reluctantly expanded out of informational space and into the physical environment. By 1904, Sears had purchased more than 40,000 square feet of office and warehouse space in Chicago alone; today, a single Amazon distribution center averages 100,000 square feet, and there are many hundreds of such centers in the United States alone. Eventually, Sears found itself constructing its own brick-and-mortar retail establishments to supplement its mail-order operations; recently Amazon, which allegedly triumphed over Sears because of its lack of such legacy brick-and-mortar, has begun doing the same.[2]

The degree to which Amazon is fundamentally in the business of managing the movement and storage of "stuff" (activities that our future business historians will no doubt refer to as transportation and logistics) cannot be overstated. In 2017 alone, Amazon shipped more than five billion packages via its Prime subscription service.[3] To accomplish this, Amazon has constructed more than 329 distribution centers in the United States, and another 380 worldwide.[4] These include massive, million-square-foot warehouses like that in Tracy, California, as well as smaller, more special-ized sorting and delivery stations.[5] For delivery between its various facilities, Amazon relies on fleets of company-owned or leased vehicles.[6] For the so-called last mile, it relies (for the moment, at least) on delivery services like UPS or FedEx and—on extraordinarily favorable terms—the United States Post Office.[7] In order to further reduce its costs, Amazon has been developing an Uber-like system called Amazon Flex to further "disrupt" its dependence on third-party carriers.[8] And famously (and prematurely, perhaps perpetually), Amazon has announced plans to implement entirely automated drone delivery.[9]

In its focus on the control and consolidation of transportation and distribution networks, Amazon resembles yet another of the early-twentieth-century corporate giants, namely Standard Oil (see fig. 1.1).[10] Although Standard Oil's dominance of the oil industry was due in part to its monopolistic consolidation of refineries, it was equally enabled by the firm's secret manipulation of the railroad network. Like Jeff Bezos, John D. Rockefeller recognized the value of vertical integration and the necessity of access to and control over critical infrastructure. Such integration is only ever in part a technological accomplishment, and it requires social, political, and financial innovation. In this respect, the continuity between the industrial-era giants and the "Big Five" tech firms (Alphabet, Amazon, Apple, Facebook, and Microsoft) is all the more apparent. When we consider the digital economy in general, and electronic commerce in particular, it seems that success is also dependent on access to infrastructure—proximity to key transportation networks like roads, bridges, and highways; the employment of large numbers of appropriately skilled (but reasonably inexpensive) labor; the ability to construct and maintain (or at least lease) physi-cal plants and other facilities; and, of course, access to the large amounts of capi-tal, credit, and political influence required to secure the aforementioned resources. This perhaps explains in part why, despite the emphasis in the digital economy on

Figure 1.1 Existing and projected Amazon small sortable fulfillment centers in the United States.

light, flexible startups, many sectors of that economy are controlled by an increasingly small number of large and established incumbents. The growing belief that the United States is in the midst of a modern Gilded Age is about more than concern about wealth inequity.[11]

Given the perceived shift in recent decades (in the Western world, at least) from an industrial to a postindustrial society, the continued dependence of information-economy firms like Amazon on material infrastructure and the manipulation of physical objects is surprising, if not paradoxical. Despite repeated claims that the defining characteristic of the information society is "the displacement in our economy of materials by information," as *Wired* magazine editor Kevin Kelly has described it—or, in the even more succinct and memorable words of MIT professor Nicholas Negroponte, the inevitable shift "from atoms to bits"—what has in fact occurred is a massive *increase* in our interaction with our physical environment.[12] Information technologies allow humans to visualize, explore, and exploit our environment more efficiently. We travel more (and more broadly), consume more (and more globally), pollute more (and more pervasively). The amount of material moving around the planet has increased exponentially in recent years, arguably as a direct consequence of the digital revolution.[13] In fact, this increase is not only enabled by information technology but *required* by it.

Consider, for example, the one aspect of Amazon's business model that is truly different from that of its historical counterparts in the industrial-era retail economy: namely, its integration of sophisticated computational technologies at every level of the firm, from customer-facing web interfaces to back-end databases to global positioning systems. It is because of its use of these technologies that we think of Amazon as a key player in the digital economy in the first place. And, indeed, Amazon's implementation of these technologies was so successful that the company soon decided to package them for sale as commodity computational services and infrastructure. Unlike Amazon's retail operations, the provision of these services and infrastructure are highly lucrative, bringing in more than $17 billion in revenue annually and comprising the majority of the company's overall profits.[14] Within the computer industry, these products are known collectively as Amazon Web Services. Colloquially, the commodity computational infrastructure that these services comprise is known simply as "the Cloud." Of all of the elements of the contemporary digital ecosystem, none is more associated with the claims of present or imminent technological, economic, and political revolution than the Cloud.[15] If trucks and warehouses are the legacy technologies that ground e-commerce companies like Amazon to materiality and geography, the invisible and ethereal infrastructure of the Cloud seems to point the way toward a truly postindustrial and entirely digital economy.

What exactly is the Cloud? At its most basic, the Cloud is simply a set of computational resources that can be accessed remotely via the internet. These resources are generally associated with particular services, such as web hosting, server-based applications, database access, or data warehousing. The value of these resources is that they are available as discrete and idealized abstractions: when the user purchases access to a Cloud-based photo-sharing service, for example, they need know nothing about how that service is provided. They do not need to purchase a computer, install an operating system, purchase and install applications, or worry about software maintenance, hardware failures, power outages, or data backup. All of this equipment and labor is located and performed elsewhere, and as a result is rendered effectively invisible to the end user. In fact, it is this quality of seamless invisibility that most defines the Cloud as a form of infrastructure; as Susan Leigh Star reminds us, the whole point of an infrastructure is that you never really have to worry about what makes it all possible.[16] No one gives much thought as to how their electricity is generated, or where, or by whom; we simply expect that when we plug in our appliances or devices that the required electrons will be available. We only notice the massive size and complexity of the underlying electrical grid when it is broken or otherwise unavailable. The same is true of all infrastructure, from sewer systems to roads and bridges to our freshwater supply—and, increasingly, the internet and the Cloud.

But despite its relative invisibility, the Cloud is nevertheless profoundly physical. As with all infrastructure, somewhere someone has to build, operate, and maintain its component systems. This requires resources, energy, and labor. This is no less true simply because we think of the services that the Cloud provides as being virtual. They are nevertheless very real, and ultimately very material. For example, a typical large data center of the kind that Amazon operates draws between 350 and 500 megawatts of power; collectively, such data centers consumed 70 billion kilowatt-hours of electricity in 2016 in the United States alone.[17] This represents close to 2 percent of the nation's entire electricity consumption—roughly the equivalent to the output of eight nuclear power plants. Considered globally, the amount of power used by data centers approaches 1.4 trillion kilowatt-hours. And while some of this electricity is no doubt provided by renewable resources, much of it derives from sources that are so old-fashioned as to be prehistorical, such as coal, oil, natural gas, and uranium. According to a 2014 Greenpeace report, if the Cloud were a country, it would be the sixth largest consumer of electricity on the planet.[18] As these resources are consumed, they return carbon back into the atmosphere—something on the order of 159 million metric tons annually—and so the Cloud is also one of the world's largest polluters.[19]

Given its insatiable demand for electricity, there is at least one sense in which the Cloud is more than a metaphor. Cooling a typical data center requires roughly 400,000 gallons of fresh water daily. A very large center might require as much as 1.7 million gallons.[20] This is independent of the massive amount of clean, fresh water that is required to manufacture the data center's computer equipment in the first place. The Cloud is a heat machine designed to circulate cool air and moisture, creating its own carefully controlled microclimate and contributing to climate change in the larger environment.

Heat, air, and water are only a few of the material resources that the Cloud hungrily devours. Also present in these computers and their associated display screens are dozens of elements, some of them rare, some of them dangerous, all of which must be painstakingly mined, purified, transported, and manufactured into finished products—processes that also involve material resources, human labor, and multiple layers of additional infrastructure, many of which are controlled by some of the least stable and most exploitive political regimes on the planet.[21] All of which is to say that just as Amazon's e-commerce operations are revealed to rely to a remarkable degree on traditional, decidedly nondigital technologies like trucks and warehouses, so too are even its most high-tech and allegedly virtual services ultimately constructed around industrial-era systems, processes, and practices.

Which brings me to the main provocation of this chapter: namely, the claim that the Cloud is a kind of factory. In making this claim, my goal is to explore the potential benefits, analytically, politically, and otherwise, of resituating the history of computing within the larger context of the history of industrialization. In the early decades of the digital economy, the material dimensions of our emerging informational computational infrastructure were captured in the concept of the "computer utility."[22] Today, the metaphor of the Cloud erases all connection between computing services and traditional material infrastructure (as well as the long history of public governance of infrastructural resources). As a result, the computer industry has largely succeeded in declaring itself outside of this history, and therefore independent of the political, social, and environmental controls that have been developed to constrain and mediate industrialization.[23] By describing itself as an e-commerce entrepreneur and not simply an email order company, Amazon was awarded a decades-long tax subsidy that allowed it to decimate its traditional competitors.[24] In claiming to be an internet service provider and not a telecommunications carrier, Comcast can circumvent the rules and regulations intended to prevent monopolies.[25] By transforming its drivers from employees into contractors, Uber can avoid paying Social Security benefits.[26] In rendering invisible the material infrastructure that makes possible the digital economy, the metaphor of the Cloud allows the computer industry to conceal and externalize a whole host of problems, from energy costs to e-waste pollution. But the reality is the world is burning. The Cloud is a factory. Let us bring back to earth this deliberately ambiguous and ethereal metaphor by grounding it in a larger history of technology, labor, and the built environment—before it is too late.

To begin our interrogation of the claim that the Cloud is a factory, let us return for a moment to the earliest of the information organizations that I have thus far identified, namely, the Sears, Roebuck company. Of the many industrial-era corporations with which we might compare Amazon and other Silicon Valley tech firms, Sears stands out as the most relevant: not only did it share a business model with Amazon, but it survived long enough into the twenty-first century to be a competitor. Like electronic commerce today, the mail-order-catalog industry of a century ago reveals the essential continuities between the industrial and informational economies.

Sears was not the first of the mail-order-catalog companies: that honor goes to Montgomery Ward, whose founder, Aaron Montgomery Ward, issued in 1872 a one-page catalog that listed some items for sale and provided information on how to order them. But the company that Richard Sears and his business partner, Alvah Roebuck, founded in 1891 quickly emerged as a leading competitor to Ward and was, by 1897, delivering a 500-page catalog to 300,000 American homes, offering up everything from bicycles to bonnets to bedroom furniture to two-bedroom homes. By 1913, Sears was issuing more than twenty-six million catalogs annually and on any given day was able to fulfill more than 40,000 orders and process 90,000 items of correspondence. While this is not even close to contemporary Amazon volume, it is nevertheless significant. It is certainly indisputable that Sears circa 1913 was a full-fledged information organization. They had solved all of the key challenges facing an essentially virtual corporation—or a corporation that was at least as virtual as any contemporary e-commerce company. How they solved these challenges is illustrative, and suggests further questions to ask of the Cloud-as-factory hypothesis.[27]

One of the key problems facing all retailers is the problem of trust. Once the scale of the market economy has increased to the extent that consumers no longer have a direct connection to producers (that is to say, they are not personally familiar with the local butcher, baker, or candlestick maker), it can be difficult for them to evaluate the quality of goods that they are purchasing. In traditional retail, the problem of trust is in large part solved by the physical presence of a local intermediary. The buyer might not know the farmer who grew the corn that was turned into the flour that was baked into the bread that she bought at the grocery store, but at least she could see the product before she purchased it, had a long-term relationship with the grocer who was selling it, and had someone and somewhere to return the product if it turned out to be unsatisfactory. Convincing that same consumer to send her money in the mail to a retail agent she had never met located in a city she had never visited for a product she had never seen in person made the need for trust even more apparent.

There are many ways to solve the problem of trust. The establishment of brand identity—made possible in large part by the technology of advertising—was one way, as were responsive customer service departments. The latter solution not only generates much more data to be processed but also requires human intervention. In the early years of Sears customer service workers would not only have to enter customer correspondence data into a form that could be processed by the information systems that the company used to manage its internal databases but would also then copy their responses by hand as a means of establishing a more personal relationship to their otherwise unknown and invisible consumers. A century later Amazon would solve the same problem using human call center operators, many of them originally hired out of local Seattle-area coffee shops in order to provide a more recognizably "authentic" interaction.[28] Even in the era of online feedback and user ratings, the human element required to establish and maintain trust remains a necessary—and extremely expensive—component of even the most highly automated high-tech operations.[29] Amazon was notorious in its early years for the ruthless efficiency with which it ran its customer service operations. Using techniques developed for the assembly line of the early twentieth century (already, by the 1930s, the subject of scathing social critique by Charlie Chaplin in his film *Modern Times*), Amazon monitored, measured, and regimented every interaction and movement of its call center workers, from how long they spent with each customer to how many minutes they spent in the lavatory.[30]

Essential to the establishment of trust in mail-based (or, for that matter, online) retail is the ability to leverage the trustworthiness of other networks and institutions. What made the early mail-order companies viable was the emergence, in the middle of the nineteenth century, of an inexpensive, universal, and reliable postal network.[31] Both buyer and seller could be confident that any money or products that they sent through the mail would arrive on time and untampered with. If this trustworthy communications and transportation infrastructure had not yet been established, Sears would have had to construct it, which would have been cost-prohibitive. The same is true of Amazon, which relies heavily on the government-established (and publicly subsidized) United States Postal Service to provide timely, inexpensive, and ubiquitous delivery service.[32] And of course the postal network is itself dependent on other infrastructures (particularly transportation and communication networks) to maintain its own high standards of reliability.[33] Equally essential were trustworthy infrastructures for handling remote financial transactions, from telegraph-enabled electronic transfers to modern credit-card processors. In the low-margin world of mass-market retail, it is hard to imagine either Sears or Amazon being able to construct and maintain these critical infrastructures *ex nihilo*.[34] From advertising to finance to customer support to supplier relations, once you start unraveling the layers of material infrastructure that make supposedly "immaterial" information economy possible, it turns out to be turtles upon turtles, all the way down . . .

In addition to solving the problem of trust, Sears also had to solve the problem of data management. Although they would not have referred to their solution to the data management problem as a "computer" (though the term was already widely used by the early twentieth century), they did call it "data processing." And, in fact, the technology that today we refer to as a computer was originally described as a mechanism for performing "electronic data processing," a direct reference to the continuity between its intended function and the systems developed decades earlier at information organizations like Sears. As was mentioned earlier, by the first decade of the twentieth century, the Sears data-processing division processed hundreds of thousands of data-related operations every day. They accomplished this remarkable throughput by organizing into an efficient assembly line a hybrid system of information-processing technologies and human operators that can unambiguously be identified as an "information factory." It is with the establishment of such information factories that the information revolution of the twentieth century truly begins: without reference to such factories, the history of computing is incomplete and perhaps inexplicable.

For anyone familiar with the popular history of computing, the claim that there was computing before there were computers might seem ridiculous. Such histories are typically told in terms of a series of inventions (or innovations, as the most recent bestseller in this genre would describe it). The focus is generally on the development of the first electronic digital computers of the mid-twentieth century, although the authors of such histories will often allow for the inclusion of some earlier "proto-computer" curiosities. But the emphasis is always on inventions that most closely resemble the modern understanding of what constitutes a computer and on inventors who most conform to the popular narrative of the heroic "computer nerd turned accidental billionaire." Such stories are almost too good not to be true, and they provide clear and simple answers to the question about how the computer so quickly and profoundly has come to define our modern information society.

But a closer look at how pre-electronic computing but nevertheless information-centric organizations like Sears solve their data-processing problems provides a radically different interpretation of the history of computing that focuses less on specific technological innovations and more on larger social, political, and economic developments. In such explanations, terms like "industrial" define not a particular historical era or economic sector but rather an approach to the organization of work that emerges out of very specific historical context but would soon become (and remains to this day) the dominant method for approaching problems involving large-scale production or processing.

It is important to note that although we often think of the classic Industrial Revolution that reshaped Western society in the early modern period as being driven by mechanization (with the machines themselves being driven by new forms of power), in fact, industrialization is better understood as a combination of mechanization, organization, and labor. An industrial textile mill, for example, differs from its predecessor in terms of how machines are used (and not necessarily in terms of the presence or absence of machines), how those machines are organized, and who does the labor. The paradigmatic textile worker in Britain in the preindustrial period was a male artisan who worked with hand-powered and general-purpose machines to transform raw materials into finished products. The typical worker in an industrial textile factory was a woman who operated a highly specialized machine to perform *one* specific task within a rigidly organized division of labor. The new machines did not replace human workers; they created new forms of work that required (or at least enabled) the mobilization of new types and categories of labor. Whether it was the new machines that drove the search for new labor or the availability of new labor that encouraged the development of new machines is not relevant. The elements of the new industrial order were dependent on one another. That is what industrialization meant: the recombination of new machines, new organizational forms, and new forms of labor.[35]

For a variety of reasons, some economic, others social and political, industrialization emerged in the early seventeenth century as a compelling approach to large-scale production and manufacturing challenges. This included the production and manufacturing of data. For example, when the Emperor Napoleon charged the mathematician Gaspard de Prony with overhauling the tax system in France along scientific (and metric principles), de Prony adopted an industrial approach to solving the massive computational problem posed by the need to produce in a timely fashion entire volumes of new logarithm tables. At the time, the cutting edge of industrial practice involved the division of labor proposed by Adam Smith in his 1776 classic, *Wealth of Nations*. De Prony duly constructed a method for dividing up the cognitive work associated with computing logarithms, known as the difference method, and mobilized the labor of recently unemployed (and therefore inexpensive) hairdressers (whose aristocratic patrons had been lucky to escape the recent revolution with their heads intact, much less their fancy hairstyles). This was perhaps the first industrial-era information factory, but it was a harbinger of subsequent developments to come.[36]

Several decades after de Prony, the English mathematician and astronomer Charles Babbage, faced with a similar need to quickly and efficiently generate large numbers of mathematical tables, also turned to contemporary industrial manufacturing practices. After making an extended tour of European industrial centers, he published *On the Economy of Machinery and Manufactures*, the most comprehensive study of industrialization to date. By that point, the focus of industrial development had turned from the division of labor to water-driven mechanization. Babbage adapted de Prony's method of differences to this new industrial regime and in the latter half of the 1820s designed his Difference Engine, which was explicitly modeled after a contemporary industrial granary. Like the granary, it had a mill, a store, and a central shaft that could be driven by a waterwheel. The fact that it would mill mathematical tables instead of flour was irrelevant. The two problems were seen as essentially similar.

Babbage never got around to actually constructing his Difference Engine, nor its intended successor, the Analytical Engine. Because of its conceptual similarities to a modern computing device, the Analytical Engine is often identified as an important precursor to the modern computer revolution. This it almost certainly was not, but as a reflection of the interrelationship between industrialization and computation, it is highly significant. De Prony designed his information factory in the style of the early Industrial Revolution, Babbage according to the fashion of a later era. But they shared the impulse to industrialize, as would later innovators.

By the end of the nineteenth century, contemporary industrial manufacturing practices had begun to incorporate electricity. In 1888, the head of the United States Census Bureau, faced with the impossible task of enumerating a large and growing population using existing methods of data processing, held a competition to stimulate innovation in this area. The winner was a young engineer named Herman Hollerith, who created a new type of machine (the punch card tabulator), a new form of encoding information (the digital punch card), and a new system of organizing and automating these cards. As with most industrial systems, then and now, the work was not fully automated, and so Hollerith also created a novel form of clerical worker, the punch card tabulator operator. The company that Hollerith founded and the technology he created would, in the 1920s, form the basis of the International Business Machines Corporation. By the 1930s, IBM had already become a globally dominant information-technology company—several decades before it would produce anything remotely similar to a modern electronic digital computer. Once again, Hollerith innovated by industrializing information processing, inventing not only new machines but also new forms of labor and organization.

These are only three examples of the larger pattern that played out throughout the late nineteenth and early twentieth century, as the management of large amounts of information became a central feature of science, business, and government. In almost every case, the best model for understanding how to address such informational challenges is not the modern digital computer but the ongoing practice of adapting industrial methods and organizations to complex problems of almost every description. And for the most part, this process of industrialization involved a combination of mechanization, organization, and new forms of labor. As with industrialization more generally, very often these new forms of labor were women. The first factory workers in the United States were women, and so were the first information factory workers. Consider the typewriter, for example, which allowed for the mechanization of document production by combining technical innovation (the typewriter) with the division of labor (the separation of the cognitive labor of authorship from the routine clerical labor of transcription). As the work of the head (authorship) was divorced from the labor of the hand (typing), the job of clerk was fundamentally transformed, becoming at once low-skill, low-wage, and almost entirely feminized. As with many industrial processes, an increasing level of mechanization almost inevitably implied a corresponding reduction in skill, and workers with other options (which in this historical period generally meant men) would explore new opportunities. The typewriter was simultaneously a machine, a person, and a new job category.

It is in this period that we can identify the early origins of what would become the computer revolution. The industrial organization of informational work, when it was found in the corporation, was generally referred to as data processing. In science, it was called computing. And while it is true that the nature of the problems in these two domains differed in significant ways (data processing often involved the manipulation of words, and scientific computation focused mainly on numbers), the actual practices and techniques involved were generally quite similar. The informational task would be organized and divided in such a way as to allow large numbers of inexpensive (female) laborers to perform machine-assisted calculations or manipulations. These machines might be typewriters, punch card tabulators, adding machines, or calculators, depending on the context, but the basic approach was identical. By the early twentieth century, data-processing work had become almost entirely feminized, and the word "computer" was universally understood as referring

Figure 1.2 Sears, Roebuck Company data division, ca. 1908.

to a female mechanical calculator operator. The origins of the computer industry can only be understood in terms of the larger history of industrialization; otherwise, the large number of women workers and the particular organization of labor are inexplicable.

It was the industrialization of information processing in the late nineteenth century that allowed Sears to compete economically with traditional retail. Photographs from this period of the Sears data division reveal the obviously factory-like nature of the contemporary information enterprise: row upon row of identical (and interchangeable) female machine operators tending highly specialized technologies, the entire operation intended to standardize, routinize, and automate as much as possible tasks that had previously required time-consuming and expensive cognitive labor. To the degree that the Sears data division performed the same function for which Amazon today relies on the Cloud, this early version of the Cloud was clearly a factory (see fig. 1.2).

When in the 1930s the looming threat of war inspired the United States military to invest in the latest generation of industrial technology, namely electronics (not to be confused with the earlier use of electrification), they modeled the first generation of electronic "computers" after their human equivalents. John Mauchly, head of the ENIAC project at the University of Pennsylvania, quite explicitly described his project as an "automated form of hand computation."[37] It is no coincidence, therefore, that the first operators of these new machines—what today we would call programmers—were women recruited directly from the human computing department. The centrality of women in early computing was neither an accident nor a wartime exigency. The first electronic computers were electronic information factories, and the female computer programmers were their first factory workers. As I have written elsewhere, it would be several decades before the work of computer programming was made masculine and elevated to its current status as the epitome of (generally male) cognitive labor.[38]

All of this is to establish that it is impossible to understand the emergence of the modern information society without reference to the larger history of industrialization. Why is this significant? Because industrialization is fundamentally as much a social and political project as it is technological or economic. The ostensible driving force behind industrialization is the pursuit of efficiency, but the actual history of how, when, and why certain economic sectors chose to industrialize suggests otherwise. New techniques and technologies do not emerge out of nothing to revolutionize work practices; they are designed explicitly to do so. Machines are designed by humans to accomplish human agendas, and as such it is essential to always ask why industrialization is happening, to what ends, and for what purposes. This is particularly true in the history of computing. It is quite clear from the business literature of the 1950s what the new technology of electronic computing was intended to do. It was meant to do for white-collar labor what the assembly line had done for the automobile industry: namely, to transform a system in which skilled human labor was central into one in which low-wage machine operators could accomplish the same basic objective.[39]

And so let us return again to the central conceit of this historical thought experiment: what happens when we consider the Cloud as a factory, and not as a disembodied computational device?

1. We restore a sense of place to our understanding of the information economy. Despite repeated claims that "distance is dead," "the world is flat," and that geography (and therefore the nation-state) is irrelevant, cyberspace is surprisingly local.[40] Ironically, this is perhaps most true in Silicon Valley, the place that makes the technologies that ostensibly make location irrelevant, and yet where geographical proximity is so obviously essential that firms and individuals will go to great expense and inconvenience to live there. When Amazon recently encouraged cities to bid for the privilege of hosting their "second headquarters," they were clearly pushing for

those cities with well-established physical and social infrastructures: housing, high-ways, schools, restaurants, and recreational facilities. When Microsoft or Facebook looks to locate a new data center, they require easy access to inexpensive electricity, a plentiful water supply, and an appropriately skilled labor force.[41] It is any surprise that these data centers are often located in the same places that housed industrial-era factories just a generation ago?

2. Closely associated with the recognition of the significance of space and place is an appreciation of the importance of infrastructure. When it is made clear that despite the ethereal implications of its defining metaphor, the Cloud is actually a ravenous consumer of earth, air, fire, and water, the essential materiality of the vir-tual becomes undeniable. If within a few years of its invention, the Cloud is already the sixth largest consumer of electricity on the planet, what might we imagine about the implications for the future? In the face of climate change driven by human-kind's industrial activity, can we continue to ignore and externalize the environmen-tal costs of our online activities? Given the looming global shortage of clean, fresh water, ought we not reevaluate our allocation of this precious resource to a data stor-age facility? At the very least, no matter how much of our activities seem to relocate into cyberspace, we will need to continue to invest in and maintain our traditional physical infrastructure. It turns out the Cloud needs roads and bridges and sewer systems just as much as humans do.

3. It is also essential that we recognize the fundamental interconnectedness (and interdependencies) of all of our infrastructures, including our virtual infrastructures. One of the most currently overhyped technologies in the computer industry is the virtual and distributed trust infrastructure known as the blockchain. This technology is attracting a massive amount of attention (and a slightly less massive amount of investment capital), and its financial and technological viability is entirely depen-dent on the mistaken assumption that the computational resources provided by the Cloud are essentially free—or will eventually be free in some unspecified and indeterminate future. This ignores the fact that the only significant implementation of the blockchain, which is the virtual cryptocurrency Bitcoin, is deliberately and irredeemably energy-inefficient. By design it is an almost infinite sink for computer power and, by extension, coal, oil, water, and uranium.[42] Already the Bitcoin net-work, which does not and cannot provide even basic functional financial services, is one of the largest consumers of computer power on the planet, with an annual appetite for electricity approaching that of the entire nation of Denmark. There are multiple ways to implement blockchain technology, of which the proof-of-work

algorithm used by Bitcoin is by far the least desirable, at least from an environmental point of view. For anyone cognizant of the relationship between virtual and physical infrastructure, the fact that Bitcoin is not only not regulated but rather actively encouraged is astonishing.

4. From infrastructure our attention moves naturally to the supply chain. The computing devices that comprise the Cloud are truly global commodities, containing among other elements lithium from South America, tin from Indonesia, cobalt from the Democratic Republic of Congo, and a variety of rare earths whose supply is almost exclusively controlled by China. Each one of these resources and resource chains represents a set of stories to be told about global politics, international trade, worker safety, and environmental consequences. Cobalt is a conflict mineral; tin is deadly to humans and animals alike; China has already declared its monopoly over rare earths to be even more economically and geopolitically significant than that of the Middle East over oil. The need for companies like Tesla to secure access to South American supplies of lithium invokes the specter of a similar history of corporate meddling by the United Fruit Company or US Steel. But, in any case, following the supply chain that enables the Cloud as factory is a reminder that the digital economy is a global phenomenon, whether or not the actors involved in that economy are consciously aware of it. Seen from this perspective, lithium miners in Bolivia and e-waste recyclers in Ghana are as much a part of the digital economy as software developers in Silicon Valley.

5. Although we often associate factories with jobs, historically speaking human labor is only one component of industrialization. Some factories create work for humans; others eliminate it. Some machines enhance worker productivity, autonomy, and creativity, but this is the exception and not the rule. At the very least, industrialization changes work and the composition of the workforce. As we imagine the Cloud as a factory, we must ask what kind of factory it is intended to be. As in more traditional manufacturing, the Cloud as factory consumes local resources and pollutes the local environment. But compared to traditional manufacturing, does having such a factory in one's town provide compensatory benefits in terms of jobs, tax income, or the development of new infrastructure? In the industrial era, social and political mechanisms were developed for the negotiation between private and public interests. Do such mechanisms still apply to the information economy? Are they even available as a resource to governments and citizens? In addition to thinking about what might be gained by positioning the Cloud as a factory, we might consider what opportunities we have lost in not doing so.

6. In addition to thinking about the work that happens in and around the Cloud facility itself, we might also consider the changes to work that the Cloud enables in other industries. For example, automated vehicles are made possible in no small part by the computational activities that happen in the Cloud. In this sense, the Cloud is an element of the larger technological environment in which autonomous vehicles operate. Are they all part of the same factory? And if so, what does it mean for the trucking industry—and for the truck drivers whose jobs will soon be automated out of existence by this new technology? In thirty of the fifty United States, the single most common occupation for men is truck driver.[43] What are the social and economic ramifications of the industrialization and computerization of such an industry?

It is clear from the comparative histories of Sears and Amazon that despite the latter's high-tech veneer, the fundamental business model of the two firms is surprisingly similar. Does this make Sears an early predecessor of the information economy or Amazon a lingering relic of the industrial era, with its focus on the movement of materials and the construction and maintenance of physical capital? Is this even a useful question to ask, or is it an artifact of the artificial distinctions that are often drawn between the old and new economy? My argument has been that by focusing on the similarity between the two firms, and the continuity across different economic epochs, we can ask new and provocative questions about the history of modern computing, including questions of political economy, labor history, and the history of capitalism. Because it is clear that the Cloud is more than just a technical term or even a series of overlapping infrastructures. It is a metaphor, an ideology, and an agenda, which means that it is a tool for both understanding the past and present as well as for shaping the future. The Cloud is a factory. But a factory for what, and for whom, and for what purposes?

NOTES

1. Paul La Monica, "Tech's Top Five Now Worth More than $3 Trillion," *CNN Money* (October 31, 2017), https://money.cnn.com/2017/10/31/investing/apple-google-alphabet-microsoft-amazon -facebook-tech/index.html.

2. Anna Schaverien, "Five Reasons Why Amazon Is Moving into Bricks-and-Mortar Retail," *Forbes* (December 29, 2018), https://www.forbes.com/sites/annaschaverien/2018/12/29/amazon-online -offline-store-retail/#23b1d1f55128.

3. Ashley Carman, "Amazon Shipped over 5 Billion Items Worldwide through Prime in 2017," *The Verge* (January 2, 2018), https://www.theverge.com/2018/1/2/16841786/amazon-prime-2017 -users-ship-five-billion.

4. "Amazon Global Fulfillment Center Network," MWPVL International, accessed August 21, 2018, http://www.mwpvl.com/html/amazon_com.html.

5. Marcus Wohlsen, "Amazon Sets Up (Really Big) Shop to Get You Your Stuff Faster," *Wired* (January 23, 2013), https://www.wired.com/2013/01/amazon-distribution-centers/.

6. Michael Lierow, Sebastian Jannsen, and Joris D'Inca, "Amazon Is Using Logistics to Lead a Retail Revolution," *Forbes* (February 21, 2016), https://www.forbes.com/sites/oliverwyman/2016/02/18/amazon-is-using-logistics-to-lead-a-retail-revolution/.

7. Brian McNicoll, "For Every Amazon Package It Delivers, the Postal Service Loses $1.46," *Washington Examiner* (September 1, 2017), https://www.washingtonexaminer.com/for-every-amazon-package-it-delivers-the-postal-service-loses-146.

8. Rachel Nielsen, "Amazon Begins to Hand Off Last-Mile Delivery Service to Contract Drivers," *Puget Sound Business Journal* (February 18, 2016), https://www.bizjournals.com/seattle/blog/techflash/2016/02/amazon-begins-to-hand-off-last-mile-delivery.html; Laura Cox, "Bye-Bye FedEx, DHL, UPS . . . Here Comes Amazon Logistics," Disruption Hub (January 11, 2017), https://disruptionhub.com/amazons-next-target-logistics/.

9. Jeff Desjardins, "Amazon and UPS Are Betting Big on Drone Delivery," *Business Insider* (March 11, 2018), https://www.businessinsider.com/amazon-and-ups-are-betting-big-on-drone-delivery-2018-3.

10. In a possible postscript to our yet-unwritten chapter on Amazon, accusations of monopolistic practices might also serve as a connection between these two corporations.

11. Timothy Wu, *The Curse of Bigness: Antitrust in the New Gilded Age* (New York: Columbia Global Reports, 2018); Susan Crawford, *Captive Audience: The Telecom Industry and Monopoly Power in the New Gilded Age* (New Haven: Yale University Press, 2013).

12. Nicholas Negroponte, *Being Digital* (New York: Knopf Doubleday, 1996); Kevin Kelly, *New Rules for the New Economy: 10 Radical Strategies for a Connected World* (New York: Penguin Books, 1999).

13. Joshua Ganz, "Inside the Black Box: A Look at the Container," *Prometheus* 13, no. 2 (1995), 169–183; Alexander Klose, *The Container Principle: How a Box Changes the Way We Think* (Cambridge, MA: MIT Press, 2015).

14. Jordan Novet, "Amazon Cloud Revenue Jumps 45 Percent in Fourth Quarter," *CNBC* (February 1, 2018), https://www.cnbc.com/2018/02/01/aws-earnings-q4-2017.html.

15. John Durham Peters, "Cloud," in *Digital Keywords: A Vocabulary of Information Society and Culture*, ed. Ben Peters (Princeton: Princeton University Press, 2016).

16. Susan Leigh Star, "The Ethnography of Infrastructure," *American Behavioral Scientist* 43, no. 3 (1999), 377–391.

17. Yevgeniy Sverdlik, "Here's How Much Energy All US Data Centers Consume," *Data Center Knowledge* (June 27, 2016), accessed August 21, 2018, https://www.datacenterknowledge.com/archives/2016/06/27/heres-how-much-energy-all-us-data-centers-consume; Christopher Helman, "Berkeley Lab: It Takes 70 Billion Kilowatt Hours a Year to Run the Internet," *Forbes* (June 28, 2016), https://www.forbes.com/sites/christopherhelman/2016/06/28/how-much-electricity-does-it-take-to-run-the-internet.

18. Gary Cook et al., *Clicking Clean: How Companies Are Creating the Green Internet* (Washington, DC: Greenpeace, 2014).

19. Bryan Walsh, "Your Data Is Dirty: The Carbon Price of Cloud Computing," *Time* (April 2, 2014), http://time.com/46777/your-data-is-dirty-the-carbon-price-of-cloud-computing/.

20. Mél Hogan, "Data Flows and Water Woes: The Utah Data Center," *Big Data & Society* 2, no. 2 (December 27, 2015), 1–12; Mél Hogan, *The Big Thirst: The Secret Life and Turbulent Future of Water* (Old Saybrook, CT: Tantor Media, 2011).

21. Roughly 1,500 kilograms per computer. See Ruediger Kuehr and Eric Williams, *Computers and the Environment: Understanding and Managing Their Impacts* (New York: Springer, 2012); and Annie Callaway, "Demand the Supply: Ranking Consumer Electronics and Jewelry Retail Companies on Their Efforts to Develop Conflict-Free Minerals Supply Chains from Congo," Enough Project (November 16, 2017), accessed August 7, 2018: https://enoughproject.org/reports/demand-the-supply.

22. Joy Lisi Rankin, *A People's History of Computing in the United States* (Cambridge, MA: Harvard University Press, 2018).

23. Nathan Ensmenger, "The Environmental History of Computing," *Technology and Culture* 59, no. 4S (2018), S7–33.

24. Alan Pyke, "How Amazon Built Its Empire on One Tax Loophole," *Think Progress* (April 26, 2014), https://thinkprogress.org/how-amazon-built-its-empire-on-one-tax-loophole-49732e358856/.

25. See Kavita Philip, "The Internet Will Be Decolonized," and Paul N. Edwards, "Platforms Are Infrastructures on Fire," both in this volume.

26. Hubert Horan, "Uber's Path of Destruction," *American Affairs* III, no. 2 (Summer 2019).

27. Daniel Raff and Peter Temin, "Sears, Roebuck in the Twentieth Century: Competition, Complementarities, and the Problem of Wasting Assets," in *Learning by Doing in Markets, Firms, and Countries*, ed. Naomi R. Lamoreaux, Peter Temin, and Daniel Raff (Chicago: University of Chicago Press, 1998).

28. Richard Howard, "How I Escaped from Amazon.cult," *Seattle Weekly* (October 9, 2006), https://www.seattleweekly.com/news/how-i-escaped-from-amazon-cult/.

29. Mary Gray and Siddharth Suri, *Ghost Work: How to Stop Silicon Valley from Building a New Global Underclass* (Boston: Houghton Mifflin Harcourt, 2019); Sarah Roberts, *Behind the Screen: Content Moderation in the Shadows of Social Media* (New Haven: Yale University Press, 2019).

30. Simon Head, "Worse than Wal-Mart: Amazon's Sick Brutality and Secret History of Ruthlessly Intimidating Workers," *Salon* (February 23, 2014), https://www.salon.com/control/2014/02/23/worse_than_wal_mart_amazons_sick_brutality_and_secret_history_of_ruthlessly_intimidating_workers/; Jody Kantor and David Streitfield, "Inside Amazon: Wrestling Big Ideas in a Bruising Workplace," *New York Times* (August 15, 2015), https://www.nytimes.com/2015/08/16/technology/inside-amazon-wrestling-big-ideas-in-a-bruising-workplace.html.

31. Richard R. John, "Recasting the Information Infrastructure for the Industrial Age," in *A Nation Transformed By Information: How Information Has Shaped the United States from Colonial Times to the Present*, ed. James Cortada and Alfred Chandler (New York: Oxford University Press, 2000), 55–105; David Henkin, *The Postal Age: The Emergence of Modern Communications in Nineteenth-Century America* (Chicago: University of Chicago Press, 2006).

32. Jake Bittle, "Postal-Service Workers Are Shouldering the Burden for Amazon," *The Nation* (February 21, 2018) https://www.thenation.com/article/postal-service-workers-are-shouldering-the-burden-for-amazon/; Brendan O'Connor, "Confessions of a U.S. Postal Worker: 'We deliver Amazon packages until we drop dead,'" *GEN* (October 31, 2018), https://gen.medium.com/confessions-of-a-u-s-postal-worker-we-deliver-amazon-packages-until-we-drop-dead-a6e96f125126.

33. Richard R. John, *Network Nation: Inventing American Telecommunications* (Cambridge, MA: Belknap Press of Harvard University Press, 2010); Shane Greenstein, *How the Internet Became Commercial: Innovation, Privatization, and the Birth of a New Network* (Princeton, NJ: Princeton University Press, 2015).

34. Richard White, *The Republic for Which It Stands: The United States during Reconstruction and the Gilded Age, 1865–1896* (New York: Oxford University Press, 2017).

35. Brooke Hindle and Steven Lubar, *Engines of Change: The American Industrial Revolution, 1790–1860* (Washington, DC: Smithsonian Institution Press, 1986).

36. Martin Campbell-Kelly, William Aspray, Nathan Ensmenger, and Jeffrey R. Yost, *Computer: A History of the Information Machine* (Boulder, CO: Westview Press, 2014).

37. David Allan Grier, "The ENIAC, the Verb to Program and the Emergence of Digital Computers," *Annals of the History of Computing* 18, no. 1 (1996), 51–55.

38. Nathan Ensmenger, "Making Programming Masculine," in *Gender Codes*, ed. Thomas Misa (Hoboken, NJ: John Wiley, 2010), 115–141; Mar Hicks, *Programmed Inequality: How Britain Discarded Women Technologists and Lost Its Edge in Computing* (Cambridge, MA: MIT Press, 2016).

39. Nathan Ensmenger, *The Computer Boys Take Over: Computers, Programmers, and the Politics of Technical Expertise* (Cambridge, MA: MIT Press, 2010).

40. Frances Cairncross, *The Death of Distance: How the Communications Revolution Is Changing Our Lives* (Boston: Harvard Business Review Press, 2001); Thomas L. Friedman, *The World Is Flat: A Brief History of the Twenty-First Century* (New York: Farrar, Straus and Giroux, 2006); Ray Kurzweil, *The Singularity Is Near: When Humans Transcend Biology* (New York: Viking, 2016); John Mark Newman, "The Myth of Free," *George Washington Law Review* 86, no. 2 (October 2016), 513–586.

41. James Glanz, "The Cloud Factories: Data Barns in a Farm Town, Gobbling Power and Flexing Muscle," *New York Times* (September 2012).

42. Karl O'Dwyer and David Malone, "Bitcoin Mining and Its Energy Footprint," *Irish Signals & Systems Conference* (2014), 280–285.

43. "Map: The Most Common Job in Every State," NPR.org, February 5, 2015, https://www.npr.org/sections/money/2015/02/05/382664837/map-the-most-common-job-in-every-state; Jennifer Cheeseman Day and Andrew W. Hait, "America Keeps on Trucking: Number of Truckers at All-Time High," US Census Bureau, June 6, 2019, https://www.census.gov/library/stories/2019/06/america-keeps-on-trucking.html.

2

YOUR AI IS A HUMAN

Sarah T. Roberts

INTELLIGENCE: WHAT IT IS AND HOW TO AUTOMATE IT

This chapter starts out with a polemic in a volume devoted to them (and we are therefore in good company):

Your artificial intelligence on your social media site of choice—what you may have imagined or even described as being "the algorithm," if you have gotten that far in your thinking about the inner machinations of your preferred platform—is a human.

I say this to provoke debate and to speak metaphorically to a certain extent, but I also mean this literally: in the cases that I will describe herein, the tools and processes that you may believe to be computational, automated, and mechanized are, in actuality, the product of human intervention, action, and decision-making.

Before we go further, your belief that AI, automation, autonomous computation, and algorithms are everywhere is not because you are a naïf or a technological ingenue. It is because you have been led to believe that this is the case, tacitly and overtly, for many years. This is because to fully automate, mimic, outsource, and replace human intelligence at scale is, for many firms, a fundamental aspiration and a goal they have pursued, publicly and otherwise, but have yet to—and may never—achieve. And it *is* true that algorithmically informed automated processes are present in many aspects of the social media user experience, from serving you up the next video to be viewed, to curating (or restricting) the advertising directed at you, to offering you a constant stream of potential friends to . . . *friend*.

Just what constitutes AI is slippery and difficult to pin down; the definition tends to be a circular one that repeats or invokes, first and foremost, human intelligence—something that itself has eluded simple definition. In a 1969 paper by John McCarthy and Patrick Hayes considered foundational to the field of artificial intelligence, the authors describe the conundrum by saying:

> work on artificial intelligence, especially general intelligence, will be improved by a clearer idea of what intelligence is. One way is to give a purely behavioural or black-box definition. In this case we have to say that a machine is intelligent if it solves certain classes of problems requiring intelligence in humans, or survives in an intellectually demanding environment. This definition seems vague; perhaps it can be made somewhat more precise without departing from behavioural terms, but we shall not try to do so.[1]

Forty years on, and computer scientists, engineers, and, in the quotation that follows, senior executives at esteemed tech publisher and Silicon Valley brain trust O'Reilly Media are still struggling to adequately define AI without backtracking into tautology and the thorny philosophical problem of what constitutes nonartificial, or human, intelligence in the first place: "Defining artificial intelligence isn't just difficult; it's impossible," they say, "not the least because we don't really understand human intelligence. Paradoxically, advances in AI will help more to define what human intelligence isn't than what artificial intelligence is."[2]

Perhaps, therefore, more easily apprehended than a definitive description of what constitutes AI are the reasons behind its seemingly universal appeal and constant invocation among tech and social media firms. This is particularly the case for those firms that desire to develop or use it to replace human intelligence, and human beings, with something that can serve as a reasonable stand-in for it in a routinized, mechanical, and ostensibly highly controllable and predictable way.

Some of these reasons are practical and pragmatic—business decisions, if you will. But there is also an underlying presupposition almost always at play that suggests, tacitly and otherwise, that the dehumanized and anonymous decision-making done by computers in a way that mimics—*but replaces*—that of human actors is somehow more just or fair.

Yet if we can all agree that humans are fallible, why is artificial intelligence based on human input, values, and judgments, then applied at scale and with little to no means of accountability, a better option? In many cases, we can find the justification for this preference for replacing human decision-making and, therefore, the human employees who perform such labor, by properly identifying it as an ideological predisposition fundamental to the very orientation to the world of the firms behind the major social media platforms.

The goal of this ideological orientation is a so-called "lean" workforce both to match the expectations of venture capital (VC) investors and shareholders to keep labor costs low and to align with the fundamental beliefs to which tech firms are adherents: that there is greater efficiency, capacity for scale, and cost savings associated with fewer human employees. And the means to this end is, conveniently, the firms' stock in trade: technological innovation and automation using computational power and AI, itself a compelling reinforcement of their *raison d'être*. The myriad benefits of such workforce reduction, from a firm's or investor's perspective, are laid bare in this 2016 piece from *Forbes* magazine, mouthpiece for the investor class:

> The benefits many employees enjoy today, like health insurance, unemployment insurance, pensions, guaranteed wage increases, paid overtime, vacation and parental leave have increased the cost of labor significantly. With new work models emerging, these benefits create an incentive for companies to minimize their headcount.

> Thanks to technological advancements, it is much easier and efficient today to manage distributed teams, remote workers and contractors. What is the point in housing thousands of workers, often in some of the most expensive areas of our cities, when communication is mostly done through email, phone or Slack anyways?[3]

But the cost savings don't stop simply at the reduction in need for office space and health insurance. A cynic such as myself might also note that having fewer direct employees might be seen as a means to control the likelihood of labor organizing or collective action, or at least greatly reduce its impact by virtue of smaller numbers of direct employees and less chance of them encountering each other at a shared work site.

Even when humanlike decision-making can be successfully programmed and turned over to machines, humans are still in the picture. This is because behind a decision, a process, or an action on your platform of choice lies the brainpower of unnamed humans who informed the algorithm, machine learning, or other AI process in the first place. Taking a larger step back, the problems that firms decide to tackle (and those they do not) in this way are also a site of the demonstration of very human values and priorities. In other words, what kinds of problems can, should, and will be turned over to and tackled by automated decision-making is a fundamentally value-laden human decision.

In this way, it becomes difficult to make claims for purely machine-driven processes at all. It may be that such computational purity of perfected AI at a scale not only cannot exist but ought not to be taken for granted as the panacea it is often touted as being. Why is this the case? Because reality calls for a much more complex sociotechnological assemblage of human and machine, procedure and process, decision-making and policy-making that goes well beyond the imagined closed system flow-chart logic solution of algorithms and computational decision-making and interventions operating at scale.[4]

THE CASE OF COMMERCIAL CONTENT MODERATION AND THE HUMANS BEHIND IT

In 2018 Facebook found itself having a tough year. A sharp decline in stock value in late July, resulting in the largest one-day value loss—a jaw-dropping $119 billion—to date in the history of Wall Street, left analysts and industry watchers searching for answers as to what might be happening at the firm.[5] While some contended that market saturation in North America and Europe were behind the poor second-quarter performance leading to the sell-off, there were likely deeper problems afoot that pushed investor lack of confidence, including major scandals related to election influence, fake political accounts and Astroturf campaigns, and a seeming inability for the platform to gain control of the user-generated content (UGC) disseminated via its properties, some of it with deadly consequences.[6]

It is worth noting that Facebook was not alone in seeing loss of value due to flagging consumer confidence; the much more volatile Twitter (having its own moment, of sorts, due to being the favorite bully pulpit of President Donald Trump) has also seen a bumpy ride in its share price as it has worked to purge fake and bot accounts many suspect of unduly influencing electoral politics, thereby vastly reducing its user numbers.[7] More recently, YouTube has (finally?) received increased ire due to its recommendation algorithms, favoritism toward high-visibility accounts, and bizarre content involving children.[8]

Throughout the many episodes of criticism that major, mainstream social media tech firms have faced in recent years, there has been one interesting, if not always noticed, constant: a twofold response of firms—from Alphabet (Google) to Facebook—to their gaffes, scandals, and disasters has been to invoke new AI-based tools at the same time they commit to increasing the number of humans who serve as the primary gatekeepers to guard against bad user behavior in the form of UGC-based abuse. In some cases, these increases have been exponential, with particular firms committing to doubling the total number of human employees evaluating content in various forms on their platforms while at the same time assuring the public that

Figure 2.1 YouTube to World: YouTube's early slogan, "Broadcast Yourself!," suggested an unencumbered, uninterrupted relationship of user to platform to world. The reality has proven to be more complex, in the form of automation, regulation, and humans.

soon computational tools using AI will be able to bridge any remaining gaps between what humans are capable of and what went undone.

Perhaps what is most interesting about these developments is that they serve as an acknowledgment of sorts on the part of the firms that, until this recent period, was often absent: the major social media platforms are incapable of governing their properties with computational, AI-based means alone. Facebook and its ilk operate on a user imagination of providing a forum for unfettered self-expression, an extension of participatory democracy that provides users a mechanism to move seamlessly from self to platform to world. As YouTube once famously exclaimed, "Broadcast yourself!" and the world has responded at unprecedented scale and scope (see fig. 2.1).

That scope and scale has been, until recently, the premise and the promise of the social media platforms that have connected us to friends, family, high school classmates, exes, work colleagues, former work colleagues, and people we met once, friended, and forgot about yet remain connected to, all of whose self-expression has been made available to us, seemingly without intervention.

Yet the case is not that simple for user-generated content (UGC), the primary currency of mainstream social media platforms and the material upon which they rely for our continued engagement—and perhaps it never was. As a researcher, I have been attempting to locate the contours and parameters of human intervention and gatekeeping of UGC since 2010, after reading a small but incredibly important article in the *New York Times* that unmasked a call center in rural Iowa specializing in the practice.[9] Immediately upon reading the article, I—then seventeen years into being a daily social internet user, computing professional, and current grad student in those areas—asked myself a two-part question: "Why have I never thought of the cease-less uploading of user content on a 24/7/365 basis from anyone in the world *as a major problem for these platforms* before?" followed quickly by, "Don't computers deal with that?"

What I discovered, through subsequent research of the state of computing (including various aspects of AI, such as machine vision and algorithmic intervention), the state of the social media industry, and the state of its outsourced, globalized low-wage and low-status commercial content moderation (CCM) workforce[10] was that, indeed, computers did *not* do that work of evaluation, adjudication, or gatekeeping of online UGC—at least hardly by themselves. In fact, it has been largely down to humans to undertake this critical role of brand protection and legal compliance on the part of social media firms, often as contractors or other kinds of third-party employees (think "lean workforce" here again).

While I have taken up the issue of what constitutes this commercial, industrialized practice of online content moderation as a central point of my research agenda over the past decade and in many contexts,[11] what remained a possibly equally fascinating and powerful open question around commercial content moderation was why I had jumped to the conclusion in the first place that AI was behind any content evaluation that may have been going on.

And I wasn't the only one who jumped to that conclusion. In my early days of investigation, I did a great deal of informal polling of classmates, colleagues, professors, and researchers about the need for such work and how, to their best guess, it might be undertaken.

In each case, without fail, the people that I asked—hardly technological neophytes—responded with the same one-two reaction I had: first, that they had never thought about it in any kind of significant way, and second, that, in any case, computers likely handled such work. I was not alone in my computational overconfidence, and the willingness and ease with which both my colleagues and I jumped to the conclusion that any intermediary activity going on between our own UGC uploads and their subsequent broadcasting to everyone else was fueled by AI engines was a fascinating insight in its own right. In other words, we wanted to believe.

To be fair, since I made those first inquiries (and certainly even more so in recent years), much has changed, and AI tools used in commercial content moderation have grown exponentially in their sophistication, hastened by concomitant exponential and continuous growth in computational performance, power, and storage (this latter observation a rough paraphrasing of Moore's Law[12]). Furthermore, a variety of mechanisms that could be considered to be blushes of AI have long been in use alongside human moderation, likely predating that practice's industrialization at scale. These include but are not limited to:

- A variety of what I would describe as "first-order," more rudimentary, blunt tools that are long-standing and widely adopted, such as keyword ban lists for content and user profiles, URL and content filtering, IP blocking, and other user-identifying mechanisms;[13]
- More sophisticated automated tools such as hashing technologies used in products like PhotoDNA (used to automate the identification and removal of child sexual exploitation content; other engines based on this same technology do the same with regard to terroristic material, the definitions of which are the province of the system's owners);[14]
- Higher-order AI tools and strategies for content moderation and management at scale, examples of which might include:
  - Sentiment analysis and forecasting tools based on natural language processing that can identify when a comment thread has gone bad or, even more impressive, when it is in danger of doing so;[15]
  - AI speech-recognition technology that provides automatic, automated captioning of video content;[16]
  - Pixel analysis (to identify, for example, when an image or a video likely contains nudity);[17]
  - Machine learning and computer vision-based tools deployed toward a variety of other predictive outcomes (such as judging potential for virality or recognizing and predicting potentially inappropriate content).[18]

Computer vision was in its infancy when I began my research on commercial content moderation. When I queried a computer scientist who was a researcher in a major R&D site at a prominent university about the state of that art some years ago and how it might be applied to deal with user-generated social media content at scale, he gestured at a static piece of furniture sitting inside a dark visualization chamber and remarked, "Right now, we're working on making the computer know that that table is a table."

To be sure, the research in this area has advanced eons beyond where it was in 2010 (when the conversation took place), but the fundamental problem of how to make a computer "know" anything (or whether this is possible at all) remains. What has improved is the ability of computers, through algorithms and AI, to make good guesses through pattern recognition and learning from a database of material. Yet when you catch a computer scientist or a member of industry in an unguarded moment, you may learn that:

- All the tools and techniques described above are subject to failure due to either being overbroad, blunt, or unable to reliably identify nuanced or ambiguous material;
- They are also subject to and largely a representation of the quality and reliability of the source material used to train the algorithms and AI, and the inputs that have informed the tools (that is, the work of humans on training sets). The old computer science adage of "garbage in, garbage out," or GIGO, applies here—and we are once again back to pulling back a curtain to find, at key points of origin, decidedly human perspectives in every aspect of the creation of these tools, whether speccing their functionality or actually training them, via data sets, to replicate decisions that individuals have made and that are then aggregated;[19]
- The likelihood of these tools and processes ever being so sophisticated as to fully take over without any human input or oversight is close to zero. I have never heard anyone claim otherwise, and I stand by this claim, based on almost a decade of asking about the possibility and evaluating the state of a variety of AI applications for CCM work—from industry members as well as academic computer scientists and engineers.

There are a number of factors that play into the last statement, but two of the most overarching and most intractable, in terms of all automation all the time, are scale, on the one hand, and liability, on the other. These two factors are intertwined and interrelated, and worthy of their own discussion alongside a third factor: regulation. All of them go to the true heart of the matter, which is to question what social media platforms are, what they do, and who gets to decide. These are matters that I take up in the next section.

SOCIAL MEDIA'S IDENTITY CRISIS

Now able to count a vast portion of the global populace among the ranks of their user base (a scale unprecedented for most other media, past or present), the mainstream social media firms emanating from Silicon Valley face an identity crisis that is not entirely self-driven. Rather, questions from *outside* the firms have provoked a reevaluation of how they are regarded. This reevaluation is far from trivial and has social, technological, and legal impacts on what the firms are able to do, what claims they are able to make, and to whom they are beholden.

To a large extent, they have been able to self-define, over and over again in the past decade and a half (see Nathan Ensmenger's extensive treatment, in this volume, of Amazon's positioning of itself as an excellent case in point). Typically, when allowed to do so (or when queried), they invariably self-describe as "tech firms" first and foremost. Perhaps this is where free speech is best implemented by the platforms, in their own self-identification practices. What they are sure to claim, in almost every case, is that they are *not* media companies. In fact, on the very rare occasions in which Facebook founder and CEO Mark Zuckerberg has said that Facebook might bear some similarity to a media company,[20] the utterance has been so unusual as to be newsworthy. At other points, Zuckerberg has referred to Facebook as "the service," harkening back to the era of internet service providers, or ISPs, which mostly predate Facebook itself.

Being "not a media company" goes beyond social media firms' wish to distance themselves from dinosaur old-media antecedents (see fig. 2.2). Indeed, there are a number of significant and very pragmatic reasons that Silicon Valley–based social media firms do *not* want to be seen as media companies, at least not in the traditional broadcast media sense. There are two key regulatory and legal reasons for this. The first is the ability of these self-identified not-media companies to avoid regulation and limits on, and responsibility for, content placed on broadcast television or radio companies by the US Federal Communications Commission.[21] The late comedian George Carlin made an entire career of lampooning, and loudly using in his standup act, the words banned from the television airwaves by the FCC.

Social media platforms are greater than the sum of their technological parts. They are more than just their tool set, more than just their content, and more than just their engineering infrastructure and code. They are a dynamic assemblage of functionality, relationship creation and maintenance, policies, and governance that are always in flux and seldom visible in their totality to the user. They aggregate data, track behavior, determine, and even influence tastes and habits, capture attention, and package all of these things as commodities for advertisers. Zuckerberg may be right that this is not a "traditional" media company in the vein of television or film

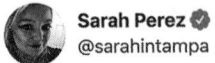

Sarah Perez ✔
@sarahintampa

'We're not a media company!' —> celebs holding
Facebook mics

3:48 PM · Jan 7, 2018 · Twitter for iPhone

391 Retweets   **1.1K** Likes

Figure 2.2 "Not a media company": An identity crisis is more than just what's in a name or how platforms self-define in the market, as this Twitter user lampoons (https://t.co/KkpMWKMgKl).

studios, for example, in that they rely largely on the production of material from average people—but even that isn't the whole story. That is where the second key reason for steadfast avoidance of the "media" label comes in.

Unlike television and radio broadcasters in the United States, who can be held liable and fined for the material they allow to pass over their channels, social media firms have been traditionally afforded a very specific legal status that both largely grants them immunity in terms of responsibility for the content they host and disseminate and also affords them the discretion to intervene upon that content as they see fit.[22] This particular legal status of "internet intermediary" dates to the Communications Decency Act of 1996 and to its Section 230, in particular.[23] Specifically, this portion of the CDA has set the tone for how internet intermediaries such as ISPs, search engines, and now social media platforms have been defined in the law for liability purposes. The Electronic Frontier Foundation, a major supporter of Section 230, describes its history as such:

Worried about the future of free speech online and responding directly to *Stratton Oakmont*, Representatives Chris Cox (R-CA) and Ron Wyden (D-OR) introduced an amendment to the Communications Decency Act that would end up becoming Section 230. The amendment specifically made sure that "providers of an interactive computer service" would not be treated as publishers of third-party content. Unlike publications like newspapers that are accountable for the content they print, online services would be relieved of this liability. Section 230 had two purposes: the first was to "encourage the unfettered and unregulated development of free speech on the Internet," as one judge put it; the other was to allow online services to implement their own standards for policing content and provide for child safety.[24]

Yet not all agree that Section 230 has been a harmless tool for self-expression. As legal scholar Mary Graw Leary describes, just who receives the greatest benefit from Section 230 may depend on whom you ask or who you are. She contextualizes Section 230 as having primarily benefited the tech companies themselves, explaining, "In Reno v. ACLU the Supreme Court struck down as vague some of the more controversial criminal provisions of the CDA, such as the prohibition on the transmission of 'indecent material.' However, §230 was not challenged, and this protection remains effective law to this day."[25]

Indeed, this outcome has largely been seen as an unqualified win for contemporary tech companies and, assuredly, those to come in the future—in fact, Section 230 may well have been key to their founding and growth. Says Leary, "Tech companies arguably achieved the best of both worlds. After Reno, much of the CDA that tech companies opposed was eliminated, but the provision that was designed to protect them remained. Thus, when the dust settled, tech companies enjoyed increased protections without the regulations."[26] The resulting power over speech resting with the platforms and their parent firms led law professor Kate Klonick to describe them as nothing short of "the new governors."[27]

Despite being legally immune for what they transmit over their platforms, most prominent mainstream firms have expressed, at various times, a moral obligation in many cases to protect people from harmful content, even beyond what US law may have mandated. But there is another compulsion that drives the need for gatekeeping that is less tangible and also difficult to separate, and it is related to brand protection practices, curation, and the platform's desire to create and maintain its user base—by providing that group with whatever content it is that keeps them engaged and coming back. The firms therefore rely on a never-ending, massive influx of UGC to keep existing users engaged and to draw new ones to the platforms, with the details of what needs to be removed sitting in second place after the race to always have new content to supply. The adjudication and removal falls in part to users, who report troubling material when they encounter it,[28] and certainly to CCM workers, with or without their AI tools to assist.

That social media platforms function legally more like libraries than newspapers, as an oft-repeated legal analogy goes, has assuredly given rise to the user base's comfort to upload and transmit its personal content, and has in turn created and supported notions of "free speech" that are synonymous with these acts. In this regard, Section 230 has done an excellent job of ensuring that one particular kind of speech, in the form of social media UGC on American-based platforms, flourishes. This is not incidental to the firms themselves, who have taken up that UGC as their stock in trade, monetized, and extracted value from it in many different ways. Without a user sensibility that an uploaded bit of UGC constitutes something other than one's own free self-expression first and foremost, it is unlikely that this model would have proliferated to the lucrative extent that it has.

The global business model of many of these major firms is now causing friction: as many social media giants are discovering, "free speech" in the form of users' social media sharing serves as a compelling enticement to current and potential users but is largely a principle that cannot be successfully operationalized on a global scale, considering that Section 230 and the CDA in its entirety is technically only pertinent to US jurisdiction (a fact divorced from the global reality of the internet's materiality, user base, and politics, as vividly described by Kavita Philip later in this book). Perhaps even more problematic, however, is that "free speech" is not a universally held principle, nor is it interpreted in the same way around the world.

Where nation-state gatekeepers demand compliance with restrictive speech laws to enter a marketplace, many firms quickly and coldly capitulate; one 2018 report describes where Facebook censors the most "illegal content," and Turkey heads the list—because Turkey has some of the most expansive, draconian laws on what constitutes "illegal content" in the first place.[29] Whereas Section 230 had long been the principle under which Silicon Valley–based social media firms approached their content obligations to the world, this privilege is no longer afforded to them in a variety of contexts. For social media firms wishing to enter a given marketplace, demonstrating a willingness and ability to respond to local norms, mores, and legal obligations when deciding what is allowed to stand and what is removed has become important, and not doing so can come at a significant financial cost. More and more, state actors are asserting their expectation that their local jurisdictional authority be recognized, and they are willing to impose fines, restrict business, and even block an entire platform or service as they see fit.

Whether the purpose is to restrict the speech of political dissidents in Turkey or to stop the flow of Nazi glorification in Germany,[30] these local laws are straining the firms' capacity to continue the status quo: to operate as they wish, aligned with their ideologies and with those enshrined in Section 230. Otherwise deeply regulation-avoidant, social media firms have developed intensely complex frameworks of internal policies (Klonick's "new governance"[31]) and the processes, technological mechanisms, and labor pools to enact them—an assemblage that, critically, has operationalized ideologies, large and small. These ideologies are identifiable on a Silicon Valley–wide scale but also reflect the particular flavor of a given platform and its permissiveness, restrictiveness, and branding relative to its peers.

The resulting enmeshments of policy, tech, and operations have largely grown up over time and internally to the firms, which have simultaneously worked to render outward-facing traces of such human values and interventions invisible on its social media platforms and sites. I have described this phenomenon elsewhere as a "logic of opacity":

> Obfuscation and secrecy work together to form an operating logic of opacity around social media moderation. It is a logic that has left users typically giving little thought to the mechanisms that may or may not be in place to govern what appears, and what is removed, on a social media site. The lack of clarity around policies, procedures and the values that inform them can lead users to have wildly different interpretations of the user experience on the same site.[32]

The invocation of AI as a means to resolve legal and regulatory demands is therefore an additional mechanism that firms use to promise compliance with these new standards while creating another layer of obfuscation, confusion, and technological complication that leaves users filling in the blanks, if they know they're there at all.[33] Communication scholar Sarah Myers West demonstrated this in her recent study on internet takedowns and how users who experience them tend to develop "folk theories" surrounding both the reasons for and the mechanisms of their content's deletion.[34]

If AI tools, which I hope by now to have convinced you are best thought of as some form of human-computer hybrid, continue to be invoked as the primary solution to social media's increasing content problems, then we will see much more reference to and reliance upon them going forward. Yet the black box nature surrounding

Figure 2.3 Fake AI tweet: As reported in the *Guardian* (Olivia Solon, "The Rise of 'Pseudo-AI'"), a Twitter user's clever take on how to start an AI-reliant firm.

the creation of tools themselves[35]—how they are built and what they do, and where human engagement with them begins and ends—will assuredly foster the kinds of folk theories unveiled by Myers West that will deliver us this *artificial* artificial intelligence while simultaneously denying any human behind the curtain, pulling levers and appealing to our collective assumptions that tech firms tech, never fake it, and that we are all the better for it.[36] Ironically, it may be that the proliferation of such computational interventions will be the very thing that leads to generalized, unquestioning acceptance of their encroachment into our collective experience of digital technology, a normalization process bolstering and bolstered by a fantasy sold to us by an industry that fundamentally benefits from it (see fig. 2.3).

HUMANS AND AI IN SYMBIOSIS AND CONFLICT

First, what we *can* know: AI is presently and will always be both made up of and reliant upon human intelligence and intervention. This is the case whether at the outset, at the identification of a problem to be solved computationally; a bit later on, with the creation of a tool based in algorithmic or machine-learning techniques; when that tool is informed by, improved upon, or recalibrated over time based on more or different human inputs and intervention;[37] or whether it is simply used alongside other extant human processes to aid, ease, speed, or influence human decision-making and action-taking, in symbiosis or perhaps, when at odds with a human decision, in conflict.

This sometimes-symbiosis-sometimes-conflict matters when it comes to social media. In addition to regulatory and legal concerns that vary greatly around the globe and are delineated by jurisdiction (as well as by less tangible but deeply important cultural and political norms), the speech that flows across social media platforms in the form of images, text postings, videos, and livestreams is subject to a vast patchwork of internal rules and other opaque policies decided by mid- and high-level employees but often implemented by a global workforce of precarious low-paid laborers. When algorithms and deep machine-learning-based tools may be present at some points in the system, even they are informed by the platform policies and the workers' decisions, captured and then redeployed as AI. In short, human beings and their traces are present across the social media production chain. Where humans stop and machines start is not always clear, and it is not always even possible to know. But that the human agents have been largely disavowed and rendered invisible by the platforms for which they labor has been no accident as firms continue to appeal to the public's assumptions and to avoid, or respond to, regulation by invoking AI.

A major challenge over the next five years or so will therefore be to recognize, acknowledge, and map the contours of these relationships, with particular attention to the dangers of overemphasizing their abilities or benefits, minimizing their shortcomings, or disavowing the value-ladenness of the tools themselves and their computational outcomes. Langdon Winner, in his field-defining 1980 essay, "Do Artifacts Have Politics?," demonstrated that technological innovation is hardly always synonymous with progress and is frequently not in everyone's best interest, distributing its benefits and harms unevenly and unequally.[38] These observations have been taken up in the internet era by numerous scholars (many of whom are present in this very volume) and by a more recent assertion that platforms, too, have politics.[39]

What, then, are the values of an algorithm? What are the politics of AI mechanisms? How can we know? To put it plainly: AI reflects, and endeavors to replicate or improve upon, human problem-solving in human contexts. When these goals and hoped-for outcomes are made plain, what becomes clear is that all human problems and all human contexts do not lend themselves equally successfully to computational interventions, and to apply such tools to some of these processes would not only be inappropriate, the outcomes could even be dangerous—often more so to those already impacted by structural or historical discrimination.[40]

Even if one were to stipulate that such problem-solving could be rendered value-free, surely the reentry of the results of such processes into their social, cultural, and political contexts make such claims difficult to take seriously. Yet this cautionary approach to AI and its potential for biases is rarely met with concern from those who celebrate it as a solution to a panoply of human ills—or build toward those ostensible goals. Beyond the case of social media UGC and commercial content moderation, examples of this would be the call to quickly resolve a multitude of social problems with technical solutions in which AI is often invoked in the context of the latter.

Predictive policing, and the tools being built around it, are being programmed using data and assuming a relationship to police that is the same for all people, under the best of circumstances; what happens when inaccurate, outdated, or racially biased information is fed into the predictive engine?[41] Tools designed to remove "terroristic content" using hashing algorithms do a great job of seeking and matching for known bad material, but who decides what constitutes terrorism and how can we know?[42] And what person or persons should be held responsible for the algorithmically determined Google search engine results that lead users to racist and false information, such as that which influenced Dylann Roof, the killer of nine African-American parishioners in a Charleston church in 2015?[43]

Rather than reflecting what is either technologically possible or socially ideal at present or in the near- to midterm, this solutionist disposition toward AI everywhere is *aspirational* at its core. The abstractions that algorithms and other AI tools represent suggest a means to quickly and easily solve for a solution and bypass the more difficult and troubling possibility of having to cast a larger net or to acknowledge the even more frightening reality that perhaps not every human dilemma or social problem can be solved in this way.

What is the goal of the AI turn in . . . everything? Certainly, in the case of social media platforms, the legal implications of being unable to invoke one or even a group of humans as being responsible for decisions rendered is one powerful benefit, if not the point in and of itself. How does one hold AI tools to account? The hope may be that no one can, and that is a position considerably balanced in favor of the firms who create and control them.

Movements to demand fairness, accountability, and transparency in algorithmic endeavors are gaining ground, not only in important works from social scientists who have traditionally critiqued social media platforms and digital life and its unequal distribution of power[44] but also from others in disciplines such as computer science and engineering—the province of the makers of the tools themselves. Entire conferences,[45] academic working groups,[46] and cross-disciplinary and cross-sector institutes joining academe and industry[47] have been constituted to tackle these issues and demand that they be addressed. The AI Now Institute's 2017 report issued a clarion call to this effect, saying:

> AI systems are now being adopted across multiple sectors, and the social effects are already being felt: so far, the benefits and risks are unevenly distributed. Too often, those effects simply happen, without public understanding or deliberation, led by technology companies and governments that are yet to understand the broader implications of their technologies once they are released into complex social systems. We urgently need rigorous research that incorporates diverse disciplines and perspectives to help us measure and understand the short and long-term effects of AI across our core social and economic institutions.[48]

The ability of regulators, advocates, and the public to meaningfully engage with AI tools, processes, and mechanisms will be predicated on the ability to see them, however, and to know when they are, and are not, invoked. To give these actors the ability to see through the AI illusion will be the first, and possibly most difficult, task of all.

NOTES

1. John McCarthy and Patrick J. Hayes, "Some Philosophical Problems from the Standpoint of Artificial Intelligence," in *Machine Intelligence 4*, ed. B. Meltzer and D. Michie (Edinburgh: Edinburgh University Press, 1969), 463–502, https://doi.org/10.1016/B978-0-934613-03-3.50033-7.

2. Mike Lorica and Ben Loukides, "What Is Artificial Intelligence?," OReilly.com (June 29, 2016), https://www.oreilly.com/ideas/what-is-artificial-intelligence.

3. Martin Strutz, "Freelancers and Technology Are Leading the Workforce Revolution," *Forbes* (November 10, 2016), https://www.forbes.com/sites/berlinschoolofcreativeleadership/2016/11/10/free-lancers-and-technology-are-leading-the-workforce-revolution/.

4. Arnold Pacey, *The Culture of Technology* (Cambridge, MA: MIT Press, 1985).

5. Aimee Picchi, "Facebook Stock Suffers Largest One-Day Drop in History, Shedding $119 Billion," *CBS Moneywatch* (July 26, 2108), https://www.cbsnews.com/news/facebook-stock-price-plummets-largest-stock-market-drop-in-history/.

6. Elizabeth Dwoskin and Annie Gowen, "On WhatsApp, Fake News Is Fast—and Can Be Fatal," *Washington Post* (July 23, 2018), https://www.washingtonpost.com/business/economy/on-whatsapp-fake-news-is-fast--and-can-be-fatal/2018/07/23/a2dd7112-8ebf-11e8-bcd5-9d911c784c38_story.html; Lauren Frayer, "Viral WhatsApp Messages Are Triggering Mob Killings in India," *NPR* (July 18, 2018), https://www.npr.org/2018/07/18/629731693/fake-news-turns-deadly-in-india.

7. Reuters staff, "Twitter Shares Fall after Report Says Account Suspensions to Cause . . . ," *Reuters* (July 9, 2018), https://www.reuters.com/article/us-twitter-stocks/twitter-shares-fall-after-report-says-account-suspensions-to-cause-user-decline-idUSKBN1JZ20V.

8. Zeynep Tufekci, "YouTube Has a Video for That," *Scientific American* (April 2019), https://doi.org/10.1038/scientificamerican0419-77; Charlie Warzel and Remy Smidt, "YouTubers Made Hundreds

of Thousands Off of Bizarre And Disturbing Child Content," *BuzzFeed News* (December 11, 2017), https://www.buzzfeednews.com/article/charliewarzel/youtubers-made-hundreds-of-thousands-off-of-bizarre-and.

9. Brad Stone, "Concern for Those Who Screen the Web for Barbarity," *New York Times* (July 18, 2010), http://www.nytimes.com/2010/07/19/technology/19screen.html?_r=1.

10. See Sreela Sarkar's chapter in this volume, "Skills Will Not Set You Free," describing a New Delhi–based computer "skills" class aimed at low-income Muslim women that may be preparing them more for the deskilling jobs outsourced from the Global North's IT sector, if that at all.

11. Sarah T. Roberts, *Behind the Screen: Content Moderation in the Shadows of Social Media* (New Haven: Yale University Press, 2019).

12. Chris A. Mack, "Fifty Years of Moore's Law," *IEEE Transactions on Semiconductor Manufacturing* 24, no. 2 (May 2011): 202–207, https://doi.org/10.1109/TSM.2010.2096437; Ethan Mollick, "Establishing Moore's Law," *IEEE Annals of the History of Computing* 28, no. 3 (July 2006): 62–75, https://doi.org/10.1109/MAHC.2006.45.

13. Ronald Deibert, *Access Denied: The Practice and Policy of Global Internet Filtering* (Cambridge, MA: MIT Press, 2008); Steve Silberman, "We're Teen, We're Queer, and We've Got E-Mail," *Wired* (November 1, 1994), https://www.wired.com/1994/11/gay-teen/; Nart Villeneuve, "The Filtering Matrix: Integrated Mechanisms of Information Control and the Demarcation of Borders in Cyberspace," *First Monday* 11, no. 1 (January 2, 2006), https://doi.org/10.5210/fm.v11i1.1307.

14. Hany Farid, "Reining in Online Abuses," *Technology & Innovation* 19, no. 3 (February 9, 2018): 593–599, https://doi.org/10.21300/19.3.2018.593.

15. Ping Liu et al., "Forecasting the Presence and Intensity of Hostility on Instagram Using Linguistic and Social Features," in *Twelfth International AAAI Conference on Web and Social Media* (2018), https://www.aaai.org/ocs/index.php/ICWSM/ICWSM18/paper/view/17875.

16. Hank Liao, Erik McDermott, and Andrew Senior, "Large Scale Deep Neural Network Acoustic Modeling with Semi-Supervised Training Data for YouTube Video Transcription," in *2013 IEEE Workshop on Automatic Speech Recognition and Understanding* (2013), 368–373, https://doi.org/10.1109/ASRU.2013.6707758.

17. A. A. Zaidan et al., "On the Multi-Agent Learning Neural and Bayesian Methods in Skin Detector and Pornography Classifier: An Automated Anti-Pornography System," *Neurocomputing* 131 (May 5, 2014): 397–418, https://doi.org/10.1016/j.neucom.2013.10.003; Vasile Buzuloiu et al., Automated detection of pornographic images, United States US7103215B2, filed May 7, 2004, and issued September 5, 2006, https://patents.google.com/patent/US7103215B2/en.

18. Sitaram Asur and Bernardo A. Huberman, "Predicting the Future with Social Media," in *Proceedings of the 2010 IEEE/WIC/ACM International Conference on Web Intelligence and Intelligent Agent Technology—Volume 01*, WI-IAT '10 (Washington, DC: IEEE Computer Society, 2010), 492–499, https://doi.org/10.1109/WI-IAT.2010.63; Lilian Weng, Filippo Menczer, and Yong-Yeol Ahn, "Predicting Successful Memes Using Network and Community Structure," in *Eighth International AAAI Conference on Weblogs and Social Media* (2014), https://www.aaai.org/ocs/index.php/ICWSM/ICWSM14/paper/view/8081.

19. Reuben Binns et al., "Like Trainer, Like Bot? Inheritance of Bias in Algorithmic Content Moderation," ArXiv:1707.01477 [Cs] 10540 (2017): 405–415, https://doi.org/10.1007/978-3-319-67256-4_32.

20. John Constine, "Zuckerberg Implies Facebook Is a Media Company, Just 'Not a Traditional Media Company,'" *TechCrunch* (December 21, 2016), http://social.techcrunch.com/2016/12/21/fbonc/

21. Federal Communications Commission, "The Public and Broadcasting (July 2008 Edition)" (December 7, 2015), https://www.fcc.gov/media/radio/public-and-broadcasting.

22. This immunity is typically known as freedom from "intermediary liability" in legal terms, but it is not all-encompassing, particularly in the context of child sexual exploitation material. See Mitali Thakor's chapter in this volume, "Capture Is Pleasure," for a more in-depth discussion of the parameters and significance of laws governing such material on social media platforms and firms' responsibility to contend with it.

23. Jeff Kosseff, *The Twenty-Six Words That Created the Internet* (Ithaca, NY: Cornell University Press, 2019). For an extended treatment of the origin and impact of Section 230, see Kosseff's monograph on the subject, which he calls a "biography" of the statute.

24. Electronic Frontier Foundation, "CDA 230: Legislative History" (September 18, 2012), https://www.eff.org/issues/cda230/legislative-history.

25. Mary Graw Leary, "The Indecency and Injustice of Section 230 of the Communications Decency Act," *Harvard Journal of Law & Public Policy* 41, no. 2 (2018): 559.

26. Leary, "The Indecency and Injustice of Section 230 of the Communications Decency Act," 559.

27. Kate Klonick, "The New Governors: The People, Rules, and Processes Governing Online Speech," *Harvard Law Review* 131 (2018): 1598–1670.

28. Kate Crawford and Tarleton Gillespie, "What Is a Flag For? Social Media Reporting Tools and the Vocabulary of Complaint," *New Media & Society* 18, no. 3 (2016): 410–428.

29. Hanna Kozlowska, "These Are the Countries Where Facebook Censors the Most Illegal Content," *Quartz* (May 16, 2018), https://qz.com/1279549/facebook-censors-the-most-illegal-content-in-turkey/

30. "Reckless Social Media Law Threatens Freedom of Expression in Germany," *EDRi* (April 5, 2017), https://edri.org/reckless-social-media-law-threatens-freedom-expression-germany/.

31. Klonick, "The New Governors."

32. Sarah T. Roberts, "Digital Detritus: 'Error' and the Logic of Opacity in Social Media Content Moderation," *First Monday* 23, no. 3 (March 1, 2018), http://firstmonday.org/ojs/index.php/fm/article/view/8283.

33. Jenna Burrell, "How the Machine 'Thinks': Understanding Opacity in Machine Learning Algorithms," *Big Data & Society* 3, no. 1 (June 1, 2016), https://doi.org/10.1177/2053951715622512.

34. Sarah Myers West, "Censored, Suspended, Shadowbanned: User Interpretations of Content Moderation on Social Media Platforms," *New Media & Society* 20, no. 11 (2018), 4366–4383, https://doi.org/10.1177/1461444818773059.

35. Frank Pasquale, *The Black Box Society: The Secret Algorithms That Control Money and Information* (Cambridge, MA: Harvard University Press, 2015).

36. Olivia Solon, "The Rise of 'Pseudo-AI': How Tech Firms Quietly Use Humans to Do Bots' Work," *Guardian* (July 6, 2018, sec. Technology), http://www.theguardian.com/technology/2018/jul/06/artificial-intelligence-ai-humans-bots-tech-companies; Maya Kosoff, "Uh, Did Google Fake Its Big A.I. Demo?," *Vanity Fair: The Hive* (May 17, 2018), https://www.vanityfair.com/news/2018/05/uh-did-google-fake-its-big-ai-demo.

37. Cade Metz, "A.I. Is Learning from Humans. Many Humans," *New York Times* (August 16, 2019), https://www.nytimes.com/2019/08/16/technology/ai-humans.html.

38. Langdon Winner, "Do Artifacts Have Politics?," *Daedalus* 109, no. 1 (1980): 121–136.

39. Tarleton Gillespie, "The Politics of 'Platforms,'" *New Media & Society* 12, no. 3 (May 1, 2010): 347–364, https://doi.org/10.1177/1461444809342738.

40. See Safiya Noble's discussion of digital redlining in her chapter in this volume, and the work of Chris Gilliard on this subject.

41. Ali Winston and Ingrid Burrington, "A Pioneer in Predictive Policing Is Starting a Troubling New Project," *The Verge* (April 26, 2018), https://www.theverge.com/2018/4/26/17285058/predictive-policing-predpol-pentagon-ai-racial-bias.

42. "How CEP's EGLYPH Technology Works," Counter Extremism Project (December 8, 2016), https://www.counterextremism.com/video/how-ceps-eglyph-technology-works.

43. Safiya Umoja Noble, *Algorithms of Oppression: How Search Engines Reinforce Racism* (New York: NYU Press, 2018).

44. Siva Vaidhyanathan, *Antisocial Media: How Facebook Disconnects Us and Undermines Democracy* (New York: Oxford University Press, 2018); Virginia Eubanks, *Automating Inequality: How High-Tech Tools Profile, Police, and Punish the Poor* (New York: St. Martin's Press, 2018).

45. "ACM FAT*," accessed August 1, 2018, https://fatconference.org/.

46. "People—Algorithmic Fairness and Opacity Working Group," accessed August 1, 2018, http://afog.berkeley.edu/people/.

47. "AI Now Institute," accessed August 1, 2018, https://ainowinstitute.org/.

48. Alex Campolo et al., "AI Now 2017 Report" (AI Now, 2017), https://ainowinstitute.org/AI_Now_2017_Report.pdf.

3

A NETWORK IS NOT A NETWORK

Benjamin Peters

The map is not the territory.
—Polish American philosopher Alfred Korzybski

INTRODUCTION

Computer networks cannot save us—nor can we save them. Often, network designers and users do not even know what networks are: a network is *not* what we think of as a network. A network is not its map.

"It's not that we didn't think about security," admitted internet pioneer David D. Clark to the *Washington Post* about the early days of the internet: "We knew that there were untrustworthy people out there, and we thought we could exclude them."[1] In the fifty years since 1969, not only has the internet not excluded bad actors, it has grown from connecting a few dozen scientists to encompassing over four billion users, more than half the world's population. Hundreds of millions more are likely to go online in Africa and Asia in subsequent years; ambitious predictions even hold that almost every living person and far more things (flip-flops?) will be online by 2030. The internet is an unparalleled access machine. But towering successes cast long shadows: what is this unparalleled growth for? Access to what and for whom? Whom does it all serve? It should not surprise us that no other class of actor has benefited more from the enormous global reach, speed, and capacity of the internet than the organizations that prop it up. Not the users. Not the visionary network designers. Other networked organizations—from the undersea cable owners to service providers to big data brokers to data aggregators—have gained the most. While surely not all organizations intend to exploit people, it remains the historical rule that at least a few will, and in so prospering just a few bad actors may normalize exploitative behavior that serves networked organizations in general: networks now let us exchange data but at the cost of rendering ourselves products on the network subject to network surveillance, discrimination, gatekeeping, platforms rigged to reward addictive behavior and groupthink, toxic attention economies, and much else. Looking back, one cannot help but counterfactually wish that a more secure design at the dawn of computer networking might have paid dividends now, half a century on.

But that would be sloppy thinking, for it is precisely such design-oriented solutionism about networks that led to the current state of computer networks in the first place. It is shortsighted analysis (that often imagines itself to be structural, fundamental, and far-reaching) that justifies a network based on the values embedded in its original design. Why are the values we embed in the design of a network not enough? It is simple: a network is not what we think it looks like. Networks may enjoy attractive visual design-reality homeomorphisms, but they do not mean what we think they mean. A live complex computer network—its uses, its meaning, its consequences—is *not* what it looks like on paper. A network design is defined visually and spatially through architectures, assemblies, gaps, links, maps, nodes, points, topologies (such as centralized spoke-and-hub, decentralized, distributed mesh, free scale, daisy chain), etc. By contrast, the internet *in situ* is more than the sum of its complex uses and users: thousands of thousands of thousands of users shuffling data between organizations for as many reasons—or no reason at all—at any given time.

How should we make sense of the internet then? Philosophers have long bristled at the fact that technology is often understood only instrumentally—by its use value. Yet network analysis often does not even get that far. Consider the analogy to money. Money and networks are both immensely useful, but by reducing their value into a single price or an ideal design, analysts multiply their potential abuses. When it comes to money, few ignore the difference between what money can be exchanged for (products for a price) and what it can be used for (products for use). By contrast, for fifty years, popular networks have extrapolated from what a network looks like to what it can be used for. Observers scowl at centralized network schematics (one hub with all spokes), judging them good for nothing but maybe centralized control and surveillance situations like a panopticon. Distributed networks (each node connects to its neighbor nodes), meanwhile, appear to share liberal democracy's commitment to all nodes being equal, locally connected, and robustly integrated. Given those two ideological endpoints, how, exactly, did the world's most distributed network end up a panopticon?

Modern-day design analogies go wrong *not* because network visualization tools—graphs, dynamic models, schematics, etc.—are not yet good enough to represent network nodes and their gaps (even if the roughly fifty billion nodes on the internet stretch even machine abstraction). Rather, it is that such network design visuals are themselves the gaps in the modern understanding of networks.[2] To twist that old line often attributed to the mathematician and early information age polymath Norbert Wiener, the best model of a cat is a cat, preferably the same cat: so too is the most reliable model of a complex large-scale network that same network at work in the world. A network on paper or screen is no computer network, and often computer network models obscure what embedded organizations might actually *use* the network for.

No historian will be surprised by the assertion that technology does not behave as it is initially designed: it is common sense that technologies often develop unforeseen affordances. For example, when the telephone was introduced into the US in the late nineteenth century, it was sold to middle-class businessmen with the promise of globalizing their companies; however, once they adopted it, women turned to the telephone in greater numbers than businessmen to connect with local friends and contacts. Only a fool clings to the stories we tell ourselves about new media. No one today insists that telephones are *inherently* global business tools: telephone sets have largely vanished, although I'm sure we could arrange a call to discuss it more. By analogy, there is no reason to imagine networks as fundamentally democratic, and yet we do all the time.

Network designs come with a second-order recognition problem that stems from the fact that all sufficiently large networks are necessarily the products of multiparty collaborations. (It takes a lot of groups to make a network.) Therefore, it is impossible to view a network from any single perspective. Instead, we insist on seeing too much in their network designs. By believing that we embed values into the universal design of networks, we wrongly attribute explanatory power to such design values.

An analogy obtains in corporate social responsibility. As media and organization scholar Siva Vaidhyanathan points out, blue-chip corporations have come to embed lofty values directly into their mission statement and operations in order to make a positive difference in the world while also gaining internal and public buy-in for their brand. No matter how genuine in intent, the more responsible and desirable these values appear, the more corporations or networks can counterintuitively deflect, defer, and delay the normal correcting hand of history—the checks and balances of industry self-criticism and state regulation meant to ensure socially responsible behavior. Everyone should want and demand more socially responsible network corporations in practice, of course. But, for the same reason that my planning to be a more caring neighbor does not make me so, designing values into our corporations or networks offers no guarantee they will behave accordingly.

This essay critically considers and compares early attempts to design large-scale computer networks, including the internet. In the 1960s and 1970s, engineers designed distributed packet-switching computer networks after the American values at the time of surviving a nuclear war, modeling democracy, market institutions, and collective smarts. Network companies continue to fashion themselves after the image of the society they reflect: perhaps no other company does this more than Facebook, a self-professed "community" of nearly two and a half billion active users that promises (like much in your cable company mailer) to increase your contact with loved ones, float you in an ocean of entertainment, amplify your voice on forums, and flatten your access to goods and services. And yet the same network companies have shaped the experience of the internet in ways their designers did not intend for reasons they are not trained to identify. Of course, not all network companies are malicious masterminds with mind-control rays hell-bent on speeding society suicide; most successful ones are, by all accounts, disciplined to well-intentioned design values that invite both our celebration as well as vigilant checks. For example, network neutrality policies are noble and vitally needed, but nothing about network neutrality should be mistaken as actually being *neutral*: in resonance with Sarah Roberts's chapter in this volume, the public claims about design values—the forward-thinking solutionism of artificial intelligence (AI) and machine learning is another one—obscure, not disclose, the causes and effects of actual network behavior.

A proposition: no one today knows exactly how or why actually existing, large-scale networks behave in the ways they do, and in the meanwhile, surveillance has become the quiet coin of the realm: the logics of surveillance, by contrast, are well understood. Surveillance online today is not some benign anonymized monitoring that feeds and informs public understanding about our own behavior on networks. No, our networks exist in an age when surveillance services compete to buy and sell information about you, without your wishing it, and every other "you" to the highest bidder. Such "smart" surveillance has made idiots out of everyone and every-thing that plugs in in the original Greek sense of *private persons*: it privatized every action. Online we are not even persons—we are *personas*. Forget Russian bots: *you* are the internet's favorite bot.

Critical interventions into our networked worlds, such as this and Kavita Philip's essay in this volume, are sure to prove both necessary and insufficient: this essay calls attention to the many variables that have shaped our global network history without informing network designs or discourse; in particular, it attends to how institutional behavior has shaped large-scale networks in ways that quietly under-mine their designs' publicly celebrated values. No one can know yet what a network is or does. This yawning gap—strung out here in an attempt to compare the history of three national computer network projects—remains pressingly open.

Nothing less than the future of networks hangs in the balance.

The remainder of this essay compares the stated design values and actual insti-tutional practices across the histories of three national computer network proj-ects since 1959: the Soviet OGAS project, the Chilean Project Cybersyn, and the American ARPANET. First, on the eastern side of the iron curtain at the height of the Cold War tech race, Soviets were designing major computer network projects dedicated to explicitly civilian purposes: namely, Soviet scientists and statesmen sought to upgrade the entire paper-based command economy to the speed of elec-tronic socialism, but what happened next simply does not follow the high-noon scripts for Cold War showdowns. The complex institutional conditions in which these Soviet network projects tried to take root tell a different story that scram-bles and refreshes mainstream network thinking since the Cold War.[3] Second, as eloquently chronicled in Eden Medina's pathbreaking book *Cybernetic Revolu-tionaries*, the relevance of broader political, organizational, and cultural variables becomes even more obvious in another major socialist cybernetic national net-work. Namely, the Chilean President Salvador Allende's computer network proj-ect, Project Cybersyn, sought to democratize and socialize by computer network parts of the Chilean national economy between 1971 and 1973. In the process, the network was designed to hand the reins of control over to worker-citizens and Allende's statesmen.[4] Here, too, the historical arc of Project Cybersyn reminds us how poorly design thinking predicts realities. Our third vignette, on the US ARPA-NET, likewise bears none of the neural or military design imprints that inspired it then or now: instead, the ARPANET, which went online in 1969, sped scientific exchange and, once paired with the personal computer revolution, grew into the contemporary internet.[5] (If pressed to speculate, I would bet that future historians will remember the globalization of the internet primarily for its uneven consolidation of knowledge and power among populations that write in world empire languages and code, but no doubt the future will unsettle such predictions at least as much as does reflection on the past!) These historical surprises have generated countless new innovations as well as cooptations, including some of the internet's most celebrated design values—such as antihierarchical packet-switching and democratizing distribution designs. This essay turns against visual network-design values and toward the explanatory weight that complex institutional behavior has had and will continue to have in shaping a less straightforward, more critical, and more incendiary history of how computer networks came to encircle and then enchain the world.

THE SOVIET OGAS NETWORK: NEITHER NERVOUS SYSTEM NOR HIERARCHY

In the fall of 1959, the young star military researcher Anatoly Kitov wrote General Secretary Nikita Khrushchev with an apparently original thought: why not connect the nation's military computers into one giant network that could benefit civilian problems, like economic planning? It was the height of the Cold War technology race, and the Soviets had all the mathematics, motivation, and military might to build the contemporary of the Semi-Automatic Ground Environment project in the 1950s or the ARPANET in the 1960s. At that very moment, Soviet scientists were expanding nuclear power, launching satellites and spacecraft into orbit, and making huge advances across the theoretical sciences, including innovations in an indigenous computer industry into the mids1960s. Why, then—after thirty years of attempts—was there never a "Soviet internet"?[6]

Between 1959 and 1989, a group of leading Soviet scientists and statesmen and women agreed there should be a national computer network to manage and digitize the nation's paper-based command economy, thus ushering in an era of electronic socialism. By circumventing all the cumbersome delays in paperwork and subsequent perpetual backlogs in the supply chain, every factory worker and economic planner with access to the proposed national network would be able to interact, exchange reports, and make real-time adjustments to their contribution to the command economy—all at the sublime speed of electricity. The result, they hoped, would be no less than what Francis Spufford has called, in the title of his delightful historical novel, the age of Red Plenty.[7]

Under the direction of Soviet cyberneticist Viktor M. Glushkov, the All-State Automated System of Management (*obshchee-gosudarstvennaya avtomitizirovannaya sistema upravleniya*)—or the OGAS project—sought to fulfill such a socioeconomic vision while sharing few of the design values that internet enthusiasts would recognize today: no packet-switching protocols, no distributed design, no peer-to-peer populism. Rather, in order to solve time-sharing and interoperability functions while appearing politically viable to the Soviet state, the OGAS network design was modeled directly after the three-tiered hierarchical structure of the planned economy. As the secretary of the institute in charge of the optimal economic models or software for managing the networked economy once put it, the decision was made to "build the country's unified net hierarchically—just as the economy was planned in those days."[8] Thus, at the top, the network would link one central computing center in Moscow to a second tier with as many as 200 computer centers in prominent city centers, which would in turn connect to a third tier with as many as 20,000 computer terminals distributed throughout the industrial base of the Eurasian continent. In order to pass as a politically feasible state hero project, the network was designed to be neither centralized nor distributed. It should be, like the state, a *decentralized* hierarchy, meaning that any user would be able to connect directly with any other user on the network, provided that Moscow had first granted clearance to do so. Worker feedback, criticism, suggestions, and updates could be submitted directly from the factory floor to anyone deemed relevant by a supervisor. All this activity would also, in turn, conveniently permit the supervising state to collect tens of thousands of dossiers on individuals and organizations in the national network.[9]

Like the Chilean and American network designers discussed here, Glushkov and his team were inspired by the grandiose cybernetic design analogs between information systems in vogue then (as well as today, although the odd label "cybernetics" has been shed in favor of new ones): Glushkov, whom the *New York Times* once called "the King of Soviet cybernetics," imagined the OGAS project to be even more corporeal than "brainlike" (*mozgopodobnyi*). He saw networks as fully embodying the material economy and thus the nation: his national network was designed to function as the nation's informational nervous system for electrifying and animating the sluggish print-based organic body of the national economy.[10] Cyberneticists and design theorists such as Glushkov, Anatoly Kitov, Nikolai Fedorenko, and many others did not make the design mistake popular in the West of imagining information to be immaterial or virtual (see Nathan Ensmenger's essay in this volume on the Cloud in cloud computing). Instead, they planned the state economic network as if it were a living machine fully embodied according to (yet another visually obvious) cybernetic design value: if the Soviet economy was the national body of workers, it needed not just the state to be its central information processor and brain. Both the body (the nation of worker) and the brain (Soviet state) needed a live nervous system—a real-time national computer network.[11]

It was a beautiful design in principle. But this is the stubborn point: none of these stated designs—socialist economic reform, decentralized hierarchical states, mass state surveillance, or cyber-network corporeal collectivism—had any bearing on the outcome of the OGAS and other early attempts to build Soviet networks. Unanticipated obstacles interfered: even though Glushkov often had the ear of the top brass in the Soviet state (once almost winning over the Politburo to the project in 1970, a handful of years after the project had been under internal review and one year after the ARPANET had gone online), very few others in the Soviet economic planning bureaucracy wanted a computer network's help in reforming the economy. That was *their* job, not a computer's, the general sentiment went. Despite the internal boasts and the complaints from capitalist critics, the heart of the Soviet economic bureaucracy was in practice no hierarchy. So middle-level bureaucratic resistance mattered in ways that no top-down decision could every control for. The number of institutions—and bureaucratic ministries—that would have to agree to implement the OGAS project underwent such a continuous "treadmill of reforms" that the resulting traffic jam of competing, conflicted, and self-interested ministers proved impossible to navigate well through the 1980s, when the economy was teetering toward internal collapse.[12] For example, immediately after Stalin's death in 1953, the number of ministries were consolidated to fewer than a dozen; a decade later, after the uneasy decentralization of Khrushchev, the relevant economic ministries would balloon to over thirty-five, with a continuous and uneven pendulum swing back and forth thereafter.[13] The resulting institutional ambiguities benefited status quo–inclined economic planners with enough moxie and strategic self-interest to hold onto their positions in such tumultuous internal reforms. The resulting mismatch between stated design values and the actual practices proved enough to shipwreck many efforts to reform the economy, including the OGAS project, against the rocks of competitive self-interest. Informal resistance reverberated up and down the chain of command: the occasional worker would sabotage prototype machines, fearing their own replacement, while ministers in the Politburo, the second highest state committee, apparently mutinied against Prime Minister Aleksei Kosygin rather than cooperate on the network.[14] Not only was the cybernetic analogy of the nation as the economic body misplaced, it could not materialize in the unfettered and turbulent interinstitutional competition over power and resources.

This Soviet network story amounts to the signal contradiction that my most recent book begins with: the US ARPANET took shape thanks to collaborative research environments and state funding, while the Soviet attempts to build civilian computer networks were stalemated by its own unregulated competition and internal conflicts of self-interest. The first global computer network took shape, in other words, thanks to capitalists behaving like cooperative socialists, not socialists behaving like competitive capitalists.[15]

THE CHILEAN PROJECT CYBERSYN: A POLITICAL REVOLUTION IN MINIATURE

At the same time that the Soviets were stumbling to fund the OGAS project, an uneasy political regime was taking place in Latin America, and with it a fascinating case study of another significant socialist network project—the Chilean Project Cybersyn, designed by the British cyberneticist Stafford Beer under Salvador Allende between 1971 and 1973, now rigorously told in Eden Medina's signal history *Cybernetic Revolutionaries: Technology and Politics in Allende's Chile*.[16] In her account, the young engineer Fernando Flores reached out to Stafford Beer in 1971 for help building and managing the 150 enterprises the new Allende regime had under its direction, including twelve of the twenty largest Chilean companies. Together, Beer and Flores design a nationwide computer network that would link together through many feedback loops a single "viable system" of government officials, industry managers, and worker-citizens.[17] Like the Soviet case, a networked democratic socialism promised to serve its citizens' economic needs with additional information, processing power, and economic predictions. Even more explicitly than in the Soviet model, in the Chilean project Beer designed the state as the head of the economic plan, with many sensory inputs making it sensitive to the changing needs of the proletariat classes; unlike the Soviet case (which was developed almost a half century after the Russian Revolution), Project Cybersyn was steeped in the revolutionary design values of a fresh regime change, when internal hopes ran high that technocratic solutions could assuage and ease brewing class tensions. That hope proved short-lived: after surviving a trucker strike, the Allende regime—and Project Cybersyn with it—collapsed on September 11, 1973, after a revolt of middle-class businessmen and a CIA-backed military coup led by Augusto Pinochet.[18]

Here, too, Beer's stated design values of democratic socialism and cybernetic viable systems tell us much, although (again) nothing that either the designers anticipated or that correlates with the network's actual development. Consider, for example, the design choices informing the futuristic Cybersyn "operations room" at the center of the state national network of telex machines. The room held "seven chairs arranged in a circle within a hexagonal room," deliberately introducing an odd number to prevent voting ties, with walls lined with screens that displayed brightly colored graphic designs signaling economic data and warnings to state officials.[19] On the armrest of each chair, a series of large, legible buttons awaited the command of both state officials and citizen-workers—although these designs reveal something about the gendered future of their networked Chile: these "large geometric buttons" were "big hand" buttons designed to be "thumped" when emphasizing a point.[20] However, nowhere in the room, in Medina's careful reading, does one encounter a keyboard, the gateway to the female secretarial classes at the time (and to incisive analyses by Mullaney and Stanton in this volume). The future was designed for male workers and officials who do not know how to type. Introducing a keyboard would "insinuate a girl between themselves and the machinery," Beer claimed, "when it is vital that the occupants interact directly with the machine, and each other."[21] As Medina holds, "Such design decisions were not neutral. They reflected who the design team believed would hold power in Chile's revolutionary context and enforced that vision. Male factory workers and government bureaucrats would have decision-making power. Other kinds of workers, such as clerical workers, and women, would not."[22]

The historical design values—democratic socialism, socioeconomic justice, and male decision-making classes—reveal volumes about the time and place in which technologies merge with and emerged in society, including a sensitivity to the pressing political and social class issues of the day. However—and here is my critical point—these design values tell us very little about the actual conditions, causes, and complications shaping the network itself. Unlike the Soviet and American case studies, the explicit network design values in the Chile *were* oriented toward managing the very historical forces that undid it—oriented, that is, *in vain*: Project Cybersyn addressed but, facing broader political and military opposition, ultimately could not resolve the conditions of class unrest and revolt that toppled the Allende regime.

The Chilean case offers a vital reminder that network technologies, even when oriented toward relevant social values, are not enough to single-handedly change the values at work in society, for (of course) no technology alone can "fix" society since no technology ever contains society, no matter how smart we may imagine its social values to be. The point is not simply that technologies are political (imagine claiming otherwise in the wake of the Russian hacking of the 2016 US elections): it is that network projects are twice political for how they, first, surprise and betray their designers and, second, require actual institution building and collaborative realities far richer than any design—even that of the most self-reflexive cyberneticist—can account for. As the Chilean case shows, every network history begins with a history of the wider world. Once network observers set aside our visual design-value biases, the possibilities previously unseen behind this realization should delight, not disappoint.

THE US ARPANET, OR HOW NOT TO BUILD A SURVIVABLE CYBERNETIC BRAIN

The United States government is a prime example of a complex organization. While intervening against socialist regimes in Chile and the Soviet Union, it also aggressively subsidized the cross-institutional collaborations necessary to jump-start the ARPANET, the first packet-switching distributed network that would, in time, seed the contemporary internet. Except perhaps scientific exchange, none of the stated motivations for key ARPANET innovations—national defense, commerce, and communication—have been fully realized in it or its network descendants. Optimistically, this may mean the underlying purpose of the internet is still wide open—and often generatively so. More realistically, this also means the internet remains open to cooptation by increasingly large organizations (corporations and states alike) keen to seize and wield private control over our network relations. Perhaps we can say that if the Soviets never agreed on *how* to build the OGAS network, no one has yet agreed on *why* the US built the internet—and there is much to learn from both disagreements.[23] The contradictory design values of the mixed American economy are so great—well-managed state subsidy coinciding with anti-socialist, anti-statist ideologies—that we can now see in retrospect that the pathway leading to the internet is marked by historical accidents.[24] The internet has never been what we think it is. It will continue to surprise.

As it long has. A few of those surprise twists and turns follow: the ARPANET went online for the first time on October 29, 1969, before growing into a network of other networks for scientific exchange such as the NSFNET, CSNET, and EUNET, even as the main motivation and financial support for several key ARPANET innovations came from the military.[25] Indeed, the communication tool most heralded as flattening global communication inequalities emerged out of the paragon example of command-and-control hierarchy: the US Department of Defense. The Advanced Research Project Agency (ARPA), a special research arm of the Department of Defense, was funded in February 1958, five months after the launch of Sputnik, with the initial goal of militarizing space, although soon thereafter its space mission was reassigned to its contemporary civilian agency, the National Aeronautics and Space Agency (NASA). The ARPA instead focused on basic computer research and information systems fit for the age of nuclear-tipped Sputniks.[26] Even as its missions molted, however, its institutional funding paths continued to straddle military defense spending and essential nonmilitary basic research funding—at least in the 1960s, before the Vietnam War pressured Congress to limit ARPA spending to specifically military projects, driving a brain drain to Xerox PARC, among other private-sector companies.[27] In the 1960s, many network innovations—distributed network design, packet-switching protocols, time-sharing, interface message processor—came together into one computer network. This period, which was perhaps the most generative in US history, stands out for the significant cross-agency collaboration supported by virtually socialist-levels of state funding for nonmilitary basic scientific research. Well-funded collaboration sped the birth of national computer networking.

For example, after some bumps, computer network engineer Paul Baran's packet-switching distributed networks grew out of the dashes in the military-industrial-academic complex. In 1960, Baran, then a researcher at RAND—a private think tank spun off from the US Air Force in 1956—was tasked with designing a "survivable" communication network: "there was a clear but not formally stated understanding," he remarked later, "that a survivable communication is needed to stop, as well as to help avoid, a war."[28] He had the strained strategies of mutually assured destruction in mind: in order to maintain peace, the nuclear superpowers would need to enter a perpetual standoff with one another, legitimately armed for the apocalypse at all times. To make good on such a threat, both superpowers would need long-distance communication network infrastructure that could withstand multiple nuclear blasts.

Military survival logics combined to inform Baran's network design values with another source of unlikely inspiration: the self-repairing human brain in the work of the founding cyberneticist and neurophysiologist Warren McCulloch.[29] "McCulloch in particular inspired me," Baran said, continuing, "He described how he could excise a part of the brain, and the function in that part would move over to another part."[30] In other words, for Baran (like Glushkov and Beer), the biological brain and the social network could become mirror models of survival and adaptation: just as a damaged brain can rewrite itself, so too could a message be rerouted in real time through what he then called "hot potato routing" and what we later called "packet-switching" and more recently "smart" protocols.[31] A dynamic distributed network design, he thought, could survive the destruction of many nodes.

This survivalist logic is a good example of a larger truth: just because the networks he designed worked does not mean that networks work today because of those design values. As in the Soviet and Chilean cases, none of Baran's stated design values (neither a survivable national security communication platform nor cybernetic neural-national networks) cash out in obvious historical consequences in the long run. Instead, the institutions at hand played a huge role in shaping—and in part *delaying*—the actual development of the ARPANET. Baran faced resistance from upper management that balked at the cross-agency collaborations such a network would require. For example, Baran's supervisors in government and industry ignored his 1964 research report "On Distributed Communication" for several years. It appears likely that Baran's supervisors deprioritized his work because it did not have weight of the top-secret imprimatur: if so, it is bitter comment on how interagency collaboration often bloomed in secrecy, not open, cultures, since Baran deliberately chose to publish his research publicly in the hopes his designs would reach the Soviets and ensure mutual competitiveness. (Perhaps the limits of contemporary open-source politics—such as the radical transparency of WikiLeaks and the private cipher that is Julian Assange—follow the limits of Cold War mutual intelligibility.) AT&T, the near analogue to a state network monopoly at the time, also rejected Baran's proposal to develop digital networks, fearing digital networks would bring them into competition against their own services.[32]

And so Baran's research languished until an Englishman named Donald Davies, with the support of the UK post office, independently developed network innovations similar to Baran.[33] Only then, in 1966, under the pressure of an outside organization to which the national government felt compelled to respond (an ally nation backed by another hefty state media monopoly), did Baran's superiors feel pressured to revisit and reclaim his team's network research for ARPA under J. C. R. Licklider. The fact that we know networks now as "packet switching" (Davies's term), not "block switching" (Baran's term), reminds us that the realities of collaboration take shape independent of, and even in spite of, their design values. Accident and contingency coauthored the internet.

COMPARING NETWORK STATES

Network history takes shape *apart from* its most cherished design values. As the Soviet and Chilean insights add to the well-known American story, the conditions for building and operating networks rest on the tricky politics of convincing institutions to collaborate and make vital infrastructural investments. The design values that retroactively appear to justify the American case should not do so. Large-scale institutions—like states and global corporations—must do the hard work of coming to agree on how to build vital infrastructure projects, including networks—but what a network looks like on paper is not among them. For the most part, Cold War states and corporations have collaborated when pressured into imitating their enemies (who are often already imitating them). The historical fickleness of institutional collaboration is not encouraging news for any modern romance with etwork design.

Consider two values that network history cannot be attributed to. First, the belief in distributed network design—that networks should connect each node to its neighboring nodes in a democratic ideal free of bottlenecks—appears to be one of the most enduring and popular network myths. As of writing, the term "distributed network," for instance, garners ten times more hits on Google than "decentralized networks," despite the startlingly simple empirical observation that *large-scale distributed networks do not, in fact, exist in the real world.* No significant computer network today is distributed, and the internet has *never* been distributed: when the ARPANET had only four nodes, one node (in Utah) had only one severable link to the other three nodes (in California). It was never evenly connected to all its peer nodes.

Some nodes matter more than others. In practice, all existing major computer networks today are unevenly decentralized, meaning that all large-scale computer networks come with consequential (but often invisible) pinch points, veto sites, and obligatory passage points and clusters. In 2011, Egyptian President Hosni Mubarak, for example, needed to pressure only a handful of internet service providers to turn off the internet in Cairo during the Arab Spring.[34] A single anchor dragged along an ocean floor pushed Somalia off the grid; the shovel of a seventy-five-year-old woman scrounging for spare copper caused an internet blackout across Armenia. The uneven material realities behind communication networks today make mincemeat of the democratic design ideals baked into increasingly global technology. Or rather, and perhaps more humbling still, networks show that, like actually existing democracy, our network realities are far from ideal and require continuous maintenance and attention.[35]

Second, internet packet-switching protocol, once heralded by its innovators as a solution to the problem of hierarchy, no longer seems the solution it once did. These protocols have been celebrated for smartening peer-to-peer, end-to-end communication across dumb networks for more than thirty years—that's not necessarily wrong, although in the same period the protocols have grown *too* smart: the consequences of fragile bottlenecks in decentralized networks pale in comparison to the fact that now *every* node in a network can be leveraged and exploited through smart backdoors built into recent editions of the internet protocol. The internet protocol works by letting each node make microdecisions about how to reroute packets in real time; however, as media scholar Florian Sprenger has argued, the microdecisions protocols have let corporate and state spies introduce deep-packet inspection and off-the-record access points to each node, shepherding in mass network surveillance.[36] Network technologies now permit the simultaneously location and addressing of mobile devices and their users worldwide. This means that surveillance networks are not the territory they map, nor, as these histories show, are networks the map itself.[37] The fact issues a stirring wake-up call: the protocol once celebrated as solving the problem of hierarchical control has proven the very vehicle for ushering in the current era of private network surveillance. Internet protocol, which Baran once diminished as just "a few housekeeping functions such as standards setting," now display, as in the Roman Saturnalia festivals where slaves would become masters, the hidden power of the housekeeper.[38]

CONCLUSION: WHY NETWORKS DEFY THEIR DESIGNS

The historical life cycle of global computer networks is perhaps in its adolescence (and, as I suggest elsewhere in this volume, midlife crisis). In its childhood, many tech visionaries called networked new media fundamentally different from conventional mass media of the twentieth century; yet future historians will likely frame the mass-mediated twentieth century as a prelude to, not a break from, the mass-networked twenty-first. Most users and technologists would do well to check rose-colored visions of our ever-brighter network future with a balance of self-critical reflection and risk-adverse maturation. This essay has taken a step in that direction by outlining three design-defiant early national computer networks projects—the Soviet OGAS, the Chilean Cybersyn, and the US ARPANET. Throughout I have argued that every history of a computer network is first a history of the organizations that tried to build it—and only secondarily a reminder of our collective failed romance

Table 3.1 Stated Design Values and Institutional Practices in Three Early Computer Networks

	Stated Design Values	Institutional Practices
Soviet OGAS	Socialist economic reform	Official consensus on purpose
	Nation as cybernetic body	Competitive in-fighting
	Three-tiered hierarchy plan	No centralized funding
	Explicit state surveillance	Bureaucratic heterarchy
Chilean Cybersyn	Socialist economic reform	Official consensus on purpose
	Nation as cybernetic body	External conflict and revolt
	"Viable system" design	Unviable political environment
		Class equality through male power
US ARPANET	Surviving a nuclear attack	Ambiguity and conflicting purposes
	Nation as cybernetic brain	Collaborative research
	Scientific data exchange	State funding of nondefense research
		Global mass surveillance

with their design values. Table 3.1 summarizes the differences between what network designers *thought* they were doing (stated design values, left column) and what their institutional practices made happen (right column).

In all three cases, the stated design values in the left column approximate how these network projects are remembered, while the right column better describes how the networks actually developed (or not) over time. Several other observations arise. Midcentury cybernetic system analogies inspired all three networks, although never to a predictable effect. So while one may be tempted to, say, celebrate or critique the ARPANET for its design commitment to incorporating individual citizens as thinking nodes in a national brain, not working cogs in the (Soviet or Chilean) national body, these easy glosses fall short of the mark: network history followed neither. The beautiful design symmetries between technology and social values are more likely to mislead than to be realized. Meanwhile, the features of network history—from the technical affordances of network protocols that permit surveillance backdoors to the historical unrest of middle-class revolutionaries, to funding squabbles, to military-industry stonewalling—will continue to unfold with a forceful reality no design symmetry can imagine or manage. Modern global networks arrived on the backs of institutional ambiguities, nonhierarchical (or, borrowing from McCulloch, what I call "heterarchical") conflicts of interest, and the tumult of socioeconomic unrest.[39] These powerful and contingent historical forces, once accounted for, help network observers resist the Narcissus temptation to see in networks only the rippled reflections of our own favorite design values. No understanding of the organizations that build networks will cohere so long as we insist that every design is necessary, not contingent; the organizational mode we call a "network" is itself a (often too) convenient fiction built out of other organizational contingencies.

If users and designers must espouse values, let us lead with risk-averse insights into their vices, not virtues. Indeed, a healthy network public might prefer the Soviet situation in which its state explicitly sought to build a surveillance network (left column Soviet) to that of the American, which only discovered the surveillance baked into its networks after the fact (right column US). Today's euphemism "cloud computing," a public relations triumph taken straight from the playbook of immaterial media myths, lacks the punch and clarity with which the Soviet network designers described "remote-access networking"; at least the Soviets had the pluck to call cloud computing what it remains today—a network designed to monitor, track, and survey all nodes in its reach.[40]

The network vocabulary inherited from the Cold War also needs a serious update: the oppositions between *capitalism* and *socialism, private corporation* and *public state* do not adequately describe the complex institutional practices that supported computer networks in Cold War mixed economies or today's global network economy. Understanding the Cold War as an ideological battle between social freedoms and fairness, it is clearer in the wake of Edward Snowden's revelations of the NSA's global surveillance and the Russian hacking of the 2016 US elections that the very technology thought to herald the end of that conflict and to usher in an age of information freedom—global computer networks—has actually left no clear victor among our multipolar world of networked citizens and migrants.[41] Instead, the sure victors of the last half century of network history have been large organizations—telecommunication companies, surveillance states, big tech (Amazon, Apple, Facebook, Google), internet providers, cybersecurity organizations, data broker services (SAS and SalesForce), and other technologically powerful few that, fueled by data breaches and the fumes of modern privacy, sort, toll, and monitor our many network traffic bottlenecks.

Networks do not resemble their designs so much as they take after the organizational collaborations and vices that tried to build them: the first civilian computer networks ever imagined tried to reproduce and manage the institutional collaboration among Soviet economic bureaucracy—and when Soviet bureaucrats proved noninteroperable, so did their computer networks. Meanwhile, the hopes of the first Chilean national network project aspired to high cybernetic theory in design, even as they were dashed against the hard soil of class history and ideological unrest. Today, the internet still reproduces many of the institutional ambiguities that have long been baked into it: while the internet can still appear generative and open to the individual user, it has, since the Department of Defense first dreamed it up, scaled to serve unscrupulous institutions. Instead of circumventing institutional power hierarchies, computer networks have embedded themselves into, served, reconfigured, and then largely reconstituted the hierarchies of the Department of the Defense. (Even open-source cultures celebrate and elevate celebrity authorities.) The early success of the ARPANET rests on the collaborations between the dashes of the US military-industrial-academic complex in the 1960s—and now, again, the internet serves first the interests of nonpublic corporate and state surveillance. The "mother of all nets," as the patriarchs of the ARPANET dubbed it, has since seeded hundreds of thousands of closed, proprietary computer networks. So too has the internet sped a media environment that rewards paywalls, firewalls, and the walled gardens we call apps, not because that design is embedded in the technology but because such tools are fashioned in the imprint, not the image, of global capital, network power, and human greed.

Given all this, we need new kinds of network organization collaborations—perhaps not unlike fiduciaries in medicine—whose self-interests are checked and balanced and whose fundamental aim is not fulfilling imagined design values but maintaining the health of the world it networks.[42] Until then, the internet now encircling the world—which is neither distributed nor evenly decentralized—is first among many puckish and powerful muses that the industrialized West built on the fortunes of its own misbehavior. But rarely by design. Given the widening gaps in our understanding of the unruly global history of networks, it is a safe bet that, until checked, networks and their designs will double-cross our fondest wishes time and again. Until then, a network is not its map. A network is anything but a network.

NOTES

1. Craig Timberg, "Net of Insecurity: A Flaw in Design," *Washington Post* (May 30, 2015), https://www.washingtonpost.com/sf/business/2015/05/30/net-of-insecurity-part-1/.

2. Manuel Castells, *Rise of the Network Society* and *End of the Millennium*. in his trilogy *The Information Age—Economy, Society, and Culture* (Oxford: Blackwell, 1998); Bruno Latour, *Reassembling the Social: An Introduction to Actor-Network-Theory* (Oxford: Oxford University Press, 2005); Geert Lovink, *Networks without a Cause: A Critique of Social Media* (New York: Polity, 2012).

3. The Soviet case study builds on, among others, my most recent book, *How Not to Network a Nation: The Uneasy History of the Soviet Internet* (Cambridge, MA: MIT Press, 2016).

4. The Chilean case study builds on the authoritative work that is Eden Medina's *Cybernetic Revolutionaries: Technology and Politics in Allende's Chile* (Cambridge, MA: MIT Press, 2011).

5. Janet Abbate, *Inventing the Internet* (Cambridge, MA: MIT Press, 1999); Paul N. Edwards, *The Closed World: Computers and the Politics of Discourse in Cold War America* (Cambridge, MA: MIT Press, 1996); Finn Burton, *Spam: A Shadow History of the Internet* (Cambridge, MA: MIT Press, 2013); Thomas Streeter, *The Net Effect: Romanticism, Capitalism, and the Internet* (New York: NYU Press, 2011). See also Katie Hafner and Matthew Lyon, *Where Wizards Stay Up Late: The Origins of the Internet* (New York: Simon & Schuster, 1996) and Walter Isaacson, *The Innovators: How a Group of Hackers, Geniuses, and Geeks Created the Digital Revolution* (New York: Simon & Schuster, 2014).

6. Peters, *How Not to Network a Nation*, 1–13. See also Slava Gerovitch, "InterNyet: Why the Soviet Union Did Not Build a Nationwide Computer Network," *History and Technology* 24, no. 4 (2008): 335–350.

7. Francis Spufford, *Red Plenty* (Minneapolis: Greywolf Press, 2010).

8. Aleksandr Ivanovich Stavchikov, academic secretary of CEMI, "Romantika pervyikh issledovanii I proktov I ikh protivorechnaya sub'dba" [Romanticism of early research and projects and their contradictory fate], unnamed, unpublished history of Central Economic Mathematical Institute (CEMI), Moscow, read in person and returned May 2008, chapter 2, 17.

9. Peters, *How Not to Network a Nation*, 108–114.

10. Quoted in Peters, *How Not to Network a Nation*, 119.

11. Peters, *How Not to Network a Nation*, 18, 27, 55, 95–96, 100, 118–120, 202.

12. Gertrude Schroeder, "The Soviet Economy on a Treadmill of Reforms," *Soviet Economy in a Time of Change*, US Congress Joint Economic Committee (Washington, DC: USGPO, 1979).

13. Peters, *How Not to Network a Nation*, 63, see also the rest of chapter 2.

14. Peters, *How Not to Network a Nation*, 160–166.

15. Peters, *How Not to Network a Nation*, 10.

16. Medina, *Cybernetic Revolutions*.

17. Stafford Beer, *Brain of the Firm* (London: Penguin Press, 1972); Stafford Beer, *Heart of the Enterprise* (Chichester: Wiley, 1979); Stafford Beer, *Diagnosing the System for Organizations* (New York: Wiley, 1985). See also, Medina, *Cybernetic Revolutions*, 34–39, 121, 143.

18. Medina, *Cybernetic Revolutions*, 45, 93.

19. Eden Medina, "The Cybersyn Revolution: Five Lessons from a Socialist Computing Project in Salvador Allende's Chile," *Jacobin* (April 27, 2015), https://www.jacobinmag.com/2015/03/allende-chile-beer-medina-cybersyn/.

20. Medina, "The Cybersyn Revolution."

21. Medina, "The Cybersyn Revolution."

22. Medina, "The Cybersyn Revolution."

23. Peters, *How Not to Network a Nation*, 200.

24. Streeter, *The Net Effect*.

25. See Abbate, *Inventing the Internet*.

26. Robert A. Divine, *The Sputnik Challenge: Eisenhower's Response to the Soviet Satellite* (New York: Oxford University Press, 1993); Audra J. Wolfe, *Competing with the Soviets: Science, Technology, and*

the State in Cold War America (Baltimore: John Hopkins University Press, 2012); Annie Jacobsen, *The Pentagon's Brain: An Uncensored History of DARPA, America's Top Secret Military Research Agency (New York: Little, Brown, 2015)*; Peters, *How Not to Network a Nation*, 92; Stuart W. Leslie, *The Cold War and American Science: The Military-Industrial-Academic Complex* (New York: Columbia University Press, 1993), 203–231.

27. I am not aware of much sober scholarship on this particular transition from East Coast government to Silicon Valley private business, although much of the dated rhetoric that pits state against corporations can be found in popular accounts, such as Michael Hiltzik, *Dealers of Lightning: Xerox PARC and the Dawn of the Computer Age* (New York: Harper Business, 1999), and L. Gordon Crovitz, "Who Really Invented the Internet?" *Wall Street Journal* (July 22, 2012), https://www.wsj.com/articles/SB10000872396390444464304577539063008406518.

28. Judy O'Neill, "Interview with Paul Baran," Charles Babbage Institute, OH 182 (February 5, 1999), Menlo Park, CA, accessed September 15, 2017, http://www.gtnoise.net/classes/cs7001/fall_2008/readings/baran-int.pdf.

29. Tara Abraham, *Rebel Genius: Warren S. McCulloch's Transdisciplinary Life in Science* (Cambridge, MA: MIT Press, 2016).

30. Stewart Brand, "Founding Father," *Wired* 9, no. 3 (1991), http://archive.wired.com/wired/archive/9.03/baran_pr.html.

31. Paul Baran, "On Distributed Communications: II. Digital Simulation of Hot-Potato Routingin a Broadband Distributed Communications Network," RAND Publications, 1964, accessed September 15, 2017, https://www.rand.org/pubs/research_memoranda/RM3103.html.

32. Peters, *How Not to Network a Nation*, 96–97.

33. Hafner and Lyon, *Where Wizards Stay Up Late*, 64.

34. Larry Greenemeir, "How Was Egypt's Internet Access Shut Off?," *Scientific American* 28 (January 2011), https://www.scientificamerican.com/article/egypt-internet-mubarak/.

35. Tom Parfitt, "Georgian Woman Cuts Off Web Access to Whole of Armenia," *Guardian* (April 6, 2011), https://www.theguardian.com/world/2011/apr/06/georgian-woman-cuts-web-access.

36. Florian Sprenger, *The Politics of Micro-Decisions: Edward Snowden, Net Neutrality, and the Architectures of the Internet* (Lueneburg, Germany: Meson Press, 2015). See also Alex Galloway, *Protocol: How Control Exists after Decentralization* (Cambridge, MA: MIT Press, 2004); Laura Denardis, *Protocol Politics: The Globalization of Internet Governance* (Cambridge, MA: MIT Press, 2009).

37. Florian Sprenger, "The Network Is Not the Territory: On Capturing Mobile Media," *New Media & Society* 21, no. 1 (2019): 77–96.

38. Paul Baran, "Is the UHF Frequency Shortage a Self Made Problem?" Accessed September 15, 2017, http://digitallibrary.usc.edu/cdm/ref/collection/p15799coll117/id/2494; also quoted in Galloway, *Protocol*, 120.

39. David Stark, *The Sense of Dissonance: Accounts of Worth in Economic Life* (Princeton, NJ: Princeton University Press, 2009). See also Peters, *How Not to Network a Nation*, 22–24, 145, 173, 193.

40. Peters, *How Not to Network a Nation*, 109, 112. See also John Durham Peters, *The Marvelous Clouds: Toward a Philosophy of Elemental Media* (Chicago: Chicago University Press, 2015), and Vincent Moscow, *To the Cloud: Big Data in a Turbulent World* (New York: Paradigm Publishers, 2014).

41. See Mario Biagioli and Vincent Lepinay, eds., *From Russia with Code: Programming Migrations in Post-Soviet Times* (Durham, NC: Duke University Press, 2019).

42. Jack M. Balkin, "Information Fiduciaries and the First Amendment," *UC Davis Law Review* 49, no. 4 (2016); see also Jack M. Balkin and Jonathan Zittrain, "A Grand Bargain to Make Tech Companies More Trustworthy," *Atlantic* (October 3, 2016); see also Lina Khan and David Pozen, "A Skeptical View of Information Fiduciaries," *Harvard Law Review* 133 (May 25, 2019), https//ssrn.com/abstract=3341661.

4

THE INTERNET WILL BE DECOLONIZED

Kavita Philip

Policy makers and internet activists were gathering in South Africa to "decolonize the internet," a 2018 news headline announced.[1] How does the internet have anything to do with colonialism, and why do global campaigners seek to "decolonize" it (see fig. 4.1)?

Activist agitations against the "colonial" internet open up a Pandora's box of questions that this chapter attempts to excavate. The slogan "decolonize the internet" belonged to a global campaign called "Whose Knowledge?," founded in the realization of an imbalance between users and producers of online knowledge: "3/4 of the online population of the world today comes from the global South—from Asia, from Africa, from Latin America. And nearly half those online are women. Yet most public knowledge online has so far been written by white men from Europe and North America."[2] Whose Knowledge? and the Wikimedia Foundation came together to address a problem exemplified by the statistics of global internet use.

Whose Knowledge? is one of a profusion of social movements that seek to reframe the wired world through new combinations of social and technological analysis.[3] Critiques of colonialism, in particular, have been trending in technological circles. All of the ways we speak of and work with the internet are shaped by historical legacies of colonial ways of knowing, and representing, the world. Even well-meaning attempts to bring "development" to the former colonial world by bridging the digital divide or collecting data on poor societies, for example, seem inadvertently to reprise colonial models of backwardness and "catch-up strategies" of modernization. Before

Figure 4.1 Campaign graphic, Decolonize the Internet! by Tinaral for Whose Knowledge? Licensed under CC BY-SA 4.0.

we rush to know the Global South better through better connectivity or the accumulation of more precise, extensive data, we need to reflect on the questions that shape, and politicize, that data.[4]

How might we use historical knowledge not to restigmatize poor countries but to reshape the ways in which we design research programs that affect the world's poorest and least connected?[5] Rather than offering new empirical data from the Global South, then, this chapter steps back in order to reflect on the infrastructures that help create facts about the world we can't see, the part of the world we believe needs to be developed and connected. In order to redesign our ideas, perhaps we need a more useful lens through which to observe intertwined historical, social, and technical processes.

This chapter begins to construct such an analytical lens by surveying diverse social and technical narratives of the internet. The internet is literally the infrastructure that enables our knowledge and activities about the world, and if the internet is shaped by colonialism, we need to understand what that means for the knowledge we consider objective. Once studied only as a remnant of past centuries, colonialism and imperialism are now seen as structuring the ways in which internet-based technologies are built, maintained, and used. In order to understand why scholars, designers, and users of the internet are grappling with the historical legacies of colonialism, we must broaden our frame of analysis to ask why history matters in the stories we tell about the internet and the techniques by which we are "datafying" global development. We begin, then, by surveying different ways of thinking about the internet, drawing from popular discourse, technical maps, and media theory.

INFORMATION IS POWER, BUT MUST TECHNOLOGY BE NEUTRAL?

"If information is power, whoever rules the world's telecommunications system commands the world," argues historical geographer Peter Hugill.[6] He traces the ways in which the power of British colonial submarine cable networks was eclipsed by post–World War II American communications satellites. He observed in 1999: "Just as American radio communications challenged the British cable 'monopoly' in the 1920s, European satellites and fiber-optic lines and Asian satellites are now challenging American satellite and submarine telephone cable dominance."[7] Despite this historically global stage on which technology and power have played out, studies of the internet still tend to be nationally or disciplinarily segregated and often remain innocent of colonial histories. It is hard to integrate varied disciplinary and national perspectives, and even harder to know why they matter. Euro-American studies tend to take high-bandwidth infrastructural functioning for granted and focus on representational and policy issues framed by national politics. "Developing world" studies, then, focus on bridging the digital divide and the challenges to infrastructural provisioning.

There are many definitions of the internet, and that descriptive diversity reflects the many ways in which people experience the internet. The last few decades have offered up several competing narratives about the internet, none of which are strictly false. For example, activists draw attention to the ways in which representational inequities are reshaped or perpetuated by new communication technologies. Media studies scholars focus on how the technical affordances of the internet, such as speed and decentralization, shift forms of media representation and the reconstitution of audiences, publics, and producers. Materials science experts tell us that the internet is really the infrastructural backbone of cables under the sea and the satellites in the sky, and its everyday functions depend on the atomic-level properties of cable and switch materials. Network engineers talk about internet exchange points and the work it takes to make them operate seamlessly and securely. Social media researchers tell us that the internet is essentially made up of the platforms that frame users' participation in dispersed communities. Internet service providers, managing the flow of information through digital "pipes," describe business plans to sell consumers space for our data while lobbying against net neutrality protocols with government regulators. All of these stories are true; each emphasizes, from a particular professional perspective, an important aspect of the internet. Each embodies particular assumptions about individual, corporate, state, and transnational behaviors.

Rather than attempting to consolidate one static, comprehensive definition, let us try to understand how disparate stories about the internet enlisted particular audiences in the last decade of the twentieth century, and why it mattered when one internet narrative was favored over another. One persistent fault line in late-twentieth-century narratives about the internet derived from the geographic and economic location of the narrator. In the 1990s, tech visionaries in the West extolled the internet's creative virtual worlds, while visionary discourse in the Global South more commonly celebrated the engineering feats of big infrastructure projects. This may have been because infrastructure tended to recede into the background when it worked well and was dramatically foregrounded when it failed.[8] First Worlders in the mid- to late twentieth century tended to take roads, bridges, wires, and computer hardware for granted, while Third Worlders experiencing the financially strapped "development decades" were painfully aware of the poorly maintained, underfunded infrastructures of communication technology. As the new millennium began, however, the growth of emerging markets reversed some of this polarity. Globally circulating scholars began to challenge this conceptual divide in the first decades of the twenty-first century (as Paul N. Edwards's chapter in this volume documents).

But a curious narrative lag persistently haunted digital maps and models. Even as the global technological landscape grew increasingly dominated by emerging economies, popular internet stories shaped the future's imaginative frontier by reviving specters of backwardness. Cold War geopolitics and colonial metaphors of primitivism and progress freighted new communication technologies with older racialized and gendered baggage.

MAPPING THE INTERNET

Internet cartographers in the 1990s devised creative ways to represent the internet. A 1995 case study from AT&T's Bell Laboratories began with a network definition: "At its most basic level the Internet or any network consists of nodes and links."[9] The authors, Cox and Eick, reported that the main challenge in "the explosive growth" in internet communications data was the difficulty of finding models to represent large data sets. Searching for techniques to reduce clutter, offer "intuitive models" of time and navigation, and "retain geographic context," the authors reported on the latest method of using "a natural 3D metaphor" by drawing arcs between nodes positioned on a globe. The clearest way to draw glyphs and arcs, they guessed, was to favor displays that "reduce visual complexity" and produce "pleasing" and familiar images "like international airline routes."

The spatial representation of the internet has been crucial to the varied conceptualizations of "cyberspace" (as a place to explore, as a cultural site, and as a marketplace, for example).[10] The earliest internet maps were technically innovative, visually arresting, and rarely scrutinized by humanities scholars. New-media critic Terry Harpold was one of the first to synthesize historical, geographical, and technical analyses of these maps. In an award-winning 1999 article, he reviewed the images created by internet cartographers in the 1990s, observing that, in Eick and his collaborator's maps, "the lines of Internet traffic look more like beacons in the night than, say, undersea cables, satellite relays, or fiber-optic cables." Harpold argued that although genuinely innovative in their techniques of handling large data sets, these late-twentieth-century internet maps drew "on visual discourses of identity and negated identity that echo those of the European maps of colonized and colonizable space of nearly a century ago." Early maps of the internet assumed uniform national connectivity, even though in reality connectivity varied widely from coast to hinterland or clustered around internet cable landing points and major cities. These maps produced images of the world that colored in Africa as a dark continent and showed Asian presence on the network as unreliable and chaotic. It seemed that in making design choices—for example, reducing visual clutter and employing a "natural" 3-D image model—internet cartographers had risen to the challenge of representing complex communications data sets but ignored the challenge of representing complex histories.

The cluttered politics of cartography had, in the same period, been treated with sensitivity by historians and geographers such as Martin Lewis and Kären Wigen, whose 1997 book *The Myth of Continents: A Critique of Metageography* called for new cartographic representations of regions and nations. Mapping colored blocks of nations on maps, they pointed out, implied a nonexistent uniformity across that space. Noting that "in the late twentieth century the friction of distance is much less than it used to be; capital flows as much as human migrations can rapidly create and re-create profound connections between distant places," they critiqued the common-sense notions of national maps, because "some of the most powerful sociospatial aggregations of our day simply cannot be mapped as single, bounded territories."[11] Borrowing the term "metageography" from Lewis and Wigen, Harpold coined the term "internet metageographies." He demonstrated how a variety of 1990s' internet maps obscured transnational historical complexities, reified certain kinds of political hegemony, and reinscribed colonial tropes in popular network narratives. Thus, the apparently innocent use of light and dark color coding, or higher and lower planes, to represent the quantity of internet traffic relied for its intelligibility on Western notions of Africa as a dark continent or of Asia and Latin America as uniformly underdeveloped. The critique of traditional cartography came from a well-developed research field. But this historical and spatial scholarship seemed invisible or irrelevant to cartographers mapping the emergent digital domain.

"New media" narratives, seemingly taken up with the novelty of the technology, rewrote old stereotypes into renewed assumptions about viewers' common-sense. Eick, Cox, and other famous technical visualizers had, in trying to build on viewers' familiarity with airline routes and outlines of nations, inadvertently reproduced outdated assumptions about global development. Colonial stereotypes seemed to return, unchallenged, in digital discourses, despite the strong ways in which they had been challenged in predigital scholarship on media and science. Harpold concluded: "these depictions of network activities are embedded in unacknowledged and pernicious metageographies—sign systems that organize geographical knowledge into visual schemes that seem straightforward but which depend on the historically and politically-inflected misrepresentation of underlying material conditions."[12]

These "underlying material conditions" of global communications have a long history. Harpold's argument invoked a contrast between "beacons of light" and internet cables. In other words, he implied that the arcs of light used to represent internet traffic recalled a familiar narrative of enlightenment and cultural advancement, whereas a mapping of material infrastructure and cartographic histories might have helped build a different narrative.

Harpold hoped, at the time, that investigations of the internet's "materiality" would give us details that might displace colonial and national metanarratives. How do new politics accrete to the cluster points where cables transition from sea to land? How do older geographies of power shape the relations of these nodes to the less-connected hinterlands? Such grounded research would, he hoped, remove the racial, gendered, nationalist cultural assumptions that shaped a first generation of internet maps.

We did indeed get a host of stories about materiality in the decade that followed. Yet our stories about race, gender, and nation did not seem to change as radically as the technologies these stories described. The next two sections explore some influential understandings about communication infrastructures. We find that an emphasis on material infrastructure might be necessary but not sufficient to achieve technologically *and* historically accurate representations of the internet.

THE INTERNET AS A SET OF TUBES: VIRTUAL AND MATERIAL COMMUNICATIONS

In 2006, United States Senator Ted Stevens experienced an unexplained delay in receiving email. In the predigital past, mail delays might have been due to snow, sleet, or sickness—some kind of natural or human unpredictability. But such problems were supposed to be obsolete in the age of the internet.

Stevens suggested that the structure of the internet might be causing his delay: "They want to deliver vast amounts of information over the internet . . . [But] the internet is not something you just dump something on. It's not a truck. It's a series of tubes."[13]

Younger, geekier Americans erupted in hilarity and in outrage. Memes proliferated, mocking the Republican senator's industrial metaphor for the virtual world, his anachronistic image revealing, some suggested, the technological backwardness of politicians. Tech pundit Cory Doctorow blogged about Senator Stevens's "hilariously awful explanation of the Internet": "This man is so far away from having a coherent picture of the Internet's functionality, it's like hearing a caveman expound on the future of silver-birds-from-sky."[14]

The image of information tubes seemed to belong to a "caveman era" in comparison to the high-speed virtual transmission of the information era. For Doctorow, as for many technology analysts, we needed metaphors that conveyed the virtual scope, rather than the material infrastructure, of information technologies. The

Stevens event hit many nerves, intersecting with national debates about net neutrality and technology professionals' anger at a conservative agenda that was perceived as stalling the progress of technological modernity.[15]

But are pipes and tubes really such terrible images with which to begin a conversation about the operation of the internet? Tubes summon up an image not far from the actual undersea cables that really do undergird nearly speed-of-light communication. Although Stevens's tubes seemed drawn from an industrial-era playbook, fiber optics, which enabled speed-of-light transmission by the end of the twentieth century, are tube-like in their structure. Their size, unlike industrial copper cable, is on the order of 50 microns in diameter. Cables a millimeter in diameter can carry digital signals with no loss. But fiber optics are not plug-and-play modules; virtual connections are not created simply by letting light flow along them. Engineers and network designers must figure out how many strands to group together, what material they must be clad in, how to connect several cables over large distances, where to place repeaters, and how to protect them from the elements. The infrastructure of tubes remains grounded in materials science and the detailed geographies of land and sea.[16]

At the end of the twentieth century, public discourse valorized images of information's virtuality, rather than its materiality.[17] The notion of an invisible system that could move information instantaneously around the universe was thrilling to people who thought of industrial technology as slow and inefficient. The promise of the digital was in its clean, transparent, frictionless essence. Internet visionaries were inspired by the new information technologies' ability to transcend the physical constraints and linear logics of the industrial age. This was a new age; metaphors of pipes and tubes threatened to drag technological imaginations backward to a past full of cogs and grease.

Tubes also recall a century-old manner of delivering mail, when canisters literally shot through pneumatic tubes. Post-office worker Howard Connelly described New York's first pneumatic mail delivery on October 7, 1897, in *Fifty-Six Years in the New York Post Office: A Human Interest Story of Real Happenings in the Postal Service*. He reported that the first canister contained an artificial peach and the second a live cat. Megan Garber, writing for the *Atlantic*, commented: "The cat was the first animal to be pulled, dazed and probably not terribly enthused about human technological innovation, from a pneumatic tube. It would not, however, be the last."[18] In Connelly's 1931 autobiography, and at the heart of the things that fascinated his generation, we see an exuberant imagination associated with material objects and the physical constraints of transporting them.

In the shift from physical to virtual mail, tubes fell out of public imagination, even though information tubes of a different kind were proliferating around the world. The 1990s were a boom time for internet cable laying. So, too, were the years after Edward Snowden's 2013 disclosures about NSA surveillance, which spurred many nations to start ocean-spanning cable projects, hoping to circumvent US networks.

Undersea cable technology is not new; in the mid-nineteenth century, a global telegraph network depended on them. In the 1850s, undersea cables were made of copper, iron, and gutta-percha (a Malaysian tree latex introduced to the West by a colonial officer of the Indian medical service). A century later, they were coaxial cables with vacuum tube amplifier repeaters. By the late twentieth century, they were fiber optics. By 2015, 99 percent of international data traveled over undersea cables, moving information eight times faster than satellite transmission.[19] Communication infrastructure has looked rather like tubes for a century and a half.

Infrastructural narratives had been pushed to the background of media consumer imaginations in the 1990s, their earthly bounds overlooked in preference to the internet's seemingly ethereal power. Internet service providers and their advertisers attracted consumers with promises of freedom from the constraints of dial-up modems and data limits. Influential media theory explored the new imaginations and cultural possibilities of virtuality. The space of the virtual seemed to fly above the outdated meatspace problems of justice, race, and gender.

These metaphors and analyses were not limited to those who designed and used the early internet. They began to shape influential models of the economic and cultural world. Mark Poster, one of the first humanists to analyze the internet as a form of "new media," argued that the internet's virtual world inaugurated a new political economy. In the industrial era, economists had measured labor via time spent in agrarian work or assembly-line labor. But in the global, technologically mediated world, he argued in 1990, "Labor is no longer so much a physical act as a mental operation."[20] Poster pioneered the theorization of virtual labor, politics, and culture.

In the 1990s, scholarship, media, and common sense all told us that we were living in a new, virtually mediated mental landscape. To the extent that the virtual power of the internet, and thus of internet-enabled dispersed global production, appeared to fly above industrial-era pipes, tubes, cogs, and other material constraints, it appeared that internet-enabled labor was ushering in a new Enlightenment. This new technological revolution seemed to immerse Western workers in a far denser global interconnectivity than that of their eighteenth-century European ancestors, thus appearing to shape more complex networked subjectivities. But to the extent that Third Worlders still appeared to live in an industrial-era economy, their subjectivities appeared anchored to an earlier point in European history.[21] In other words, these narratives made it seem as if people in poorer economies were living in the past, and people in high-bandwidth richer societies were accelerating toward the future.

THE INTERNET'S GEOGRAPHIC IMAGINARIES

It sounds crazy, but Earth's continents are physically linked to one another through a vast network of subsea, fiber-optic cables that circumnavigate the globe.
—*CNN Money*, March 30, 2012

The financial dynamics of the global economy were radically reshaped by the seemingly instantaneous digital transmission of information. Financial trading practices experienced quantum shifts every time a faster cable technology shaved nanoseconds off a transaction. Despite the centrality of information tubes to their 2012 audience, *CNN Money* writers sought to capture everyman's incredulous response to the discovery of a material infrastructure undergirding the virtual world: "crazy!"[22] This 2012 news item suggests that the assumption that the internet was "virtual" persisted into the second decade of the new millennium. This was part of a now-common narrative convention that tended to emphasize "surprise" at finding evidence of materiality underpinning virtual experiences in cyberspace, a space conceived of as free from territorial and historical constraints.

Despite metaphors such as "information superhighway," which seemed to remediate metaphors of physical road-building that recalled the infrastructural nineteenth century, a twenty-first-century American public, including many critical humanist academics, had come to rely on a set of metaphors that emphasized virtual space. The promise of cyberspace in its most transcendent forms animated a range of Euro-American projects, from cyberpunk fiction to media studies. Virtuality was, of course, tied to cultures with robust infrastructures and high bandwidth.

One magazine that prided itself on innovative technological reporting did, however, call attention to cables and material geographies as early as the mid-1990s. In December 1996, *Wired* magazine brought its readers news about a fiber-optic link around the globe (FLAG), under construction at the time. "The Hacker Tourist Travels the World to Bring Back the Epic Story of Wiring the Planet," the cover announced, with an image of the travelogue author, superstar cyberpunk writer Neal

Stephenson, legs spread and arms crossed, astride a manhole cover in the middle of the Atlantic Ocean,

The "Hacker Tourist" epic was the longest article *Wired* had published. The FLAG cable, at 28,000 kilometers, was the longest engineering project in history at the time. The cable, Stephenson predicted, would revolutionize global telecommunications and engineering. Well-known for his science fiction writing, Neal Stephenson wrote a technopredictive narrative that was accurate in all its major predictions: fiber-optic information tubes did in fact dramatically improve internet speeds.

The essay appealed to roughly the same kind of tech-savvy audience that found Ted Stevens's metaphors caveman-like. Here was a cyberpunk writer, representing the future of the internet via metaphors of pipes and tubes. Two decades later, a leading Indian progressive magazine listed it among the "20 Greatest" magazine stories, crediting it for the revelation of "the physical underpinning of the virtual world."[23]

THIS IS A PIPE: THE INTERNET AS A TRAVELING TUBE

Stephenson introduces his "epic story" via a mapping narrative: "The floors of the ocean will be surveyed and sidescanned down to every last sand ripple and anchor scar."[24] The travelogue is filled with historical allusions. The opening page, for example, is structured in explicit mimicry of an Age of Exploration travel narrative. The essay's opening foregrounds "Mother Earth" and the cable, the "mother of all wires," and celebrates "meatspace," or embodied worlds. The grounding of the story in "Earth" appears to subvert virtuality's abstraction, and the term "meatspace" appears to invert a common dream of uploading of user consciousness to a disembodied "computer."

Like Cory Doctorow in the pipes-and-tubes episode, Stephenson draws a contrast between industrial past and digital future. Unlike Doctorow, though, Stephenson celebrates pipes and tubes, or materiality, rather than virtuality. Stephenson successfully brought attention to the internet's materiality in a decade otherwise obsessed with the virtual and drew in the history of colonial cable routes. It might seem, then, that Stephenson's essay addresses what Terry Harpold identified as the need for materialist narratives of the internet.

Known for the deep historical research he puts into his fiction writing, Stephenson ironically emulates the stylistics of early modern explorers, describing experiences among "exotic peoples" and "strange dialects." All the cable workers we see are men, but they are depicted along a familiar nineteenth-century axis of masculinity.

The European cable-laying workers are pictured as strong, active, uncowed by the scale of the oceans. A two-page photo spread emphasizes their muscled bodies and smiling, confident faces, standing in neoprene wetsuits beside their diving gear on a sunny beach. Their counterparts in Thailand and Egypt are pictured differently. Thai workers are photographed through rebar that forms a cage-like foreground to their hunched bodies and indistinct faces. A portly Egyptian engineer reclines, smoking a hookah.

The representation of non-Western workers as caged or lazy, the structure of the world map that recalls telegraph and cable maps, and the gendered metaphors of exploration and penetration are familiar to historians of empire. Scientific expeditions to the tropics in the eighteenth and nineteenth centuries collected specimens, produced maps, and accumulated resources that were key to the explosion of post-Enlightenment scientific and technological knowledge. Stephenson interweaves this knowledge with global power. Colonial administrators and armies that later used the navigational charts produced by early expeditions, and administrative experiments from policing to census policy, relied on scientific expertise. Stephenson knows, and cheekily redeploys, the historians' insight that Enlightenment science and colonial exploration were mutually constitutive. The language of eighteenth- and nineteenth-century science plays with and rigidifies dichotomies between male-gendered workers and female-gendered earth, emasculated Eastern men and tough Western men, modernity's feminizing effects and technology's remasculinizing potential.[25] These stereotypes originated in nineteenth-century European popular culture and science writing.

Harpold, a cultural critic of technology, had called for an account of materiality as a corrective to what he saw as the semiotic excess in early internet maps. But, as we see in Stephenson's "Hacker Tourist" epic, an emphasis on the internet's materiality doesn't eliminate the racial, gender, and colonial resonances from narratives about it. Rather, the sustained popularity of this hacker travelogue was bound up in the frisson of historical recognition, and approval, with which readers greeted its mash-up of nineteenth-century gendered, raced, and colonial tropes.

In the two decades after Stephenson's travelogue, infrastructure studies boomed in science and technology studies, anthropology, and media studies. These studies further reveal that infrastructure does not provide an escape from politics and history. For example, infrastructure scholars Bowker and Star described the links between infrastructure and historicity in a founding moment of this field: "Infrastructure does not grow *de novo;* it wrestles with the inertia of the installed base and inherits strengths and limitations from that base. Optical fibers run along old railroad lines, new systems are designed for backward compatibility; and failing to account for these constraints may be fatal or distorting to new development processes."[26]

Some critics of the "cultural turn" in technology studies, recoiling at the disturbing historical and political history that seem to be uncovered at every turn, yearned for a return to empirical, concrete studies. For a moment, the focus on materiality and nonhuman actors seemed to offer this.[27] But neither Neal Stephenson's deep dive to the Atlantic Ocean floor nor the rise of "infrastructure studies" yielded clean stories about matter untouched by politics. It seemed to be politics all the way down.

Even as Stephenson waxed eloquent about the new Western infrastructure that was bringing freedom to the world, US news media outlets were filled with reports of a different wave of economic globalization. There was a global shift afoot that in the next decade would alter the economic grounds of this technocolonialist discourse.

OTHER INFRASTRUCTURES

The consortium of Western telecom companies that undertook the massive, adventurous cable-laying venture Stephenson described encountered the digital downturn at the end of the 1990s, going bankrupt during the dot-com bust of 2000.[28] India's telecom giant Reliance bought up FLAG Telecom at "fire-sale prices" at the end of 2003. By 2010, there were about 1.1 million kilometers of fiber-optic cable laid around the globe. By 2012, the Indian telecommunications companies Reliance and Tata together owned "20% of the total length of fiber-optic cables on the ocean floor and about 12% of high-capacity bandwidth capacity across the globe."[29] In the year 2000, the fifty-three-year-old postcolonial nation's popular image was of an underdeveloped economy that successfully used computational technology to leapfrog over its historical legacy of underdevelopment.[30]

Many internet narratives of the 1990s—incorporating assumptions about "advanced" and "backward" stages of growth, light and dark regions of the world, actively masculine and passively feminized populations, and so on—had pointed to real disparities in access to networks, or the "digital divide." But they missed the ways in which the developing world was responding to this challenge by accelerating infrastructure development.

Another economic downturn hit the US in 2008. After Edward Snowden's 2013 allegations about NSA surveillance, Brazil's President Dilma Rousseff declared the need to produce independent internet infrastructure. The US responded with allegations that Rousseff was embarking on a socialist plan to "balkanize" the internet. *Al Jazeera* reported that Brazil was in the process of laying more undersea cable than any other country, and encouraging the domestic production of all network equipment, to preclude the hardware "backdoors" that the NSA was reported to be attaching to US products: "Brazil has created more new sites of Internet bandwidth production faster than any other country and today produces more than three-quarters of Latin America's bandwidth."[31]

In the second decade of the twenty-first century, information's pipes and tubes returned to the headlines. The "BRICS" nations—Brazil, India, China, and South Africa—were jumping over stages of growth, confounding the predictions of development experts. They were increasingly hailed as the superheroes of a future technological world, rather than disparaged as examples of the dismal failures of development and decolonization.

The rise of Asian technology manufacture and services and the shifting domain of economic competition regularly makes headlines and bestseller lists. The 1990s had brought Indian data-entry workers to the global stage through the Y2K crisis. Western corporations, finding skilled computer professionals who fixed a programming glitch for a fraction of the cost of Western labor rates, did not want to give up the valuable technological labor sources they had discovered. Software "body shopping" and outsourced technical labor grew significantly in the decade following Y2K. By 2015, Asian countries had taken the global lead by most indicators of economic progress, particularly in the technical educational readiness of workforces and in infrastructural and technical support for continued development. Yet global political and economic shifts in technology had not permeated popular Western images of information transmission.

There is a representational lag in Western writing about the state of non-Western technological practice. Despite the shifts in technological expertise, ownership, and markets, and the undeniable force of the former colonial world in the technological economy, Western representations of global information and telecommunications systems repeatedly get stuck in an anachronistic discursive regime.

TECHNOLOGY IS PEOPLE'S LABOR

Despite the shift in the ownership of cable and the skilled labor force, Euro-American representations of non-Western workers still recall the hunched-over, lazy colonial subjects that we encountered in Neal Stephenson's ironic replay of nineteenth-century travelogues. By the 2010s, the "material infrastructure" versus "virtual superstruc-ture" split was morphing into debates over "rote/automatable tasks" versus "creative thinking."

Informatics scholar Lilly Irani has argued that "claims about automation are fre-quently claims about kinds of people."[32] She calls our attention to the historical conjunctures in 2006, when Daniel Pink's bestseller, *A Whole New Mind,* summoned Americans to a different way of imagining the future of technological work. In the process, he identified "Asia" as one of the key challenges facing Western nations.[33] Pink, comparing a computer programmer's wage in India with the US ($14,000 versus $70,000), suggests that we abandon boring technical jobs to the uncreative minds of the developing world while keeping the more creative, imaginative aspects of technol-ogy design in Western hands. The problem of Asia, he notes, is in its overproduction of drone-like engineers: "India's universities are cranking out over 350,000 engineering graduates each year." Predicting that "at least 3.3 million white-collar jobs and $136 billion in wages will shift from the US to low-cost Asian countries by 2015," Pink calls for Western workers to abandon mechanical, automaton-like labor and seek what he called "right brain" creative work. In this separation, we see the bold, active imagi-nations of creative thinkers contrasted with the automaton-like engineer steeped in equations and arithmetic. Irani reminds us of the historical context for this policy-oriented project: "The growth of the Internet . . . had opened American workers up to competition in programming and call center work. . . . Pink's book *A Whole New Mind* was one example of a book that stepped in to repair the ideological rupture."[34]

The binaries and geographies associated with Pink's project have shaped policies and plans for design education and infrastructure development in the US. How are Pink's predictions about "creative" tech design dependent on the narratives that excite him (imaginative, creative "design thinking" about technology) and the images that frighten him (hordes of technological experts speaking unknown lan-guages, hunched over keyboards)? What implications do his visions of information design and creativity have for global infrastructure development and technical train-ing? How do Pink's fears draw on two centuries of anxiety about colonial subjects and their appropriation of Western science?[35]

THIS IS NOT A PIPE: ALTERNATIVE MATERIALITIES

Let us return to pipes, tubes, and those now-creaky but still-utopian internet imaginaries.

In an ethnographic study of the internet that uncovers rather different human and technical figures from Neal Stephenson's travelogue, engineer/informatics scholar Ashwin Jacob Mathew finds that the internet is kept alive by a variety of social and technical relationships, built on a foundation of trust and technique.[36] Mathew's story retains Stephenson's excitement about the internet but sets it in a decentered yet politically nuanced narrative, accounting for differences between the US and South Asia, for example, through the details of technical and political resources, rather than via racial essences and gendered stereotypes. In his 2014 study titled "Where in the World Is the Internet?," he describes the social and political relationships required to bring order to the interdomain routing system: we find established technical communities of network operators, organized around regional, cross-national lines; nonprofit, corporate, and state-run entities that maintain internet exchange points; and massive scales across which transnational politics and financial support make the continual, apparently free, exchange of information possible. He tells a dynamic human and technical story of maintaining the minute-by-minute capacities of the internet, undergirded by massive uncertainties but also by cross-national, mixed corporate-state-community relationships of trust.

> Risk and uncertainty in interdomain routing provide part of the justification for trust relationships, which can cut across the organizational boundaries formed by economic and political interests. However, trust relationships are more than just a response to risk and uncertainty. They are also the means through which technical communities maintain the embeddedness of the markets and centralized institutions involved in producing the interdomain routing system. In doing so, technical communities produce themselves as actors able to act "for the good of the internet," in relation to political and economic interests.[37]

Infrastructures, Mathew reminds us, are relations, not things.[38] Drawing on feminist informatics scholar Susan Leigh Star, and using his ethnography of network engineers to articulate how those relations are always embedded in human history and sociality, Mathew distinguishes his analysis from that encapsulated in John Perry Barlow's 1996 *A Declaration of the Independence of Cyberspace*. Avowing that the internet does offer radical new social possibilities, he suggests that these are not because it transcends human politics but rather because it is "actively produced in the ongoing efforts and struggles of the social formations and technologies involved in the distributed governance of Internet infrastructure."[39]

Mathew's aim is "to uncover the internal logic of the production of the virtual space of the Internet, and the manner in which the production of virtual space opposes and reconciles itself with the production of the spaces of the nation state, and the spaces of capital."[40] His study begins with a methodological question as original as Stephenson's. While Stephenson follows a cable ethnographically, Mathew asks: how can one study a protocol ethnographically? In his exploration of the Border Gateway Protocol and the maintenance labor that goes into internet exchange points, he offers us a more complex narrative of the internet. Both authors follow a technological object that cannot be pinned down to one point in space and time. Stephenson circumscribes this nebulous object of study in classic 1980s cyberpunk style, but globalism has changed its face since then. We need new narratives and new questions that acknowledge the internet's non-Western geographies, and Mathew's is just one of the various global voices that depart radically from internet models of the late twentieth century.

Two other internet mappers from the changing contexts of twenty-first-century narratives offer images of the production of virtual spaces. These, it turns out, look nothing like the one-way pipes of 1990s internet maps nor the imperial telegraph cables of the nineteenth century. Internet artists have articulated refreshingly complex and refractory models of the internet in the first decade of the twenty-first century, replaying but modifying the cartographies of the 1990s.

Artist Barrett Lyon's *Opte Project* (opte.org) offers a visualization of the routing paths of the internet, begun in 2003 and continually updated, allowing a time line of dynamic images of decentered globalization patterns (fig. 4.2). Lyon describes the *Opte Project* as an attempt to "output an image of every relationship of every network on the Internet." This is rather different from Cox and Eick's 1995 aesthetic parsimony. Compared to 1990s internet visualizations, which attempted to eliminate confusion and excessive lines, Lyon's approach seems to revel in the excess of data. This design choice is largely attributable, of course, to the massive increase in computational power in the decade or so after Cox and Eick's first visualization schemas. The ability to display massive data sets to viewers fundamentally alters the historical and geographical picture brought by designers to the general public. Lyon's visualizations retain maximalist forms of route data, incorporating data dumps from the Border Gateway Protocol, which he calls "the Internet's true routing table." The project has been displayed in Boston's Museum of Science and New York City's Museum of Modern Art. Lyon reports that it "has been used an icon of what the Internet looks like in hundreds of books, in movies, museums, office buildings, educational discussions, and countless publications."[41]

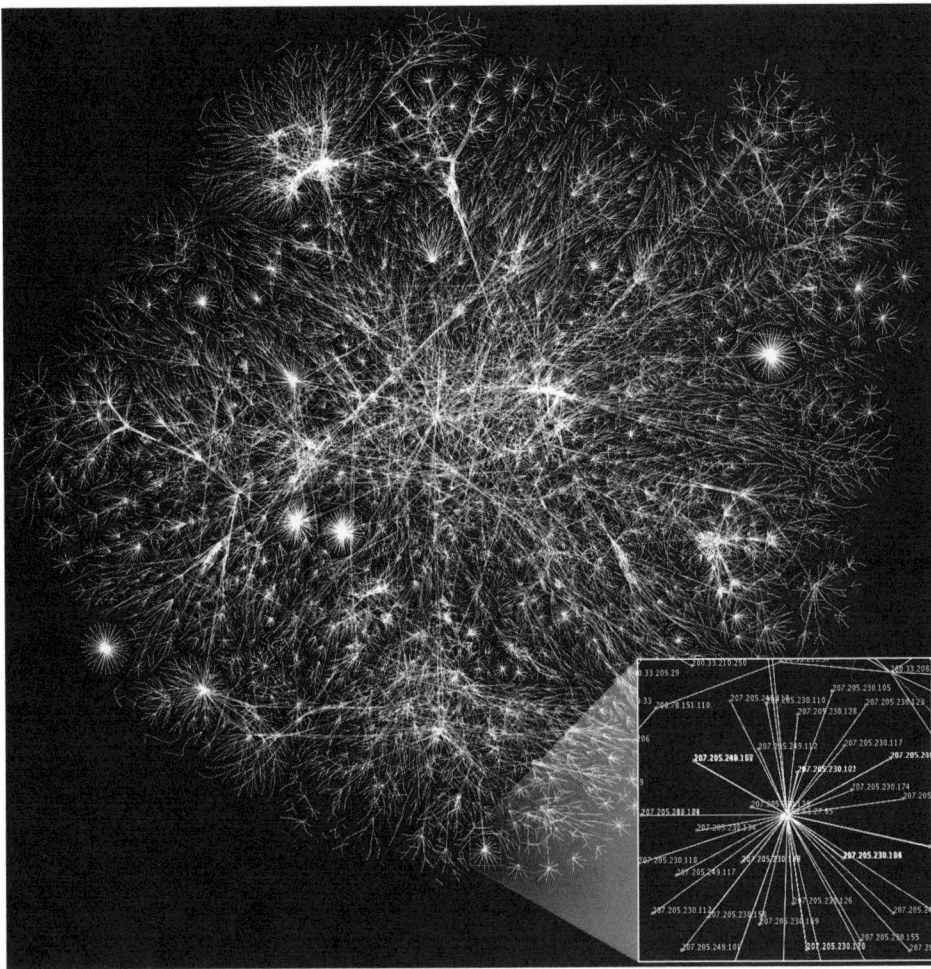

Figure 4.2   Barrett Lyon's internet map (2003).

Artist Benjamin De Kosnik offers an idiosyncratic imaging of the internet in a collaborative digital humanities project called Alpha60. Beginning with a case study mapping illegal downloads of *The Walking Dead,* he proposes, with Abigail De Kosnik, to develop Alpha60 into "a functional ratings system for what the industries label 'piracy.'" Tracking, quantifying, and mapping BitTorrent activity for a range of well-known US TV shows, De Kosnik and De Kosnik analyze images of the world that are both familiar (in that they confirm common-sense ideas about piracy and the diffusion of US popular culture) and surprising (in that they demonstrate patterns of activity that contradict the 1990s model of wired and unwired nations, bright ports and dark hinterlands, and other nation-based geographic metaphors for connectivity). This has immediate critical implications for media distribution. They write: "We do not see downloading activity appearing first in the country-of-origin of a television show . . . Rather, downloading takes place synchronously all over the world. . . . This instantaneous global demand is far out of alignment with the logics of the nation-based 'windowing' usually required by international syndication deals."[42] In a data-driven, tongue-in-cheek upending of conventional media policy, they show how "over-pirating" happens in concentrations of a "global tech elite." They overturn our assumptions about where piracy occurs, suggesting that the development and deployment of technology in the West often goes hand-in-hand with practices that disrupt inherited ethical and legal norms, but elite disruption doesn't generate the same kind of criminal stereotypes as poor people's pirating.

The internet itself was not one thing in the 1990s. Its shift away from a defense and research backbone toward a public resource via the development of the World Wide Web is a military-corporate history that explains some of the confusion among media narratives. And technological freedom did not bring freedom from history and politics. The internet looks like a pipe in many places, but it moves along a radically branching structure and it is powered by human labor at all its exchange points. In the 1990s, internet narratives gave us a picture of a pipe constructed from colonial resources and bolstered by imperialist rhetoric, but, in more recent research, it appears to be uneven, radically branched, and maintained by a world of diverse people.

The internet is constituted by entangled infrastructural and human narratives that cannot be understood via separate technological and humanistic histories. The historical forces that simultaneously shaped modern ontological assumptions about human difference and infrastructures of technological globalization are alive today. Racial, gendered, and national differences are embedded, in both familiar and surprising ways, in a space that John Perry Barlow once described as a utopia free from the tyranny of corporate and national sovereignty. With sympathy for Barlow's passionate manifesto for cyberspace's independence from the prejudices of the industrial era ("We are creating a world that all may enter without privilege or prejudice accorded by race, economic power, military force, or station of birth," he announced in Davos in 1996),[43] this chapter recognizes, also, that we cannot get there from here if we uncritically reuse the narratives and technologies thus far deployed by the internet's visionaries, bureaucrats, builders, and regulators.

Adding universalist human concerns to a globalizing internet infrastructure (as well-meaning technology planners suggest) is not sufficient. Engaged scholars and designers who wish to shape future openness in internet infrastructures and cultures need to do more than affirm an abstract humanism in the service of universal technological access. Racial, gendered, and nationalist narratives are not simply outdated vestiges of past histories that will fall away naturally. They are constantly revivified in the service of contemporary mappings of materials, labor, and geopolitics. The internet is constantly being built and remapped, via both infrastructural and narrative updates. The infrastructural internet, and our cultural stories about it, are mutually constitutive.

The internet's entangled technological and human histories carry past social orders into the present and shape our possible futures. The concrete, rubber, metal, and electrical underpinnings of the internet are not a neutral substrate that enable human politics but are themselves a sedimented, multilayered historical trace of three centuries of geopolitics, as well as of the dynamic flows of people, plants, objects, and money around the world in the era of European colonial expansion. We need new narratives about the internet, if we are to extend its global infrastructure as well as diversify the knowledge that it hosts.

But first we must examine the narratives we already inherit about the internet and about information technology. Many of us still think of information technology as a clean, technological space free from the messiness of human politics. Clean technological imaginaries, free of politics, can offer a tempting escape from history's entanglements and even facilitate creative technological design in the short run. But embracing the messiness, this chapter suggests—acknowledging the inextricability of the political and the digital—might enable better technological and human arrangements.

The ideas that excite artists like Benjamin De Kosnik and Barrett Lyon and media/technology scholars like Abigail De Kosnik and Ashwin Jacob Mathew, and the internet models that undergird this new research, are rather different from those that animate Daniel Pink's global economic forecasts, Stephenson's material-historical cyberpunk metaphors, or Cory Doctorow's optimism about virtuality. As varied as the latter Western representations were at the turn of the twentieth century, they have already been surpassed by the sheer complexity, unapologetic interdisciplinarity, and transnational political nuance that characterize a new generation of global internet analysts.

POWER'S PLUMBING

From claims about the "colonial" internet in the early twenty-first century, this chapter has led us backward in time. In order to understand why the Wikimedia Foundation found itself allied with decolonial social movements,[44] we broadened our story. Concerns about social justice and decolonization have begun to shape many of the new analyses not just about the internet and its participants but about all internet-driven data. The "decolonial" urge to democratize the production, dissemination, and ownership of knowledge has proved inextricable from material understandings of the internet. Thus we moved from statistics about identities of users and producers of online knowledge to infrastructural issues about who lays, maintains, and owns the infrastructures (servers, cables, power, and property) that enable this medium, as well as representational strategies by which we map and weave together human and technological narratives.

We have seen how forms of representation of information online are inseparable from the material histories of communication. Older, colonial-era infrastructures predated and enabled twentieth-century cable-laying projects. Geopolitics as well as undersea rubble repeatedly break cables or shift routing paths. Representational inequities cycle into policy-making about regulating access to knowledge. Once seen as a neutral technological affordance, the internet is increasingly beginning to be understood as a political domain—one in which technology can enhance some kinds of political power and suppress others, radically shifting our notions of democratic public sphere. Celebrating this new technology that set the world on fire in the late twentieth century, this chapter seeks not simply to tar it with the brush of old politics but to ask how this fire, in all its social, political, and technological aspects, can be tended in the interests of an increasingly diverse global community of users.

As in Sarah Roberts's chapter ("Your AI Is a Human"), it turns out that the infrastructural internet and the creators of its imaginaries are connected by the labor of global humans. Like Nathan Ensmenger ("The Cloud Is a Factory") and Benjamin Peters ("A Network Is Not a Network"), I turn to the history of the material, infrastructural view of the internet—a version of the internet's origin story that was once articulated primarily by infrastructural engineers.[45] But we found that there's more than infrastructure at play here. As Halcyon Lawrence ("Siri Disciplines") and Tom Mullaney ("Typing Is Dead") suggest, nonwhite users are increasingly resisting the imperial norms that shape our internet-related practices. National histories and imperial politics help select the stories we tell about the internet's past, as well as shape what we can do with its future. Seeing this requires skills in analyzing metaphor, narrative, and social history. We find that the separation of technical and cultural skills has produced knowledge that impedes the integrated improvement of technical design and use. As decolonial technological activism grows, the power dynamics involved in this historical separation of technology and culture will come under intense political and cultural scrutiny.

To revalue and resituate both technical and humanist knowledge, we need to find better ways to teach multiple kinds of skills in media, information, and computer science's educational and practical contexts. We need stories that showcase the irreducibly political realities that make up the global internet—the continual human labor, the mountains of matter displaced, its diverse material and representational habitats. Acknowledging that the political and the digital cannot be separated might serve as an antidote against the seductive ideal of technological designs untouched by the messiness of the real world. Examining the historical underpinnings of our common-sense understandings of the internet can lead us to insights about the ways in which we shape ourselves as modern, globally connected technological designers and users. Using this historical/technical lens, we are able to ask questions about the plumbing of power implicit in every internet pipe: who labored on these connections, who owns them, who represents them, who benefits from them? Tracing these things, their maintenance, as well as their shifting cultural understandings reveals the myriad ways in which the internet is already being decolonized.

NOTES

1. Julia Jaki, "Wiki Foundation Wants to 'Decolonize the Internet' with more African Contributors," *Deutsche Welle* (July 19, 2018), https://www.dw.com/en/wiki-foundation-wants-to-decolonize-the-internet-with-more-african-contributors/a-44746575.

2. Whose Knowledge?, accessed August 1, 2018, https://whoseknowledge.org/decolonizing-the-internet-conference/.

3. See, for instance, Martin Macias Jr., "Activists Call for an End to LA's Predictive Policing Program," *Courthouse News Service* (May 8, 2018), https://www.courthousenews.com/activists-call-for-an-end-to-las-predictive-policing-program/; the work of the activist "Stop LAPD Spying Coalition," https://stoplapdspying.org/; and the activist-inspired academic work of the Data Justice Lab at Cardiff, https://datajusticelab.org.

Decolonial history and theory shape data analysis in Nick Couldry and Ulises Ali Mejias, *The Costs of Connection: How Data Is Colonizing Human Life and Appropriating It for Capitalism* (Stanford, CA: Stanford University Press, 2019), accessed August 16, 2019, https://colonizedbydata.com/.

4. On the politicization of developing-country data, see Linnet Taylor and Ralph Schroeder, "Is Bigger Better? The Emergence of Big Data as a Tool for International Development Policy," *GeoJournal* 80, no. 4 (August 2015): 503–518, https://doi.org/10.1007/s10708-014-9603-5, and

Morten Jerven, *Poor Numbers: How We Are Misled by African Development Statistics and What to Do about* It (Ithaca, NY: Cornell University Press, 2013).

5. Overall, connectivity has increased in Africa in the twenty-first century, but it remains uneven and tied to national and colonial histories. International Telecommunications Union (ITU) statistics offer us a snapshot: Between 2000 and 2017, internet connectivity in Angola increased from 0.1% to 14% of the population, in Somalia from 0.02% to 2%, in South Africa from 5% to 56%, and in Zimbabwe from 0.4% to 27%. In the US, internet connectivity grew from 43% to 87% in the same period. ITU Statistics (June 2018 report on "Country ICT Data 2000–2018)," https://www.itu.int/en/ITU-D/Statistics/Pages/facts/default.aspx. Statistics cited from ITU Excel spreadsheet downloaded August 15, 2019.

6. P. J. Hugill, *Global Communications since 1844: Geopolitics and Technology* (Baltimore: Johns Hopkins University Press, 1999), 2.

7. Hugill, *Global Communications since 1844, 18.*

8. Susan Leigh Star and Geoff Bowker, *Sorting Things Out: Classification and Its Consequences* (Cambridge, MA: MIT Press, 1999).

9. Kenneth C. Cox and Stephen G. Eick, "Case Study: 3D Displays of Internet Traffic," in *Proceedings on Information Visualization* (INFOVIS '95) (1995), 129.

10. For an overview of internet maps, see "Beautiful, Intriguing, and Illegal Ways to Map the Internet," *Wired* (June 2015), https://www.wired.com/2015/06/mapping-the-internet/. Spatial/technology mapping has evolved into a tool critical to business logistics, marketing, and strategy, see e.g. https://carto.com/, accessed August 15, 2019.

11. Martin W. Lewis and Kären E. Wigen, *The Myth of Continents: A Critique of Metageography* (Berkeley: University of California Press, 1997), 200.

12. Terry Harpold, "Dark Continents: A Critique of Internet Metageographies," *Postmodern Culture* 9, no. 2 (1999), https://doi.org/10.1353/pmc.1999.0001.

13. See https://en.wikipedia.org/wiki/Series_of_tubes.

14. Cory Doctorow, "Sen. Stevens' hilariously awful explanation of the Internet," BoingBoing (July 2, 2006), https://boingboing.net/2006/07/02/sen-stevens-hilariou.html.

15. Princeton professor Ed Felten paraphrased Stevens's quote as: "The Internet doesn't have infinite capacity. It's like a series of pipes. If you try to push too much traffic through the pipes, they'll fill up and other traffic will be delayed." https://en.wikipedia.org/wiki/Series_of_tubes.

16. For the range of technical concerns that go into the cable's material and physical structure, see, e.g., https://www.globalspec.com/learnmore/optics_optical_components/fiber_optics/fiber_optic_cable.

17. The virtuality/materiality divide was vigorously debated in informatics and science and technology studies. Infrastructure analysis was pioneered by feminist scholars. See, e.g., Lucy Suchman, *Human-Machine Reconfigurations: Plans and Situated Actions* (Cambridge: Cambridge University Press, 2012); Katherine N. Hayles, *Writing Machines* (Cambridge, MA: MIT Press, 2002); Susan Leigh Star and Odense Universitet, *Misplaced Concretism and Concrete Situations: Feminism, Method and Information Technology* (Odense, Denmark: Feminist Research Network, Gender-Nature-Culture, 1994); Paul Duguid, "The Ageing of Information: From Particular to Particulate," *Journal of the History of Ideas* 76, no. 3 (2015): 347–368.

18. Howard Wallace Connelly, *Fifty-Six Years in the New York Post Office: A Human Interest Story of Real Happenings in the Postal Service* (1931), cited in Megan Garber, "That Time People Sent a Cat Through the Mail," *Atlantic* (August 2013), https://www.theatlantic.com/technology/archive/2013/08/that-time-people-sent-a-cat-through-the-mail-using-pneumatic-tubes/278629/.

19. Data from TeleGeography, https://www2.telegeography.com/our-research. TeleGeography's assessment of 99% of all international communications being carried on undersea fiber optic cables was also reported by the "Builtvisible" Team in "Messages in the Deep" (2014), https://

builtvisible.com/messages-in-the-deep/. For an ethnogeography of these cables, explicitly inspired by Neal Stephenson's cable travelogue, see Nicole Starosielski, *The Undersea Network* (Durham, NC: Duke University Press, 2015). For continuing updates on cable technology, see https://en.wikipedia.org/wiki/Submarine_communications_cable.

20. Mark Poster, *The Mode of Information: Poststructuralism and Social Context* (Chicago: University of Chicago Press, 1990), 129.

21. Johannes Fabian, *Time and the Other: How Anthropology Makes Its Object* (New York: Columbia University Press, 2014).

22. "A giant undersea cable makes the Internet a split second faster," *CNN Money* (February 30, 2012). On remediated metaphors in the digital world, see Jay David Bolter and Richard Grusin, *Remediation: Understanding New Media* (Cambridge, MA: MIT Press, 2003). For nineteenth-century metaphors in modern road-building, see *America's Highways 1776–1976: A History of the Federal Aid Program, Federal Highways Administration* (Washington, DC: US Government printing Office, 1977).

23. "The 20 Greatest Magazine Stories," *Outlook Magazine*, November 2, 2015, https://www.outlookindia.com/magazine/story/the-20-greatest-magazine-stories/295660.

24. Neal Stephenson, "The Epic Story of Wiring the Planet," *Wired* (December 1996), 160.

25. Neal Stephenson's reinvention of Enlightenment science and travel narratives is more extensively explored in "'Travel as Tripping': Technoscientific Travel with Aliens, Gorillas, and Fiber-Optics," Kavita Philip, keynote lecture, Critical Nationalisms and Counterpublics, University of British Columbia, February 2019.

26. Susan Leigh Star and Geoffrey Bowker, *Sorting Things Out: Classification and Its Consequences* (Cambridge, MA: MIT Press, 1999), 35.

27. Severin Fowles, "The Perfect Subject (Postcolonial Object Studies)," *Journal of Material Culture* 21, no. 1 (March 2016), 9–27, https://doi.org/10.1177/1359183515623818.

28. "The Dot-Com Bubble Bursts," *New York Times* (December 24, 2000), https://www.nytimes.com/2000/12/24/opinion/the-dot-com-bubble-bursts.html.

29. High-capacity bandwidth is defined as 1 million Mbps or more. All data from consulting firm Terabit, reported by *Economic Times* (April 2012), http://articles.economictimes.indiatimes.com/2012-04-15/news/31342442_1_undersea-cable-submarine-cables-fibre.

30. J. P. Singh, *Leapfrogging Development? The Political Economy of Telecommunications Restructuring* (Binghamton: State University of New York Press, 1999).

31. Bill Woodcock, "Brazil's official response to NSA spying obscures its massive Web growth challenging US dominance," *Al Jazeera* (September 20, 2013); Sascha Meinrath, "We Can't Let the Internet Become Balkanized," *Slate* (October 14, 2013), http://www.slate.com/articles/technology/future_tense/2013/10/internet_balkanization_may_be_a_side_effect_of_the_snowden_surveillance.html.

32. Lilly Irani, "'Design Thinking': Defending Silicon Valley at the Apex of Global Labor Hierarchies," *Catalyst: Feminism, Theory, Technoscience* 4, no. 1 (2018).

33. Daniel H. Pink, *A Whole New Mind: Why Right-Brainers Will Rule the Future* (New York: Riverhead Books, 2006). Pink identifies "abundance, Asia, and automation" as the three key challenges to Western dominance of the global technological market.

34. Irani, "Design Thinking."

35. The full import of these questions takes us outside the scope of this chapter's interest in internet narratives. For an exploration of these, see Lilly Irani, *Chasing Innovation: Entrepreneurial Citizenship in Modern India* (Princeton, NJ: Princeton University Press, 2019), which offers a transnationally researched analysis of computational "design thinking" as a new imperial formation.

36. Ashwin Jacob Mathew, "Where in the World Is the Internet?," PhD dissertation, School of Informatics, UC Berkeley, 2014.

37. Mathew, "Where in the World Is the Internet?"

38. Susan Leigh Star and Karen Ruhleder, "Steps toward an Ecology of Infrastructure: Design and Access for Large Information Spaces," *Information Systems Research* 7, no. 1 (1996), 111–34.

39. Mathew, "Where in the World Is the Internet?," 230.

40. Mathew, "Where in the World Is the Internet?," 231.

41. Barrett Lyon, The Opte Project, opte.org/about, retrieved February 1, 2019. Archived at https://www.moma.org/collection/works/110263 and the Internet Archive, https://web.archive.org/web/*/opte.org.

42. Abigail De Kosnik, Benjamin De Kosnik, and Jingyi Li, "A Ratings System for Piracy: Quantifying and Mapping BitTorrent Activity for 'The Walking Dead,'" unpublished manuscript, 2017.

43. John Perry Barlow, "A Declaration of the Independence of Cyberspace," Davos, Switzerland, February 8, 1996, Electronic Frontier Foundation, https://www.eff.org/cyberspace-independence.

44. The Wikimedia Foundation supports efforts to decolonize knowledge because, simply, their claim to house the sum of all knowledge in the world has been shown, by social movements from the Global South, to be woefully incomplete. Like the "World" Series in baseball, it turned out that a national tournament had been erroneously billed as global. Recognizing "a hidden crisis of our times" in the predominance of "white, male, and global North knowledge" led to the forging of the world's first conference focused on "centering marginalized knowledge online," the Whose Knowledge? conference. Thus the Wikipedia project promises to go beyond its predecessor and original inspiration, the eighteenth-century encyclopedia project. See also https://meta.wikimedia.org/wiki/Strategy/Wikimedia_movement/2017, retrieved August 1, 2018. Aspiring to create a "world in which every single human being can freely share in the sum of all knowledge," the Wikimedia Foundation underwrites human and infrastructural resources for "the largest free knowledge resource in human history."

45. The importance of the material underpinnings of virtual technologies is a thriving subfield in the humanities; see Susan Leigh Star, "The Ethnography of Infrastructure," *American Behavioral Scientist* 43, no. 3 (1999): 377–391, https://doi.org/10.1177/00027649921955326; and Tung-Hui Hu, *A Prehistory of the Cloud* (Cambridge, MA: MIT Press, 2016).

5

CAPTURE IS PLEASURE

Mitali Thakor

CAPTURING THE UNDETECTABLE

In January 2018, thousands of Facebook users were shocked by a video circulating through their feeds and messages. The video showed an adult man sexually assaulting a young girl. Viewers left comments on the video, expressing their disbelief and disgust at seeing it, tagged friends to alert them, and flagged the video to report it to Facebook. The video, originally recorded in Alabama, reached users as far away as France within hours. It remained online for two days before Facebook managed to remove it. The man in the video was not apprehended until he turned himself in to police in Montgomery, Alabama, one week after the viral video incident. Germaine Moore, a middle-aged man from Detroit, faced eleven counts of federal charges for the production and distribution of child pornography, as well as one count of first-degree rape, three counts of traveling to meet a child for an unlawful sex act, and one count of facilitating the travel of a child for an unlawful sex act in Alabama.[1] He was later sentenced to fifty years in prison for three counts of production of child pornography.[2]

While this abuse case is clearly disturbing and distressing, I want to focus on the key elements of the video's circulation and deletion from Facebook: Why did it take so long for the video to be flagged and removed? Why did users continue to circulate the video despite its clearly violent content? Ruchika Budhraja, a director for safety at Facebook, warned in a press statement that "Sharing any kind of child exploitative imagery using Facebook or Messenger is not acceptable—even to express outrage. We are and will continue to be aggressive in preventing and removing such content from our community."[3] In 2018 Facebook came under fire as users and politicians questioned the platform's role as an arbiter of content published by individual users, news media, corporations, and bots.[4] Facebook was embroiled in the Cambridge Analytica election manipulation scandal and allegations of fake accounts, fake news, and hate speech on the site; therefore, it would be easy to believe the Facebook spokesperson's comments are just one more promise to proactively police content on their platform. However, Facebook has been moderating child abuse content for one decade, nearly as long as the site has been operating. Along with other technology companies, Facebook is obligated by a federal law, the 2008 PROTECT Our Children Act,[5] to report instances of child abuse on their platforms to federal law enforcement. The legislative package of the Fight Online Sex Trafficking Act and Stop Enabling Sex Traffickers Act, passed in 2018, further penalizes technology providers for failing to flag and remove such content.[6] In order to comply with these legal mandates, a new networked infrastructure has developed to assist platforms in monitoring the photos and videos of potential violence that are uploaded and streamed every day. This new infrastructure revolves around key corporate bodies designing in-house computer algorithms to work alongside existing law enforcement and child protection specialists. Facebook has emerged at the forefront of a push to proactively police harmful images on digital platforms, sponsoring child abuse conferences and deploying its own team of artificial intelligence researchers to develop image-recognition software to identify instances of child nudity, abuse, and recognizable offender faces. As the Germaine Moore case shows, the implementation of image-recognition software on social media platforms is not without pitfalls, as Moore "failed" to be identified by current facial recognition algorithms. Some attributed this failure to Moore's race (he is African American), echoing the growing concern that facial-recognition algorithms are biased toward more accurate detection of white skin tones over people of color—a critical point I will return to later in this essay.

On a global scale, the very core of policing has fundamentally transformed under the auspices of child pornography detection across digital platforms and national jurisdictions. The first change is structural: new technical personnel are being hired within both law enforcement and technology companies to work on reviewing illicit and violent images. Policing is expanding into nonjudicial realms, that is, increasingly performed by the tech companies themselves. The second change is softer but more insidious: under legal mandates to address the most egregious crimes, tech platforms are building an infrastructure keyed toward better image recognition and data collection, one that is steadily expanding a global digital surveillance state.

We might consider the new personnel entering the policing domain as algorithmic detectives, racing to resolve abuse cases alongside more traditional law enforcement and child protection agencies. These new algorithmic detectives exemplify how policing has expanded from so-called traditional law enforcement into a diverse array of technology specialists, computer programmers, content reviewers, and social media company executives who participate in establishing shared ways of seeing child pornography. Shared ways of seeing index shared *feelings* for images, gut desires to apprehend as fast as possible in order to execute arrests. To police images of child pornography requires a peculiar contortion of visual pleasure—a way of seeing that instantly locates abuse images as evidence of crime and the perpetrator of crime as the object of attention.

To be clear, the pleasure of recognizing offenders is *not* the same as the warped gratification experienced by a viewer of child pornography. Rather, I am calling attention to the satisfaction of embodied sensing needed for algorithmic detectives to work with image-detection software to repeatedly "see" child pornography in order to pursue criminal investigation. By foregrounding embodied sensing, I pay attention to the particular pleasures of policing and apprehending offenders like Germaine Moore. An attention to this "pleasure" is the key to understanding the new digital surveillance state we already, comfortably, live with.

Image-detection forensic software is a technology of apprehension, a search tool with the purpose of expediting arrests through the generation of incriminating data. Capture is the process of making apparent, the procurement of identifying data to build a corpus of evidence. Capture[7] involves photographic recognition and image simulation, a capturing of likeness to align with a suspected offender. To know *what* to detect one must become knowledgeable about the object *sought*. Germaine Moore, the suspect in the viral video case, was virtually *un*detectable, appearing in a video rather than still image and having never had his photograph logged in a prior arrest database against which to match his face. Further still, as a person of color Moore remains less detectable by facial-recognition algorithms that continue to bias accuracy toward white male faces.[8]

APPREHENDING IS EMBODIED SENSING

To police is to apprehend—to recognize. Recognition is recall, data processing, and identification. To recognize is to bring close, to know again, to make one's own.

Recognition is a condition of proximity and closeness, or how near both beloved and despised people are presumed to be. Hateful individuals—such as child pornographers—are recognized out of the desire to keep them at a distance. The internet is ablaze with such threats, making the online child abuser seem always proximate, always lurking and then found. The implication with such public arrests is that reprehensible others must be recognized to be captured. They must be studied and "brought close" in order to be kept away.[9] In a world structured by risk and everyday fears,[10] potential abusers are assumed to always be creeping nearby, lying in wait. We live in a digital age dominated by the feeling of paranoia. Paranoia[11] is that particular signature of fear that renders us subjects, allowing us to envision ourselves as part of a larger global system, one in which we are actively clicking our participation.

The response to Germaine Moore's viral abuse video reveals our core impulses toward recognition—the desire to point out and identify the offender through bubbling frenzies of commenting and reposting the video in order to hasten apprehension. We have seen this phenomenon over and over again. Consider, for example, the rush to identify potential suspects in the aftermath of the Boston Marathon bombing of 2013. Users on Reddit, Twitter, Facebook, and other platforms scrambled to post and parse through cellphone footage of the street scenes in vigilante attempts at recognition and apprehension. The eventual capture of one suspect, and the death of another, rent the city with a piercing relief. Perhaps the offenders were innocent until proven guilty, but until then, apprehension—sleuthing, positive identification, recognition—had been made. To be clear, the digital vigilantes who claimed to be hunting the bombers did not actually make the arrest possible; if anything, the digital suggestions heightened the paranoid targeting and harassment of brown, Middle Eastern, and Muslim people in the city. But while the public had not been the linchpin in identifying the Tsaernev brothers, they had nonetheless succeeded in making the crime visible, in spectacular fashion.

People are good at playing at recognition. Digital vigilantism based on facial recognition is made banal, even playful,[12] in the proliferation of everyday software rooted in recognition, from automated photo album tagging of friends to Google's Arts and Culture app matching one's face with a portrait from (Western) art history. Recognition of one's likeness is comfortable, pleasurable. It is made easy through popular encounters with image-classification algorithms. Recognition thrills. Self-recognition offers joy; recognition of the other, however—that which is not the self or an object of admiration—elicits a spike of repulsion and malice. Recognition is pleasure for this conjoined moment: the marking and apprehension of lovable selves and reprehensible others.

Algorithms offer us abstract comforts, experienced pleasurably as they assist in image recognition for family photo albums and experienced queasily as they are operated by third-party firms to suggest advertising targets or political preferences. Many scholars have begun critiquing big data collection by corporations and governments,[13] as well as exposed issues of racial and gendered bias in how machine-learning algorithms view and classify people of color and others considered nonnormative.[14] In her study of facial-recognition systems, for example, Kelly Gates notes how militarized biometrics have seeped into widely available commercial technologies, blurring the lines between law and industry as designers and users crave improved accuracy and speed of recognition.[15] Corporations and law enforcement promote a "catch-all"[16] approach to data collection, a speculative practice that assumes that when we possess a complete and comprehensive archive, then the truth shall yield itself. However, the algorithmic sorting of data does not mean full automation, that humans are removed from any process. Far from it, especially in the case of image detection for the purpose of policing. For despite increasing calls for full automation of child pornography detection, "so that no one will ever have to see these images,"[17] this process can never be fully automated. The work of child pornography detection relies on the discursive and bodily collaboration between human and machine, as "elite content reviewers"[18] make specific decisions to escalate cases toward arrest.

This form of content review extends policing into extrajudicial realms. Content reviewers, at a certain scale, view photographs as empirical data to generate evidence in reporting cases to law enforcement. For example, in one visit to Facebook's main campus in Menlo Park, California, in the spring of 2014, I shadowed some of the staff on the Community Operations team. Community Operations is one of several entities at Facebook that deal with child exploitation issues, alongside Trust and Safety and Marketing managers. I sat at a long conference table in a sunny boardroom with Joan and Rhea, and a third manager, Mona, videoconferencing in from Dublin. The team walked me through a demo of their review process using a software program called PhotoDNA. Once images had been flagged[19] as potentially containing child abuse or nudity, the team needed to determine if other similar photos existed, and if so, if the image file needed to be further "escalated." The software would help them rapidly check for other similar photos, but it would then be up to the team to discuss if the image did in fact seem to be part of a child pornography case (as opposed to, say, an incorrectly flagged photo of a child at the beach or in the bathtub). The images would be written up into a file and reported to the National Center for Missing and Exploited Children (NCMEC), a nonprofit based near Washington, DC, that serves as the federal clearinghouse for child pornography content review. If NCMEC reported back to Facebook that the image was unknown or seemed to be a "live" case, Mona and her colleagues would escalate the case to work with law enforcement to locate other identifying information. They went on to explain that oftentimes images reported to their team were so-called "borderline" cases flagged by content moderators for further review. Large companies like Facebook, Microsoft, and others outsource image content review to contract workers often based outside of the US: Sarah Roberts and the journalist Adrian Chen[20] have extensively documented the precarity of "commercial content moderators,"[21] who perform the first pass on images that might contain child abuse or nudity. These flagged images are then sent to Community Ops to run through their internal database and report on to NCMEC and law enforcement as necessary. Borderline cases, Mona explained, might be images featuring persons of unknown age, or nudity without implied sexual or abusive activity. In these cases, the image would be discussed and reviewed in person by the team in a meeting room like the one we were sitting in to determine which forms of forensic analysis would be necessary. The viewing and classification of photos is thus a discursive, collaborative process between skilled human viewers and their image-recognition algorithms.

My argument contradicts the prevalent discourse that all labor is supposedly heading toward automation, a discourse that affords epistemological supremacy to algorithms. In fact, the indexical work of elite content reviewers hinges upon a human memory cache of previously viewed and filed images. These viewers rely on a combination of their tacit knowledge and accumulated expertise as they sift through child abuse image content to identify faces and bodies and draw on instinctual feelings on when to pursue or escalate a case.

The database against which NCMEC would match reported photos is the National Child Victim Identification System, established in 2002 along with the Child Victim Identification Program and containing images contributed by local, state, federal, and international law enforcement. Immigration and Customs Enforcement (ICE) also runs a database that contains filed images of child abuse. As of 2012, analysts had reviewed more than seventy-seven million images and videos.[22] NCMEC reports that approximately 5,400 children have been identified through the database review and matching process from 2002 to 2014.

I visited the NCMEC offices several times in 2013, shadowing an analyst with the Child Victim Identification Program. The analyst, Linda, walked me through demos of another image detection software program, NetCleanAnalyze,[23] as it matched uploaded photos with images from NCMEC's database. At various intervals in the software's detection, Linda would gesture and make side comments on what the program was doing. She scrolled down the results display and paused to point out a couple of photos. "See, the child looks a little older here, or at least larger. So this is a problem." A problem, in the sense that if the child seems to have aged during the course of the reporting, and various photos at different ages are on file, it is more likely that the child is still being abused and coerced into posing for such photos. Despite the automated matches, Linda still continued to pause and point out her own data points in the images. Her remark that "this is a problem" indicated a gut feeling—her tacit knowledge—that some images represented possible other ages and the need for further investigation. As the anthropologist Charles Goodwin has noted, gestures like Linda's pointing and verbal cues are embodied communicative modes that structure practice and indicate expertise.[24] Her expertise was interactional,[25] reflecting the dynamic relationship between software and human at a corporeal level. Image content review is perceptive, interpretive work, and such work is also trainable. Once Linda's search query results in several hundred photos, she searches through the results to disqualify any images that are very obviously—to her eye— not a match or should not belong in the set. In doing this disqualification, she helps train the software to produce more accurate results in its next search query based on the classification elements of the qualifying and disqualifying images. Such work entails skilled vision, "a capacity to look in a certain way as a result of training."[26] My understanding of apprehension, then, takes into account how "viewing"[27] becomes a skillful practice through trainings that establish shared ways of "seeing" child exploitation images. Different actors—be they law enforcement investigators, digital forensic startups, social media company reviewer teams, or the outsourced content review workers who are contracted by larger corporations to make the first reviews of potentially abusive images—learn to adapt shared ways of seeing images. Cristina Grasseni emphasizes that "one never simply looks. One learns *how* to look."[28] Such ways of seeing,[29] distributed and honed across human and computer vision, become manifest as ways of accessing the world and managing it.

SEEING LIKE AN IMAGE-RECOGNITION ALGORITHM

The image forensics software used by NCMEC searches through an in-house database of known images of child pornography to see if the new image might be similar, or even identical, to an image that has already gone through that system. If two images are exactly identical, they will have the same hash value. Hash matching is a simple form of image classification conducted at many of the agencies that are involved in child pornography detection. However, as one expert explained to me, hash matching techniques have historically prioritized detecting offenders over victims. I interviewed Special Agent Jim Cole, of US Homeland Security, to better understand law enforcement's use of hash sets. Agent Cole heads the Victim Identification Program lab at the Cyber Crimes Center housed within the Department of Homeland Security, a program of US Immigrations and Customs Enforcement (ICE), and we first met at a law enforcement training presentation at the international Crimes Against Children Conference in Dallas, Texas, in 2013. Cole explained that his lab at Homeland Security used hash values to perform "signature" matching between images. He explained, "Every image file that's seen by a computer is made of 1s and 0s. You can run an algorithm over those files. The most common in law enforcement is Message Digest 5, 'MD5,' or Secure Hash Algorithm 1, 'SHA1.'" He continued, gesturing to sample images in a PowerPoint slide used for demonstrations, "When you run the algorithm against an image file, it basically just cares about the 1s and 0s that make up that file, and so it can produce a unique identifier, or what we like to refer to as a 'digital signature' for that file because it's unique to that file. So if I have two of the exact same file, or if I make a direct copy of a file, digitally, and I don't change anything, then those two binary values, signatures, will be identical—they'll be the same. And so when we're processing suspects' machines,[30] and we have a set of the hashes that we're interested in mining, we can run those up against and find the digital duplicates in an automated way." While this matching might seem intuitively simple, Cole clarified to me that law enforcement was relying too much on binary (duplicate) hash sets, which prioritized the recognition of suspected offenders instead of victims. "Law enforcement in general is very offender-focused. It's, y'know, 'put the bad guy in jail,'" he explained. A forensics team would "run a hash set, get several thousand hits [on the offender], which would be more than enough to prosecute an individual . . . and then they'd call it good." To him and other managing officers, this quick-match technique focused on the perpetrators of abuse missed the larger picture of new child abuse image production: "If there are *newly* produced images of a *victim* in that volume in a series, you're not gonna find them. Because you're relying on a hash set and just by the very nature of being in a hash set it's already been seen before, and designated as something we're looking for. So law enforcement—*across the globe*—was for all intents and purposes shoving victims in the evidence room, never to be found."

Thus, Agent Cole's lab, and, internationally, the Virtual Global Taskforce managed by INTERPOL, aim to remedy the offender-focused image-detection process by encouraging more focus on child victims and using newer software better able to detect new child abuse images without relying exclusively on past known-hash image sets. "Look, we'll use the hash sets to get workload reduction so that you're not having to manually review all these files like you did in the past—but what we want you to do is instead of focusing on the hits, we want you to focus on the stuff you didn't hit," Agent Cole explained, "because that's where new victims will be." Cole's lab actively solicits and tests new software packages from multiple companies and countries to provide for free to local law enforcement agencies. As part of the reporting process with electronic service providers, NCMEC is also authorized to send elements related to any image reported to its CyberTipline back to the service provider if it will help them stop what they call "further transmission" of similar photos. Elements, as the law stipulates, can include "unique identifiers" associated with an image, any known website locations of the image, and "other technological elements" that can be used to identify other images of that nature. The elements for each case are, as one might guess, quite discretionary and necessitate frequent communication and close ties between the Child Victim Identification Program at NCMEC and its complementary units at electronic service providers.

In practice this works out as an extremely close relationship between NCMEC and some software companies that have a dedicated task force or unit for the investigation of child exploitation. Facebook, as mentioned, has several units that might potentially deal with child abuse content. Other companies have until recent years had no single person or unit dedicated to assessing and reporting child abuse content. Despite a legal obligation to report child exploitation image content, the reach of the federal law is limited in the extent to which companies must establish standardized internal procedures or personnel to perform reviews of potentially violent content. The cyclical process of reporting to NCMEC, and NCMEC reporting back in case there is need of further review, thus plays out very differently in each case with each company.

Facebook's director of trust and safety, Emily Vacher, discussing PhotoDNA with me, exclaimed, "We love it [PhotoDNA]," but "we want to be able to do more, to be more proactive and find child pornography without the images having already been known about." She expressed a desire to work on "hash sharing" with other technology companies but said that the current search processes varied greatly between companies. Her colleague, Michael, a manager in Facebook's eCrime division, separately made nearly the same remark to me: "We're pushing for hash sharing between tech companies. The hashes are a good signal as well." He added that PhotoDNA software, a sort of emblematic software that set the baseline for child exploitation detection but is not used as much anymore, is also symbolic of collaboration: "[PhotoDNA's] biggest piece of value is as a signal."

Vacher and her colleague's remarks on hash sharing indicate a particular form of sociality that characterizes this extension of the policing investigation network. The exchange of image data and hash sets between agencies and companies produces a reciprocal exchange in which law enforcement investigators and social media companies dissolve mistrust and establish the basis for investigative cooperation through reciprocal exchanges of data. Federal law enforcement from the US often made remarks about trust, making casual distinctions between social media companies as "good partners" or "bad partners." The networking meetings and trainings at which cooperation is discussed—not always amicably—provide space for the negotiation of trading illicit data as well as proprietary investigative software. These exchanges of data brought forth by legal necessity also establish a shared ethos for digital investigations and forensic work.

THE PLEASURES OF CAPTURE

Critically, the digital forensics software programs I discuss here are surveillance technologies that flow between corporate and law enforcement institutions. For example, Amazon is implementing its proprietary Rekognition software to perform real-time image-recognition analysis on live video footage; at the time of this writing, police departments in Florida and Oregon have purchased the software.[31] Image surveillance technologies sort through the massive amounts of text, image, and video data uploaded online every day and work in collaboration with content moderators to keep the internet secure and "clean." There is a certain pleasure in this mode of surveillance. That is, I do not intend to imply that algorithmic detectives gain pleasure from looking at child exploitation images themselves but instead a diverted, yet highly embodied, form of pleasure in seeing these images for the purpose of policing. While image-detection algorithms are increasingly built into platforms' digital infrastructure, automating, in a sense, the notion that "code is law,"[32] it is only at the moment of human revelation that apprehension is consummated. Recognition is pleasure because of this moment of revelation—of apprehending, capturing, really seeing, and cataloguing. Policing through recognition blends here with visual capture, with the moment of reproduction/apprehension/seizure so beauti-fully analyzed by scholars throughout the history of technologies of visual culture. Surveillance technologies both produce and are produced by a pleasure in looking.[33] This pleasure is justified by the pressing moral obligation and urgency to find and stop child exploitation online. Photography "furnishes evidence,"[34] transforming moments of harm into kernels of documentation. Videographic and photographic representations of child abuse become forensic data. Recognition creates juridical subjects—the victim and the one who has performed injury to produce the vic-tim. It also externalizes knowledge production for evidence, as recognition must be verified and validated across multiple institutions in a network. Image-recognition algorithms become justifiable when rooted in an imperative to *intervene* and make criminal investigations more efficient. This software has promissory value as a tool for a particular view of safety and security—for some, not all.

Policing power has been extended into the hands of extrajudicial agencies; this is an extension that is made permissible through its use in the case of child protection. The protection of the child hinges upon defining its other, that which threatens childhood. Underlying support of such projects is an assumed universal under-standing that sexual exploiters of children must be othered—located, corralled, and arrested—and are deserving of punishment. It is also understood, then, that inves-tigators who wish to locate and apprehend offenders must first become intimately knowledgeable about sexual abuse—bring it *close*, in a sense—before arresting and enacting punishment. The othering of potential offenders is a form of extension,[35] a reaching toward those who must be kept away. Carceral technologies perform this act of extension—apprehending and making proximate violent others—by extend-ing the corpus of policing power in order to seek certain bodies. We might consider them as what Sandy Stone has called a "prosthetic technological self,"[36] an extension of a punitive impulse to entrap those who might offend. The intimacy of making the (e)strange(d) familiar is evident in the care devoted to visualization of sex offender identities through forensic sketch artistry and facial recognition algorithms.

There is an intimacy as well between police agencies and the computer scientists experimenting, developing, and playing with machine-learning algorithms for facial recognition in sex offender and child exploitation cases to better detect offender and child faces across gender and racial diversity. Machine-learning software is trained against virtual databases of existing, reported child pornography known to exist online. Recent work in science and technology studies (STS) and the social study of algorithms has pointed out, critically, that algorithms have embedded values and biases, leading to potential discrimination as well as the erasure of human analysis and input. In the case study of child pornography detection, digital forensics designers use certain training sets to improve their algorithm's ability to detect faces, genders, and races. Historically, these training sets have treated the white male adult face as a default against which computer vision algorithms are trained.[37] Databases of missing and abused children held by NCMEC in the US disproportionately contain images of white children, and nonwhite children are statistically less likely to be reported as missing or have extensive case files of data. In examining how image-recognition software and content reviewers "see" abuse images, I consider how skin tone, as a category for detection, is made manifest as digital racial matter.[38] In the course of my fieldwork, computer vision researchers often remark that they earnestly want to address the "problem" of underdetectability, or even undetectability, of darker skin tones, Black and East Asian features, and younger ages. I note that these researchers' use of "problem" might differ from the social scientists' perspective on the issue of detectability.

Consider, for instance, the case of Germaine Moore discussed at the outset of this essay. Six years earlier, in November 2012, a different video was found, this time in Denmark. It was initially flagged on a social media site by Danish police, who identified the video as having potential American origins and reported the file to US Homeland Security. This video showed an adult white man sexually assaulting a child. Agent Jim Cole, the Homeland Security manager mentioned earlier, worked on this case. His team of computer vision and forensic specialists was able to zoom in on aspects of the video, focusing on the partially obscured face as well as a shot of the man's hand with fingerprints exposed and another showing a medicine cabinet in the background left ajar with prescription pill bottles showing. He described how he and his team sharpened the image focus on a finger in the first still shot, generated a spatial map capturing the unique ridges and contours of the surface of the finger and managed to extract enough fingerprint detail to run the data through their biometric database of convicted offenders. They matched this information from the first image with the second: tightening into the still image of the prescription pill bottles, using deconvoluting software to reduce motion blur and clarify the text on one of the bottles, revealing a pharmacy name, the first name "Stephen," and two letters of a last name, "K" and "E." The image and text detection of both images provided enough context for the team to locate the pharmacy in a small town just outside Savannah, Georgia. They matched the fingerprint database results against patients at the pharmacy to identify Stephen Keating as the suspect and arrested the man in his Jesup, Georgia, home on November 15, 2012. In July 2013, Keating was sentenced to 110 years in prison for multiple counts of production and distribution of child pornography. The entire arrest, from Danish origin to apprehension on US soil, transpired over just twenty-four hours. Fourteen of Keating's victims were identified over the subsequent three weeks.

In contrast, Moore was virtually undetectable, his arrest made possible outside the bounds of image recognition. Why did it take so long for Moore's video to be flagged and removed? Did Keating's whiteness, evident only through the shades of skin visible in fingertips and part of a face, help make his rapid capture possible? Moore's unrecognizability has irked safety specialists at Facebook, punctuating the need to rapidly and accurately detect a variety of skin tones. But any analysis of new image technologies must stay rooted in the knowledge that visibility has always been tied to logics of colonial control and capture.[39] Simone Browne, in her searing historical analysis of biometric surveillance, argues that the inability of computer vision software to distinguish dark skin tones is perilously paired with a hypervisibility tying the history of surveillance to the history of the surveillance of Blackness.[40] Child pornography is often seen as a "limit case" justifying expanded surveillance. But what would increased accuracy in recognizing Black and brown skin tones mean for the policing of *other* types of online content? And, critically, despite Moore's failure to be "seen" by image-detection algorithms, hundreds of human viewers participated in the impulse to find him. The viral, agitating comments and reposts of the video hastened his apprehension, exposing people's gut desires to point out and identify the offender, with or without law enforcement's intervention.

Despite recent critiques of algorithmic detection and surveillance, child abuse lingers as an exceptional case. Sex offenders are statically configured as people who are "despised and disposable"[41] or even othered "beyond the pale of humanness."[42] If the imagined offender is other than human, I argue that it shares space with the virtual child. For in the question of the undetectability of Germaine Moore lurks the question of the undetectability of the girl being abused in that same video, a child rendered completely invisible in the rush to apprehend the offender. Agent Cole's comment that hash matching favored offender detection over victim detection conjoins the two, thesis and antithesis. The identification and protection of the child hinges upon defining its other, that which threatens childhood. Algorithmic detectives across the network, from elite content reviewers at social media companies to analysts at NCMEC to special agents with US federal agencies, must be keenly attuned to the faces and bodily contours of both offenders and victims, using human and computer vision to "see" skin tones, violence, and vulnerability alike. Child pornography images, as recordings of abuse, bear traces of the original violence while also magnifying the potential for arrest. Such potentiality drives the highly embodied and kinetic skilled vision needed to feel images out, rapidly apprehend them, and participate in the collaborative experience of policing.

We must remain attuned to the significance of perception and corporeality in the practice of using representational and classificatory technologies. The programmer who designed PhotoDNA has worked on adapting his software for counterterrorism with the US government[43] in order to locate photos and videos of violence and terrorist speech, especially across multiple skin tones. Such an expansion of child pornography detection software for other purposes underscores the stakes of embodied viewing. Who determines what constitutes terrorist content? Do those images get "seen" in the same way as child pornography? By following current collaborations into the policing of child pornography, we can pay attention to the ways in which impulses to protect and punitive logic combine to form dual forms of apprehension: image-detection software for identifying and locating abuse photos and videos online, and the actual arrests of current and future exploiters made possible through partnerships between activist organizations, law enforcement, and software companies. To conclude, image-recognition technologies are expanding a surveillance state under the auspices of child protection—and furthermore, these technologies enroll nonpolice to take on surveillant roles to patrol and police digital space. Software like PhotoDNA signal to the public that digital space has been and can continue to be policed through extrajudicial and embodied measures. They signal a future of policing that is, perhaps, already here—in which not only recognition but policing itself is pleasure.

NOTES

1. Alex Horton, "Child Porn Went Viral on Facebook, and Police Say Its Creator Has Been Charged," *Washington Post* (February 6, 2018), https://www.washingtonpost.com/news/true -crime/wp/2018/02/06/child-porn-went-viral-on-facebook-and-police-say-its-creator-has-been -charged/?utm_term=.86921578a26a.

2. Department of Justice, US Attorney's Office, Middle District of Alabama, "Millbrook Man Sentenced to 50 Years in Prison for Production of Child Pornography" (December 6, 2018), https://www .justice.gov/usao-mdal/pr/millbrook-man-sentenced-50-years-prison-production-child-pornography.

3. Horton, "Child Porn Went Viral on Facebook."

4. See Sarah Roberts, "Your AI Is a Human," this volume.

5. 18 US Code § 2258A—Reporting requirements of electronic communication service providers and remote computing service providers.

6. S. 1693—Stop Enabling Sex Traffickers Act 2017.

7. The etymology of the words "capture" and "to seize" are instructive. "Capture" derives from the Latin *capere*, "to take or to seize." "To seize," in turn, comes from the medieval Latin *sacire* or *ad proprium sacire* (to "claim as one's own"). To capture is to claim a sought object as one's own, to bring close and into proximity.

8. Joy Buolamwini, "When the Robot Doesn't See Dark Skin," *New York Times* (June 21, 2018).

9. Sarah Ahmed, *Queer Phenomenology: Orientations, Objects, Others* (Durham, NC: Duke University Press, 2006).

10. Brian Massumi, *The Politics of Everyday Fear* (Minneapolis: University of Minnesota Press, 1993).

11. Sianne Ngai, "Bad Timing (A Sequel): Paranoia, Feminism, and Poetry," *differences* 12, no. 2 (1993): 1–46.

12. Ariane Ellerbrok, "Playful Biometrics: Controversial Technology through the Lens of Play," *Sociological Quarterly* 52, no. 4 (2011): 528–547.

13. cf. Mark Andrejevic, "The Big Data Divide," *International Journal of Communication* 8 (2014): 1673–1689; Kelly Gates, *Our Biometric Future: Facial Recognition Technology and the Culture of Surveillance* (New York: NYU Press, 2011); Lisa Gitelman and Virginia Jackson, "Introduction," in *Raw Data Is an Oxymoron*, ed. Lisa Gitelman, 1–14 (Cambridge, MA: MIT Press, 2013); David Lyon, "Surveillance, Snowden, and Big Data: Capacities, Consequences, Critique," *Big Data & Society* (July–December 2014): 1–14.

14. cf. Sarah Brayne, "Big Data Surveillance: The Case of Policing," *American Sociological Review* 82, no. 5 (2017): 977–1008; Simone Browne, *Dark Matters: On the Surveillance of Blackness* (Durham, NC: Duke University Press, 2015); Shoshana Amielle Magnet and Tara Rodgers, "Stripping for the State: Whole Body Imaging Technologies and the Surveillance of Othered Bodies," *Feminist Media Studies* 12, no. 1 (2011): 101–118; Safiya Umoja Noble, *Algorithms of Oppression: How Search Engines Reinforce Racism* (New York: NYU Press, 2018).

15. Gates, *Our Biometric Future*.

16. Mark Andrejevic and Kelly Gates, "Big Data Surveillance: Introduction," *Surveillance & Society* 12, no. 2 (2014): 185–196.

17. Tracy Ith, "Microsoft's PhotoDNA: Protecting Children and Businesses in the Cloud," press release (2015), https://news.microsoft.com/features/microsofts-photodna-protecting-children-and -businesses-in-the-cloud/.

18. I use the term "elite content reviewers" to denote the class (and often racial) difference between "commercial content moderators" (in the sense used by Sarah T. Roberts) and the reviewers I study on the other end, at police agencies, NCMEC, and Facebook and Microsoft's child safety teams. Sarah T. Roberts's important work on commercial content moderators has documented the intense precarity and psychological burden of CCM as outsourced, contractual labor. See Sarah T. Roberts, "Digital Detritus: 'Error' and the Logic of Opacity in Social Media Content Moderation," *First Monday* 23, no. 3 (2018), http://dx.doi.org/10.5210/fm.v23i3.8283.

19. See Roberts, "Your AI Is a Human," this volume, on initial stages of content moderation flagging.

20. Adrian Chen, "The Laborers Who Keep Dick Pics and Beheadings Out of Your Facebook Feed," *Wired* (October 2014), https://www.wired.com/2014/10/content-moderation/.

21. Cf. Roberts, "Your AI Is a Human," this volume; Sarah T. Roberts, "Digital Refuse: Canadian Garbage, Commercial Content Moderation and the Global Circulation of Social Media's Waste," *Wi: Journal of Mobile Media* 10, no. 1 (2016): 1–18, http://wi.mobilities.ca/digitalrefuse/.

22. Michael C. Seto et al., "Production and Active Trading of Child Sexual Exploitation Images Depicting Identified Victims. National Center for Missing and Exploited Children" (2018),

http://www.missingkids.com/content/dam/ncmec/en_us/Production%20and%20Active%20 Trading%20of%20CSAM_FullReport_FINAL.pdf.

23. At the time of this writing, many agencies, including NCMEC and Facebook, use some combination of NetClean, PhotoDNA, Griffeye's Child Exploitation Image Analytics, and other proprietary image forensic tools.

24. Charles Goodwin, "Pointing as Situated Practice," in *Pointing: Where Language, Culture, and Cognition Meet*, ed. Sotaro Kita (London: Psychology Press, 2003).

25. Harry Collins, "Interactional Expertise and Embodiment," in *Skillful Performance: Enacting Expertise, Competence, and Capabilities in Organizations: Perspectives on Process Organization Studies*, ed. Linda Rouleau Jorgen Sandberg, Ann Langley, and Haridimos Tsoukas (Oxford: Oxford University Press, 2016).

26. Christina Grasseni, "Skilled Visions: An Apprenticeship in Breeding Aesthetics," *Social Anthropology* 12, no. 1 (2004): 41.

27. I draw from Charles Goodwin's analysis of "the public organization of visual practice within the work-life of a profession"; Charles Goodwin, "Practices of Seeing Visual Analysis: An Ethnomethodological Approach," in *The Handbook of Visual Analysis*, ed. Theo Van Leeuwen and Carey Jewitt (London: Sage Publications, 2000), 164.

28. Grasseni, "Skilled Visions," 47.

29. John Berger, *Ways of Seeing* (New York: Penguin, 1972).

30. A US district court ruled in 2008 that running hash values on suspect files could be considered a police search protected under the Fourth Amendment. *United States v. Crist*, 2008 WL 4682806 (October 22, 2008).

31. Russell Brandom, "Amazon Is Selling Police Departments a Real-Time Facial Recognition System," *Verge* (May 22, 2018), https://www.theverge.com/2018/5/22/17379968/amazon -rekognition-facial-recognition-surveillance-aclu.

32. Lawrence Lessig, *Code and Other Laws of Cyberspace* (New York: Basic Books, 1999).

33. Magnet and Rodgers, "Stripping for the State," 103.

34. Susan Sontag, *On Photography* (New York: Penguin Books, 1977).

35. Sara Ahmed, *Queer Phenomenology: Orientations, Objects, Others* (Durham, NC: Duke University Press, 2006), 115.

36. Sandy Stone, "Split Subjects, Not Atoms; Or, How I Fell in Love with My Prosthesis," in *The Cyborg Handbook*, ed. C. H. Gray (New York: Routledge, 1991).

37. Kate Crawford, "Artificial Intelligence's White Guy Problem," *New York Times* (June 25, 2016).

38. Sara Ahmed and Jackie Stacey, eds., *Thinking through the Skin* (London: Routledge, 2001).

39. Steve Anderson, *Technologies of Vision: The War between Data and Images* (Cambridge, MA: MIT Press, 2017).

40. Simone Browne, *Dark Matters: On the Surveillance of Blackness* (Durham, NC: Duke University Press, 2015).

41. Emily Horowitz, *Protecting Our Kids? How Sex Offender Laws Are Failing Us* (Santa Barbara: Praeger, 2015).

42. John Borneman, *Cruel Attachments: The Ritual Rehab of Child Molesters in Germany* (Chicago: University of Chicago Press, 2015).

43. Julia Franz, "How We Can Use Digital 'Fingerprints' to Keep Terrorist Messaging from Spreading Online," *PRI* (February 12, 2017), https://www.pri.org/stories/2017-02-12/how-we -can-use-digital-fingerprints-keep-terrorist-messaging-spreading-online.

II

THIS IS AN EMERGENCY

6

SEXISM IS A FEATURE, NOT A BUG

Mar Hicks

In 2017, I went to Google to speak about my work on women in the history of computing. I explained how women were the field's first experts, though they weren't accorded that respect at the time, and how they were excluded when men in management saw computers becoming powerful tools for control, not of just work but of workers too. Women, of course, weren't considered management material in this era, and so as computers became aligned with management women were shown the door—oftentimes after training their (male) replacements. It was not that women lacked technical skills, or even opportunities to show their skills. It was that their very status as women precluded them from advancing in a new field whose power and prestige was swiftly growing.

In the question and answer period after my talk, I got the predictable question: wasn't it actually that women just *didn't like* computing work as much as men did? And wasn't it true that women who succeeded in computing, both then and now, were somehow less "feminine" and more "masculine"? This is a question I get often, but this time I got it from a woman engineer. She wasn't being hostile; she just appeared to really want to be reassured that she was different, that she wouldn't face the same fate as the women I'd described, and also that women weren't being held down or forced out. She seemed to need to believe that the current state of affairs was somehow the natural order, or else she might have to confront the deeply unfair past and present of her chosen line of work—and the fact that it would likely affect her own future.

This engineer's response is not all that surprising when you remember that the high-tech sector presents itself as the ultimate meritocracy. Silicon Valley is the high-tech version of the American Dream, where anyone with talent can supposedly succeed, even if they start out in a garage. Yet its makeup does not reflect this—it is deeply skewed in terms of gender, race, class, sexuality, and many other categories, toward historically privileged groups.

Gender discrimination, in particular, is a major stumbling block for the high-tech industry—so much so that even some of the most powerful, white women feel they need to admonish themselves and their peers to "lean in" to combat sexism and break through the glass ceiling.[1] This belief in the promise of a more equal future, if only women would try harder, seems alluring if one takes it on faith that Silicon Valley's goal is to measurably improve itself.

But a culture of rampant sexual harassment, persistent racial inequalities in positions of power, and pay and promotion inequalities industry-wide shows that Silicon Valley culture is mired firmly in the past, even as companies claim to be building a better future.[2] From sexist manifestos on how women's supposed intellectual inferiority disqualifies them from tech careers, to golden parachutes for serial sexual harassers at corporations that simultaneously choose not to cooperate with federal equal pay investigations, to platforms that position misogynist and racist hate speech and threats of sexual assault as just a normal part of online discourse, it is no surprise that women workers in the tech industry might internalize sexism and blame themselves—particularly because when they speak out they tend to lose their jobs.[3]

Online, Black women, and particularly Black trans women, are targeted with inordinate amounts of hatred, yet the platforms that enable it refuse to seriously address the harms they are causing. Twitter's moderation decisions routinely allow misogynoir and hate speech that specifically affects Black women, multiplying sexist and racist harms. These gendered harms are also built in to platforms at their core. One of the most highly valued companies in Silicon Valley started out as a site that stole women's pictures without their consent and asked users to rate their attractiveness.[4] Facebook now commands the attention of more than two billion users worldwide and uses its power to influence everything from minor purchasing decisions, to who gets elected, to—in extreme instances—which populations live or die.[5]

That the concept of merit structures our understanding of high technology and its successes, even as a lack of diversity at the top of Silicon Valley corporations creates an echo chamber, means that the infrastructure tech builds actually often worsens social inequalities.[6] As Safiya Noble has shown, existing problems become entrenched and magnified by profit-seeking technologies masquerading as neutral public resources.[7] Google, for example, presents itself as a neutral information tool when the company's real business model is to sell as many advertisements as possible—Google brings in more than 100 billion dollars annually in advertising revenue, the vast majority of its revenue.[8]

Such technological systems have little incentive to push back against sexism and racism and a strong profit motive to look the other way, or even to lean into them. This is not new: as something that grew out of the Second World War and the Cold War that followed, electronic computing technology has long been an abstraction of political power into machine form. Techno-optimist narratives surrounding high technology and the public good—ones that assume technology is somehow inherently progressive—rely on historical fictions and blind spots that tend to overlook how large technological systems perpetuate structures of dominance and power already in place.[9]

As the United States finds itself in the midst of a "techlash" or backlash against high tech's broken promises, the history of computing offers us a chance to reflect critically on the roots of these developments and the potential dangers that lie ahead. Because Silicon Valley seems to be pointing the way to our national and even global future, it is difficult to critique it and to imagine alternatives. The US tech sector's outsized level of power has a disproportionate influence over the terms of the conversation, particularly as social media platforms replace traditional news outlets. In this situation, comparative history is a powerful tool: by taking us out of our current moment and context and showing us a different scenario, we can better unpack the roots of what has now become normalized to the point of seeming inevitable and unchangeable.

This chapter offers an example from the United Kingdom, the nation that invented the computer. As a close historical cousin of the United States, it is a harbinger of our own current technological troubles, and a prime example of how high-tech economies rise and fall on the cultural ideals engineered into their systems. The British experience offers a compelling example of how a computing industry's failure is intimately linked with social problems that may seem ancillary or unrelated. It shows how the fiction of meritocracy can scuttle an industry and how computing has long been aligned with neocolonial projects that present fantasies not just of national but of global control as being possible through high technology.

Contrary to what we might believe or hope, this history shows us that computing purposely heightens power differences, and that those who commission and control these systems benefit from that. It also shows that this intentional program of concentrating power at the top by discriminating against certain groups paradoxically lowers the quality and output of what a nation's technology sector can produce. Ultimately this proves to be self-destructive—not only to that technology but to entire economies and democracies.

COMPUTING FIRSTS, COMPUTING FAILS

The UK context presents us with one of the most shocking failure stories in the history of computing. In 1943, Britain led the world in electronic computing, helping to ensure the Allied victory in Europe during World War II. By deploying top-secret code-breaking computers—the first digital, electronic, programmable computers in the world—the British were able to use computing to alter geopolitical events at a time when the best electronic computing technology in the US was still only in the testing phase.[10] By leveraging their groundbreaking digital methods for wartime code-breaking, the British conducted cyberwarfare for the first time in history, with great success.

After the war, British computing breakthroughs continued, matching or anticipating all of the major computing advances in the US. But by 1974, a mere thirty years after deploying the first electronic programmable computers, the British computing industry was all but extinct.

Nowhere is the story of a nation undone by a broken high-tech industry more apparent than in the case of Great Britain. Yet this swift decline of the computer industry in the nation that *invented* the computer has for too long been ignored or misunderstood. If the rising power of computing is one of the biggest stories of the twentieth century, then the failure of the nation that broke new ground in electronic computing is undoubtedly one of history's most urgent cautionary tales.

The myth of meritocracy is easier to investigate in the UK, a nation riven by class antagonisms far more obvious and long-standing than those in the US, as the unrealized ideal it has always been. The UK, still influenced by a divine-right monarchy, struggled to remove class from the equation of who could govern.

The British Civil Service was set up to be the ultimate meritocracy, with an elaborate system of exam-based hierarchies and promotions to ensure that those in charge of the country's resources derived their positions from skill rather than social standing and were suitably qualified to hold them. Nothing less than the nation's ability to function was at stake. British leaders realized that a nonmeritocratic system would result in cronyism, corruption, nepotism, graft, and ultimately the destruction of democratic civil society. A strict meritocracy was needed, because if the government could not be trusted as a fair and just instrument of the will of the people, democracy itself was in danger.

But although the Civil Service claimed to be able to transcend the unfairness of British society, describing itself as a "fair field with no favor" for all of its employees, civil servants were not judged simply according to their own talents and abilities. For most of the twentieth century, being white was a primary assumption and requirement for holding any position of power in the UK, as well as for holding what were then seen as "respectable" office jobs. Although a small minority of mostly lower-level civil servants were Britons of color, they were the exceptions that proved the rule of a system of powerful, unspoken white supremacy.[11]

More remarked upon at the time was the subordinate position of the mostly white women who had to take different entrance exams than men and for much of the twentieth century were not allowed to work in the same offices, use the same lunchroom facilities, or even enter and exit buildings by the same doors and stairwells as their male coworkers. Because of the heteronormative supposition that women would marry and leave the labor force to take care of a family, they were also either expected, or outright required, to leave their jobs upon marriage.

These women also earned less than their male peers for performing the exact same jobs. Men, it was reasoned, needed a "family wage," whereas women were supposedly working to support only themselves. White, mostly middle-class women were treated as a temporary labor force, and their lack of adequate pay or promotion prospects meant that many left upon getting married even if they wished to continue working. For the most part, women's work in the burgeoning information economy was kept separate—and seen as different in type and inferior in kind when compared to men's work. While this was true throughout industry, the inequity of the situation was most obvious within the government's "meritocracy."

It was paradoxically for this reason that women ended up being on the cutting edge of computing in the UK. Women had been the first computer operators and programmers during the war, working on the top-secret Colossus code-breaking computers that allowed the Allies to successfully land on D-Day. Their suitability for these positions was defined not simply by war's labor constraints but by the low esteem in which early computer work was held. Viewed as rote and excessively technical, early computer work was denigrated for its association with machinery rather than elevated by it. It was seen as almost akin to factory work, and the introduction of machines into office environments was often referred to as the "industrialization of the office" by managers.

Despite the fact that these women worked assembling, troubleshooting, testing, operating, and programming the computers that contributed directly to the war's positive outcome for the Allies, their work was hidden and largely unremarked upon for decades after the war. This was not simply due to the secrecy surrounding their work, although the Official Secrets Act played a role. Early computer operation and programming was such a feminized and devalued field that few people singled it out as being historically important or cared to remember and record the names, tasks, and accomplishments of these women. Once the official veil of secrecy started to lift in the 1970s, women wartime computer workers began to tell their stories, but many of them felt that their stories were not worth telling.[12]

Even prior to the term "computer" being a name given to a machine, it had been the name given to job. Originally, a computer was a person—almost always a young woman—who computed complex equations with the help of pen and paper or a desktop accounting machine. A common misperception is that women got into computing during World War II simply because men were at the front, but the gendering of computing work existed before the war, and before computers were electronic. The feminization of this work continued through and after the war, with women returning to the civilian workforce to perform computing work with electromechanical and later electronic systems—everything from programming and operation, to systems analysis, to hardware assembly.

BUILDING A BROKEN SYSTEM

As computers began to percolate out of the military and academia into industry and government more broadly, women's computing work became ever more intertwined with computers, and critical to the functioning of the economy. While women were sent into lower-level jobs, often ones that depended on proficiency with office machines like computers, men were tapped for higher level, supposedly more intellectual work that led to managerial and administrative promotion tracks where machines were nowhere in sight. The latter positions could eventually lead to policy-making roles at the highest levels of government.

In 1959, one woman programmer spent the year training two new hires with no computer experience for a critical long-term set of computing projects in the government's main computer center while simultaneously doing all of the programming, operating, and testing work as usual.[13] At the year's end, her new trainees were elevated to management roles while she was demoted into an assistantship below them, despite her longer experience and greater technical skills. In this era, the idea that women should not be in a position to hold power over men made it highly unlikely they would be promoted into management positions.

Women also continued to earn less money for doing the same jobs. For women who worked with computers, their economic worth was tied to their identity as women but largely unmoored from their proficiency on the job. One of the earliest UK civil servants to come out at work as a trans man found his pay immediately raised after his transition in the late 1950s, simply by virtue of his employer now recognizing him as a man. At the same time, another civil servant, a trans woman, was given advice by her managers to hide or delay her transition so that her pay would not go down.[14]

For decades, the government fended off increasing pressure from workers to grant women civil servants equal pay. In 1955, an equal pay plan was slowly phased in over a period of several years, but only for job categories in which men and women held identical job titles, not jobs in which they did similar work. Since the Civil Service had been so highly segregated into men's and women's jobs, this meant that the majority of women working in government would *not* actually get equal pay. The largest block of these workers were the women in the "Machine Grades," the class of workers who did the computing work of the government—at the time the nation's largest computer user.[15]

These women were explicitly excluded from the provisions of the Equal Pay Act, and as a result the Machine Grades became known as the "Excluded Grades" after the passage of equal pay. The Treasury reasoned that because so few men worked in computing, women's significantly lower pay rate had now become the market rate for the work. Because women had been doing machine-aided computation work for so long, and in such majority, their relative lesser worth in the labor market had attached to this work and lowered the jobs' *actual* worth, not just culturally but in a literal, economic sense.[16]

By the 1960s, however, the power of computing was becoming more apparent to those in charge. The woman in the earlier example was required to train her supervisors not because there were no longer enough women to do the work, but because computer systems were expanding to take over more aspects of government. Although the complexity of the work did not suddenly change, the perception of its worth skyrocketed. As such systems became recognized as more than merely technical, government ministers realized that computers would be important tools for consolidating and wielding power over workflow, workers, industrial processes, and even the shape of government itself.

With computing work becoming aligned with power, women computer workers who possessed all of the technical skills to perform the jobs found themselves increasingly squeezed out by new hiring rubrics that favored untested men trainees with no technical skills. Despite its "meritocracy," the government explicitly forbade women from the former Machine Grades from applying for the newly created class of management-aligned computer jobs designed to help management gain greater control over the mounting number of computerized processes poorly understood by those at the top levels of government and industry.[17]

Though these women could easily do the work in a technical sense, they were not allowed to occupy the positions of managerial and political power that computing jobs were suddenly becoming aligned with. At the same time, the men who were tapped for these jobs lacked the technical skills to do them and were often uninterested in computing work, in part because of its feminized past.

LOWERING STANDARDS TO CREATE AN ELITE

As a result, government and industry began a major push to recruit men into these technical positions while simultaneously grooming them for management positions. This process entailed lowering standards of technical proficiency to create an elite class of management-level computer workers, above the old Machine Grades in name and power, but beneath them in technical skill. It was supposed to result in the construction of a cadre of high-level, management-aligned "computer men"—technocrats who would be able to manage people as well as machines and make informed decisions at the highest levels about the future of computerization.

Instead, this recruitment change resulted in a devastating labor shortage. Promising young men tapped for the positions usually had little interest in derailing their management-bound careers by getting stuck in the still largely feminized "backwater" of computer work.[18] At a time when computing was tied up with ideas about the "industrialization of the office," many still saw machine work in general as unintellectual and liminally working class.

Most young men who were trained for these new computing positions, at great government expense, left to take better, noncomputing jobs within a year. Government ministers in charge of the changeover were blindsided by the results and the problems caused by hemorrhaging most of their computer staff. This trend continued throughout the 1960s and into the 1970s, even as the status of the field rose, and computing work began to professionalize. Young men who were trained for the new positions often did not want them for long, while young women who had the skills were forced out through turnover, demotion, or by being strongly encouraged to "retire"—in other words, marry and leave the workforce—before the age of thirty. Not coincidentally, thirty was roughly the age at which all promotion for a woman computer worker in the public sector ended.

FAKE IT UNTIL YOU (CAN GET A WOMAN TO) MAKE IT

As a result, the programming, systems analysis, and computer operating needs of government and industry went largely unmet. The government was the biggest computer user in the nation, and its inability to train and retain a technical labor force had national reverberations. It slowed and complicated computerization and modernization plans throughout the country, holding back Britain's economy, and even causing political problems for the UK on the international stage.[19]

One way that corporations and the government attempted to satisfy this dire need for more programmers was to turn to outsourcing the work. At this time, software was still something that normally came bundled with a mainframe, rather than being seen as a product that could be sold separately. Software either came with the machine, direct from the computer company, or was written by the employees who ran the mainframes. People who started companies in the nascent software services industry in the 1960s were making a big gamble that companies would pay for software as a standalone product after spending hundreds of thousands of pounds to buy a mainframe.

But some people who set up software companies in this period did so because they felt they had no other choice. The most famous and, eventually, most successful of these software startups was headed by Stephanie "Steve" Shirley—a woman who had previously worked for the government until the glass ceiling had made it impossible for her to advance any further. Born in Germany, Shirley had been a child refugee during World War II. She was saved from being murdered by the Nazis when she was brought to England with 10,000 other Jewish children on the Kindertransport.[20]

Shirley often credited her escape from the genocide in Europe as a primary reason for her later drive to succeed: she felt she had to make her life "worth saving."[21] After leaving school, she went to work at the prestigious Dollis Hill Post Office Research Station in the 1950s, the same government research center where Tommy Flowers had built the Colossus computers, and she worked with Flowers briefly. As a young, technical woman worker, Shirley recalled that the sexism of the time usually dictated that she follow orders without much independent thought.

Animated by her drive to succeed, Shirley went to school for her university degree at night while working full time, and sought to raise the responsibility level and difficulty of the work she was assigned. But she began to realize, as she put it, that "the more I became recognized as a serious young woman who was aiming high—whose long-term aspirations went beyond a mere subservient role—the more violently I was resented and the more implacably I was kept in my place."[22] After being passed over multiple times for a promotion she had earned, she found out that the men assigned to her promotions case were repeatedly resigning from the committee when her case came up, rather than risking having to give a woman a promotion. Her ambition was seen as a liability, even though in a man it would have been rewarded.

Realizing there would be no chance for her to get ahead in the Civil Service's false meritocracy owing to her gender, Shirley left her job and after a brief stint in industry, where she encountered the same sexist prejudice, she decided to quit and start her own computing company while raising her son. In 1962 she founded Freelance Programmers, which was unique not only because it was one of the first companies to recognize that software as a standalone product was the way of the future but also because she learned from the mistakes that government and industry were making with computer workers.

While government and industry starved themselves of programmers because they refused to hire, promote, or accommodate women technologists, Shirley scooped up this talent pool. Shirley considered her business a feminist enterprise—one that would allow women like her to continue working in computing and use their technical skills. One of her first job advertisements read, in part, "many opportunities for retired programmers (female) to work part-time at home," and described the jobs as a "wonderful chance, but hopeless for anti-feminists."[23] In other words, there was a woman boss.

By giving her employees flexible, family-friendly working hours and the ability to work from home, Shirley's business tapped into a deep well of discarded expertise. Desperate for people who could do this work, the government and major British companies hired Shirley and her growing team of mostly women programmers to do mission-critical computer programming for projects ranging from administrative software and payroll to programs designed to run industrial processes.

Initially, having a woman's name at the helm of the company prevented Shirley from getting work. But when she started using her nickname "Steve" for business purposes, her letters to potential clients were no longer ignored. From there, she could get her foot in the door long enough to impress potential clients. Shirley also took pains to hide the infrastructure of her startup. Although she worked from home with her child, often accompanied by other women employees and their children, she presented as polished and professional an image as possible, playing a recorded tape of typing in the background whenever she took a phone call in order to drown out the noise her young son might make, and instituting a strict, conservative business dress code for meetings with clients.

One of the most prestigious contracts Shirley and her team were able to win was the Concorde's black box project. The programming for the Concorde was managed and completed entirely by a remote workforce of nearly all women, programming with pencil and paper from home, before testing their software on rented mainframe time. Dubbed "Europe's version of the Space Race," the Concorde remains the only passenger supersonic jet to go into service, and its creation pushed the limits of what was possible in terms of international technological cooperation (it was a UK–French collaboration). Perhaps even more importantly, British cooperation on the project served as a political concorde as well, and effectively paved the way for the UK to finally be accepted into the European Economic Community, the forerunner to the EU, boosting the nation's flagging economic fortunes.

In the image below, one of Shirley's employees, computer programmer Ann Moffatt, sits at her kitchen table in 1966, writing code for the Concorde's black box flight recorder. Moffatt was the technical lead for the Concorde programming project and managed the project while working from home, taking care of her young child at the same time. She would later become technical director at the company, in charge of a staff of over 300 home-based programmers (fig. 6.1).

Steve Shirley's company functioned, and indeed succeeded, by taking advantage of the sexism that had been intentionally built into the field of computing. By utilizing a portion of this wasted talent—capable women computer professionals who were being excluded from contributing to the new digital economy despite having all the skills required—Shirley employed thousands of workers over the course of her company's lifetime whose skills would have otherwise been discarded. Her feminist business model allowed many technical women to fulfill their potential, and in the process also serve the nation's ever-growing need for computer programmers.

For every woman Shirley employed, however, there were always several more applicants vying for positions, looking for a place that would allow them to use their computing skills and judge them on their capabilities rather than their gender.

Figure 6.1 Computer programmer Ann Moffatt sits at her kitchen table in 1966, writing code for the black box flight recorder for the Concorde. The baby in the photograph is now over fifty years old. Photo courtesy of Ann Moffatt.

That the women Shirley could not employ were usually unlikely to have computing careers elsewhere, particularly if they needed working hours that could accommodate their family responsibilities, meant that their productivity and talent was lost to the labor market and the nation as a whole.

The irony that these women were not perceived by their male managers as good enough to keep in formal employment but at the same time were so indispensable that the government and major corporations would outsource important computing projects to them was not lost on Shirley and her workers, but the realization seemed to sail over the heads of most men in management. Her startup eventually became an international billion-dollar company and produced some of the most important software used in British business and the public sector.

Shirley's success gives some clue as to why sexism was bound to create major problems for the computing industry. Her successful business model not only highlighted the fact that sexism was part of the way the field functioned but also showed how these operating parameters hurt computing's progress. Managers in government and industry saw their sexist labor practices as a positive and necessary feature of creating the new high-tech digital landscape—as something that would ensure computerization proceeded smoothly, along the lines they intended, and with the right people in control. In point of fact, however, it was a hindrance, and it would turn out to have serious and wide-ranging negative consequences.

EMPIRE 2.0

While Shirley's company and staff of mostly women provided the know-how and programming labor to keep the nation's computers running, and keep the process of computerization expanding, another national conversation was going on. Leaders in government and industry viewed the progress of computing as being about more than just improving standards of efficiency or revolutionizing work inside the UK. As a once-powerful empire, the British government saw computing as a powerful new tool in its international political arsenal.

Where it had once dominated by gun and boat, Britain now saw that as its empire shrank, it could only revive its power abroad by dominating the informational infrastructure of other countries and their economies. Computers had started out as powerful weapons of war, funded heavily by the US and UK militaries. Even as swords turned to plowshares, computers and the government interests that largely controlled them continued to be tools for wielding power over other nations. The UK believed it could gain influence and economic power through technological exports, and even use British computers and computing expertise as a back door into the governments of other nations that were becoming politically independent from British rule.

For this reason, the UK insisted on British computers to run all UK government work. They presciently understood that foreign powers could gain a foothold through foreign computing technologies being inserted deeply into the structure of British government. American computers running the British government would present a national security issue, even though the US was an ally. Computers could give a foreign power a back door into the highest levels of the state.[24]

While the British government was determined not to allow this to happen to them, they actively planned to use these same techniques against other nations, particularly former colonial nations, in order to gain power and political influence.[25]

Narratives and images of British computing abroad confirmed this agenda. Supported by the government, British computing companies embarked on a worldwide program of aggressive expansion, attempting to sell computing technologies, and the idea of their necessity, to countries with little need for labor-saving calculating machines. While United States companies, particularly IBM, balked at selling the latest technologies to nations that they considered underdeveloped—only deigning to sell India, for instance, older models in IBM's line—British companies seized the opportunity to get as much British technology as possible into the hands of Indian companies, educational institutions, banks, and government agencies, in order to forge relationships and accustom Indian consumers to buying, using, and relying upon British computers.[26]

As part and parcel of this, the UK exported to these other nations the gender roles that it had put into place for its own labor forces. When British computing companies set up a computing installation for a company in India, for example, the gendered contours of the labor force followed those in Britain. But when Indian companies purchased a British computer and set it up for themselves, they more often staffed it with a mixed-gender model or sometimes even gave "feminized" jobs like punching to all-male workforces.[27]

The British plan to dominate through technology was symbolic as well as material: advertisements for computing extended the rhetoric of Britain's "civilizing mission" from its imperial period, and used sexualized images of exoticized "foreign" women to portray the triumphs of British technology abroad. In figure. 6.2, the obvious sexual caption plays upon a deeper preoccupation with colonial power. In these images, women often stood in as symbols of the national cultures the UK expected to dominate with British technology, extending the nation's imagined international might in a cultural as well as political sense.

These tactics laid bare the power implications of computing and used high technology to continue the logic of British imperialism, even as nations that had been subjugated by the British fought for and won their independence from British rule in a political sense. It was no coincidence that the UK put so much faith and effort into computing technology during a time when its empire was contracting and its power on the world stage was diminishing. Computing technologies, then as now, were expected to be the new lever to move the world.

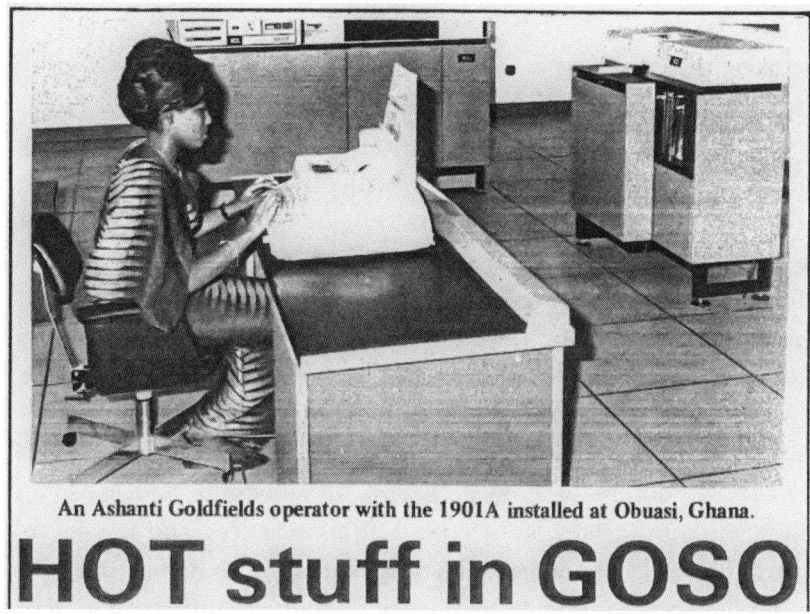

Figure 6.2 An Ashanti Goldfields computer worker in Ghana. *ICL Marketing* (February 6, 1970).

Figure 6.3 The punch room at Air India employed both men and women as operators, something British-run computing installations would not have allowed. "ICT in India," *ICT Magazine* 9 (1961).

A BUG AFTER ALL

Despite recognizing the enormous power of computing technology to reshape British fortunes at home and abroad, by the mid-1960s the UK government had allowed the nation's computer labor shortage to sharpen into a crisis that threatened to undermine all of its plans. Despite instituting career paths in the civil service for the ideal model of male, white-collar, management-oriented computer workers, the state continued to struggle to even get qualified candidates to train. Given this, its commitment to hire only management-aspirant young men began to soften, and in the mid-1960s a number of women programmers were able to get jobs in the new, higher-level technical grades of the civil service, as long as they came from the service's white-collar office worker grades and not from the "pink collar" machine grades.

But this brief wave of more egalitarian, meritocratic hiring, in terms of gender if not class or race, did not last long. By mid-1967, the labor crisis had abated just enough to allow a return to previous gendered hiring practices, and the government simultaneously decided to embark on an alternate plan for fixing its computing problems. If labor shortages could not be easily fixed, then a different part of the system would need to be changed to ensure the success of Britain's computerization.

By the late 1960s, the government's fear of losing control over the country through losing control over the technology that ran it was perilously close to becoming a reality. The power that technical workers held, given the interwoven nature of computing processes with all of the functions of the state—from the Bank of England to the Atomic Energy Authority—meant that they were becoming increasingly indispensable. A strike by keypunch staff doing data entry for the new value added tax (VAT) computer system, for instance, brought that project to a halt and sent high-level officials scrambling.[28] The computers that enabled the sprawling bureaucracies of the state to function were unreliable and consistently understaffed, unable to maintain the level of programmer and operator labor required to make them work most effectively.

It was in this context that the Ministry of Technology, backed by other government agencies, hatched a plan to reconfigure the computing infrastructure on which the government relied. Training more young men had not worked; it simply resulted in more turnover and wasted outlay, and outsourcing was only a temporary solution. Returning to a feminized computer labor force was a nonstarter given the rising power and prestige now attached to the jobs. This meant that the only solution was to re-engineer the computing systems to function with a much smaller labor force.

Managers at the top of the government therefore decided they would need ever more massive mainframes, so that all of the nation's computing operations could be centralized to the greatest extent possible. This would allow a far smaller technical labor force to control Britain's digital infrastructure. The government forced a merger of all the remaining viable British computer companies in order to literally gain more control over the shape of British computing. The merger created a single company, International Computers Limited (ICL), indebted to the government, giving government technocrats the power to demand that the company produce product lines that would best accommodate the government's needs.

The central government would fund the new merged company's research program and, even more importantly, would promise to purchase its computers exclusively. Since solving its computing problems through labor and software had not worked, the government asserted that the very design of the systems themselves would need to change. In return for government grants and contracts, ICL accepted a high degree of government control over its product line and embarked upon the project of producing the huge, technologically advanced mainframes the national government believed it needed to solve its computer labor problems.

Unfortunately, this move began right as the mainframe was losing its pride of place in business computing. Massive, expensive mainframes were on the way out, with smaller mainframes and more flexible, decentralized systems becoming the norm. By the time ICL delivered the requested product line in the mid-1970s—the highly advanced 2900 mainframe series—the British government no longer wanted these expensive and complex machines, and neither did any other potential customers. As a result, the company was effectively dead in the water. Moored to its advanced but effectively unwanted new mainframe line, ICL could not compete with the offerings of foreign companies like IBM. ICL was now—thanks to the merger—the majority of the British computer industry. Because the company had neglected the development of its smaller, and better-selling, lines of mainframes in order to focus on developing the 2900 for the government, its failure effectively took down the whole of the British computer industry.

LOOKING FORWARD BY LOOKING BACK

The discrimination that produced the technical labor shortage in UK government and industry in the twentieth century was a highly constructed and artificial feature. It was not an evolutionary or somehow natural change: nor was it a particularly logical one. The gender associated with computing jobs changed in this period because the idea that the work was appropriate only to one gender or the other was a defining element in how computing systems were organized. The sexism that made women the go-to labor force for computing early on also meant women were seen as an unreliable or inadequate group of workers once the power of computers became more apparent.

But the gendered structure of the field heavily influenced its growth and possibilities. Sexism was intended to keep power out of the hands of those deemed unworthy of wielding it, but it ended up devastating Britain's computing industry and the country's global technological standing.

This history is of crucial importance today because it is not that unusual: the British reliance on sexist hiring and firing practices, and on the association of technical work with low-status workers but powerful managerial work with high-status workers, has been a common pattern in the postindustrial West. The manner in which government leaders and industry officials worked together to standardize and codify a gendered underclass of tech workers, and then to later upskill that work once the managerial power of computers became clear, was not evolutionary or accidental. It was an intentional set of systems design parameters intended to ensure that those who held the most power in predigital society, government, and industry continued to hold that power after the "computer revolution."

It is also not a coincidence that once computing became seen as powerful and important, more than merely "technical" work, it began to exclude white women and people of color except in times of labor crises, like war or severe labor shortages in peacetime.[29] High technology is often a screen for propping up idealistic progress narratives while simultaneously torpedoing meaningful social reform with subtle and systemic sexism, classism, and racism. As Margot Lee Shetterly points out in *Hidden Figures*, Black women workers were only brought into critical jobs in NASA when Cold War tensions made their labor too valuable to ignore—too important to continue to exclude on the basis of their Blackness.[30] But even once included, people who were not seen as worthy of wielding power were deliberately engineered out of newly powerful and profitable roles, even when they had the required technical skills.[31]

From the beginning, societal biases have played a major role in constructing and deploying computer systems because high technology is not the radical agent of change that those in power often profit from presenting it as. The computer revolution was not a revolution in any true sense: it left social and political hierarchies untouched, at times even strengthening them and heightening inequalities. That the US in the twentieth century escaped the fate of the UK has more to do with the relative size of its labor force, and its relative wealth following World War Two, than with differences in sexist hiring practices. But currently the US is being forced into competition with rising superpowers like China and India, which have far greater available labor forces, and it simultaneously finds itself confronted again by Russian technological power and political warfare. In this context, the US as a fading empire mirrors the UK in the twentieth century, and we have begun to see Silicon Valley corporations, and US society as a whole, hurt by an inability—or refusal—to solve deep, structural problems with discrimination.

This is largely because this aspect is an intentional feature, not an accidental bug. Computing was designed to help those at the highest levels of government and industry wield power. Powerful technologies like this often heighten inequalities not by accident but specifically because they they are designed to protect the interests of the powerful actors who control them. The contours of this story may change in different national and temporal contexts, but the idea that technological systems normally preserve existing hierarchies and power structures, rather than being revolutionary, holds true much more often than we recognize.

Though Britain's actions backfired and had highly negative effects on its computing industry, its entire economy, and even its political standing in the wider world, the actions themselves were fully intentional.[32] At the time, UK leaders in government and industry generally believed that there could be no other way to design such systems—that this mode of operation represented a kind of natural social order, despite vast evidence to the contrary.

THE POWER OF THE STORIES WE TELL ABOUT TECHNOLOGY

Historians of technology can point out the many ways in which status in high tech is arbitrary, how categories of prestige swiftly change, and how pay and rewards are often unmoored from the specific skills the industry purports to value.[33] Yet, as a field, the history of computing has often taken for granted the ideal of meritocracy as an implicit structuring element in narratives of technological progress and the lessons drawn from those narratives.[34] Because technological success stories map to the idea that computing's history is one of social progress, even when it is not, the main narrative of meritocracy-driven progress in the history of computing needs to be rethought, with an emphasis on the important historical lessons gained from technological failure.

By privileging stories of corporate success and the ingenuity of high-status individuals, historians have often constructed narratives that seem to support the fiction of meritocracy.[35] White, heteronormative, male, and cis professionals crowd the pages of most histories of computing, while Black and white women programmers, Navajo women semiconductor manufacturers, LGBTQIA technologists and computer users, and many others have been regarded, up until recently, as interesting but marginal characters in the main narrative of computing history. Yet history shows us that these "nontraditional" actors not only were present but in fact played a much larger role in shaping the history of computing than previously understood. And through shaping computing, these groups also altered social, economic, and cultural contexts to the point of determining the fortunes of entire industries and nations.[36]

The United States today finds itself in a similar position to the UK in the twentieth century, as a technologically advanced but socially regressing late-imperial power. We would do well to look at the British example and commit to studying failure instead of focusing only on success in our technological history. Failure narratives highlight the ways we have gotten things wrong in the past, guiding us to better outcomes through deeper understanding of negative examples. If we look at the decline of British computing critically, we can recognize it as a harbinger of what is already occurring with our own technological "successes" in computing. As the CEOs of major platforms like Facebook and Twitter begin to openly struggle with how their companies have impacted, and continue to alter, the course of democratic elections and civil rights in the US, we can begin to see how tech's tendency to concentrate power in the hands of a few wealthy individuals is inimical to democracy and tends to heighten inequalities and exacerbate existing problems. For years, marginalized voices tried to raise the alarm about this important, destructive aspect of techno-optimism but were ignored. We can also see how the ahistorical technolibertarianism favored by Silicon Valley elites—most of whom come from similarly privileged backgrounds—ensures that their technologies are neither radical nor forward-thinking.[37]

By understanding the racism, sexism, and classism that formed and torpedoed our closest historical cousin's dreams of a modern, technological society, we can also begin to address how these inequalities have constructed the bedrock of our computing industry for decades, rather than being a simple mistake or an unintentional afterthought. We can begin to see that even though Silicon Valley still profits from these inequalities today, there is a clear connecting line between these features of technological economies and negative consequences for the whole of society. The power-centralizing and authoritarian tendencies of computing systems were indispensable for warfare and strong government control of massive amounts of data, but these same systems, in the hands of those determined to escape democratic oversight, become inimical to functional democracies and civil society.

As our technological systems become increasingly destructive to our professed social and political ideals, we can no longer afford to collectively fail to understand the layers and decades of intentional decisions that have led to these supposedly unforeseen consequences. The current situation shows us clearly how, as large computing and telecommunications systems have scaled, the power imbalances they foster have altered all of our social institutions, including our political process. These imbalances go far beyond the tech industry, and they help determine who holds power at every level within our society and government, altering the future of entire nations. This problem cannot be fixed without breaking up the systems in place, because these failures are not simply accidents: they are features of how the systems were designed to work and, without significant outside intervention, how they will continue to function. Our task for a better future must be to foster a willingness to subdivide and reengineer failing systems at a basic, structural level, rather than contenting ourselves to simply patch existing failures in our digital infrastructures—leaving the broader systems that created these problems largely unchanged.

NOTES

1. Sheryl Sandberg's *Lean In: Women, Work, and the Will to Lead* (New York: Alfred A. Knopf, 2013) admonished women to try to overcome structural sexism through sheer force of will and individual action, ignoring the enormous privilege that this kind of message takes for granted. The *Guardian*'s Zoe Williams described it as a "carefully inoffensive" book that was not "about how women could become more equal, but about how women can become more like Sheryl Sandberg." Zoe Williams, "Book Review: *Lean In*," *Guardian* (February 13, 2013), https://www .theguardian.com/books/2013/mar/13/lean-in-sheryl-sandberg-review. It is perhaps not surprising that Sandberg, chief operating officer of Facebook, wrote a supposedly feminist book that nonetheless endorsed and sought to maintain all of the exclusionary power structures that, on the surface, it claimed to critique.

2. Google Walkout for Real Change, "Google Employees and Contractors Participate in Global 'Walkout for Real Change,'" *Medium* (November 2, 2018), https://medium.com/@GoogleWalkout/ google-employees-and-contractors-participate-in-global-walkout-for-real-change-389c65517843; Matthew Weaver et al., "Google Walkout: Global Protests after Sexual Misconduct Allegations," *Guardian* (November 1, 2018), https://www.theguardian.com/technology/2018/nov/01/google -walkout-global-protests-employees-sexual-harassment-scandals; Susan Fowler, "Reflecting on One Very, Very Strange Year at Uber," personal blog (February 19, 2017), https://www.susanjfowler .com/blog/2017/2/19/reflecting-on-one-very-strange-year-at-uber; Maya Kosoff, "Mass Firings at Uber as Sexual Harassment Scandal Grows," *Vanity Fair* (June 6, 2017), https://www.vanityfair.com/ news/2017/06/uber-fires-20-employees-harassment-investigation.

3. For instance, many of the leaders of the Google Walkout have been forced to leave the company. See Alexia Fernández Campbell, "Google Employees Say the Company Is Punishing Them for Their Activism," *Vox* (April 23, 2019), https://www.vox.com/2019/4/23/18512542/google-employee-walkout-organizers-claim-retaliation; Julia Carrie Wong, "'I've Paid a Huge Personal Cost': Google Walkout Organizer Resigns over Alleged Retaliation," *Guardian* (June 7, 2019), https://www.theguardian.com/technology/2019/jun/07/google-walkout-organizer-claire-stapleton-resigns; Nitasha Tiku, "Most of the Google Walkout Organizers Have Left the Company," *Wired* (July 16, 2019), https://www.wired.com/story/most-google-walkout-organizers-left-company/.

4. Katharine A. Kaplan, "Facemash Creator Survives Ad Board," *Harvard Crimson* (November 19, 2003), https://www.thecrimson.com/article/2003/11/19/facemash-creator-survives-ad-board-the/; Kate Losse, "The Male Gazed: Surveillance, Power, and Gender," *Model View Culture* (January 13, 2014), https://modelviewculture.com/pieces/the-male-gazed.

5. On the last point, see Alexandra Stevenson, "Facebook Admits It Was Used to Incite Violence in Myanmar," *New York Times* (November 6, 2018), https://www.nytimes.com/2018/11/06/technology/myanmar-facebook.html. For a general overview, see Siva Vaidhyanathan, *Antisocial Media: How Facebook Disconnects Us and Undermines Democracy* (Oxford: Oxford University Press, 2018).

6. Tasneem Raja, "'Gangbang Interviews' and 'Bikini Shots': Silicon Valley's Brogrammer Problem," *Mother Jones* (April 26, 2012), https://www.motherjones.com/media/2012/04/silicon-valley-brogrammer-culture-sexist-sxsw/; Mar Hicks, "De-Brogramming the History of Computing," *IEEE Annals of the History of Computing* (January–March 2013), https://doi.org/10.1109/MAHC.2013.3. Also see, for instance, the seriousness with which the claims of the sexist "Google Memo" written by Google programmer James Damore were considered by major news outlets, prompting historians to write refutations of the memo's blatantly false premises. Daisuke Wakabayashi, "Contentious Memo Strikes Nerve, Inside Google and Out," *New York Times* (August 8, 2018), https://www.nytimes.com/2017/08/08/technology/google-engineer-fired-gender-memo.html; Abby Ohlheiser, "How James Damore Went from Google Employee to Right-Wing Internet Hero," *Washington Post* (August 12, 2017), https://www.washingtonpost.com/news/the-intersect/wp/2017/08/12/how-james-damore-went-from-google-employee-to-right-wing-internet-hero/; Mar Hicks, "What the Google Gender 'Manifesto' Really Says About Silicon Valley," *The Conversation* (August 10, 2017), https://theconversation.com/what-the-google-gender-manifesto-really-says-about-silicon-valley-82236; Mar Hicks, "Memo to the Google Memo Writer: Women Were Foundational to the Field of Computing," *Washington Post* (August 9, 2017), https://www.washingtonpost.com/opinions/memo-to-the-google-memo-writer-women-were-foundational-to-the-field-of-computing/2017/08/09/76da1886-7d0e-11e7-a669-b400c5c7e1cc_story.html.

7. Safiya Noble, *Algorithms of Oppression* (New York: NYU Press, 2018).

8. Jake Swearingen, Can Google Be More Than an Advertising Company?" *New York Intelligencer* (February 5, 2019), https://nymag.com/intelligencer/2019/02/google-earnings-show-it-needs-to-be-more-than-an-ad-company.html.

9. See, for example, Amy Slaton, *Race, Rigor and Selectivity in U.S. Engineering: The History of an Occupational Color Line* (Cambridge, MA: Harvard University Press, 2010).

10. Mar Hicks, *Programmed Inequality, How Britain Discarded Women Technologists and Lost Its Edge in Computing* (Cambridge, MA: MIT Press, 2017), chapter 1.

11. See, for example, Hazel Carby, *Imperial Intimacies* (New York: Verso, 2019) on the struggles of the "Windrush generation" to prosper and gain their civil rights in the UK.

12. In my conversations with women who worked in computing at Bletchley Park, and also after the war, I have been repeatedly greeted with responses to the effect of "Are you sure you want to interview me? I don't think what I did was that important." Part of my task in recording this history has been to convince these women that their contributions were important and that historians do value what they have to say.

13. Hicks, *Programmed Inequality*, 1–3.

14. For more on this, and a case study of one of the earliest examples of mainframe-era transphobic algorithmic bias, see Mar Hicks, "Hacking the Cis-tem: Transgender Citizens and the Early Digital State," *IEEE Annals of the History of Computing* 41, no. 1 (January–March 2019): 20–33, https://doi.org/10.1109/MAHC.2019.2897667.

15. Hicks, *Programmed Inequality*, 90–93.

16. Hicks, *Programmed Inequality*, 93–96.

17. Hicks, *Programmed Inequality*, 151.

18. Hicks, *Programmed Inequality*, 187.

19. Britain's first attempts to join the EEC (European Economic Community), for instance, were denied in part because its economy was perceived as not having adequately modernized. Hicks, *Programmed Inequality*, 190.

20. Note: Despite the UK's relatively small size, it took over ten times as many Jewish child refugees as the United States, which only accepted roughly 1,000 children.

21. Stephanie Shirley, *Let IT Go: The Memoirs of Dame Stephanie Shirley*, rev. ed. (Luton, UK: Andrews UK Limited, 2017), 6.

22. Shirley, *Let IT Go*, 58.

23. Job advertisement, "Freelance Programmers," *Times (London)* (June 26, 1964).

24. Hicks, *Programmed Inequality*, 178–179.

25. Hicks, *Programmed Inequality*, 116–122.

26. Hicks, *Programmed Inequality*, 118–121.

27. Hicks, *Programmed Inequality*, 120.

28. Hicks, *Programmed Inequality*, 204.

29. See, for example, Jennifer Light, "When Computers Were Women," *Technology and Culture* 40, no. 3 (1999): 455–483; Janet Abbate, *Recoding Gender: Women's Changing Participation in Computing* (Cambridge, MA: MIT Press, 2012); and Margot Lee Shetterly, *Hidden Figures: The American Dream and the Untold Story of the Black Women Mathematicians Who Helped Win the Space Race* (New York: William Morrow, 2016).

30. Shetterly, *Hidden Figures*.

31. See, for example, Ford, *Think Black*.

32. The UK's failed modernization plans hurt its integration into the EEC and diminished its economic and political capacities on the world stage. For more, see Hicks, *Programmed Inequality*, chapters 3, 4, and 5.

33. See, for example, Slaton, *Race, Rigor and Selectivity in U.S. Engineering*; Abbate, *Recoding Gender*; Light, "When Computers Where Women"; Hicks, *Programmed Inequality*.

34. In Meredith Broussard's *Artificial Unintelligence: How Computers Misunderstand the World* (Cambridge, MA: MIT Press, 2018), Broussard discusses how the history of computing often mirrors the technological success stories that computer scientists tell about their own work.

35. See, for example, the popular biographies of software industry personalities like Steve Jobs that fall into the "great man" history trope. Biographies of notable women in computing, like Grace Hopper, sometimes borrow from this style as well, switching great man narratives of success for "great woman" narratives. The ubiquity and popularity of such accounts gives the impression that individual actors are responsible for historical change, rather than foregrounding the complex actions of large masses of people. In general, they focus only on historical success stories as instructive. Walter Isaacson, *Steve Jobs* (New York: Simon & Schuster, 2011); Kurt Beyer, *Grace Hopper and the Invention of the Information Age* (Cambridge, MA: MIT Press, 2009).

36. Lisa Nakamura, "Indigenous Circuits: Navajo Women and the Racialization of Early Electronics Manufacture," *American Quarterly* 64, no. 4 (December 2013): 919–941; Shetterly, *Hidden Figures*; Slaton, *Race, Rigor and Selectivity in U.S. Engineering*.

37. For a fuller discussion of these issues, see David Golumbia, "Do You Oppose Bad Technology, or Democracy?," *Medium* (April 24, 2019), https://medium.com/@davidgolumbia/do-you-oppose-bad-technology-or-democracy-c8bab5e53b32, and Fred Turner, "How Digital Technology Found Utopian Ideology: Lessons from the First Hackers' Conference," in *Critical Cyberculture Studies: Current Terrains, Future Directions*, ed. David Silver and Adrienne Massanari (New York: NYU Press, forthcoming), https://fredturner.stanford.edu/wp-content/uploads/turner-ccs-hackers-conference.pdf.

7

GENDER IS A CORPORATE TOOL

Corinna Schlombs

Your company does not love you, no matter how loudly it may sing otherwise. In 1927, IBM first published a songbook to praise the company's executives and employees at corporate gatherings. Today, Silicon Valley corporations don't engage in the same overt morale-building, but they still pursue similar goals. Nowadays, tech companies are often hailed as model employers because of the perks they provide, which have evolved from the Ping-Pong tables and free pizza of the 1980s to today's free cafeterias and company shuttles, nap pods, concierge services to run employee errands, and wellness centers and free massages. But these perks hide an inconvenient truth: tech employees have no say in their companies' decisions. Indeed, the West Coast tech industry has shunned giving employees a say in company matters through unionization. Intel founder Robert Noyce supposedly said that "remaining non-union is essential for survival for most of our companies. If we had the work rules that unionized companies have, we'd all go out of business."[1] And while bus drivers and security guards in Silicon Valley have unionized, tech staff and programmers largely have not. But tech employees are beginning to speak up. In recent years, they have protested their companies' ethics—from defense contracts to the industry's handling of privacy and sexual harassment—through open letters, petitions, and walkouts. In the gaming industry, employees organize an international unionization drive to address precarious work conditions, such as long hours, frequent layoffs driven by the business cycle, and lack of health care. The fire of employee-relations issues is burning hot in the valley and in tech companies elsewhere.

Although Silicon Valley firms often decry older, more conservative models of regulation and employee relations, their strategy of perks-for-unionization has a very old model: IBM, the once-bellwether of the tech universe. At IBM, employees in the 1950s literally sang the praise of their longtime chief executive, Thomas J. Watson Sr., the "man of men," a "sterling president," and "friend so true" who made his "boys" (emasculating grown men as corporate sons) "smile, smile, smile."[2] Mr. Watson himself saw IBM as his family. With his wife, Jeanette Kittredge Watson, by his side, Mr. Watson presented himself as the family father in a corporate culture that emphasized mutual understanding and respect over labor organization and conflict.[3] But Mr. Watson's frequent talk of the "IBM family" was indicative in some possibly unintended ways: many families, of course, are not egalitarian but have clear hierarchies of age and gender, between spouses, and between parents and kids. It also obfuscated that the company's generous benefits and pay served a larger goal: to maintain Mr. Watson's control over a docile workforce, free from union interference. After all, who would unionize against their own father?

IBM back then was an up-and-coming tech company with a competitive share in the punch-card-machine market, and one of a few successful companies positioning themselves in the emerging electronic computer market. It also was a typical welfare capitalist company, that is a company offering above-average pay and benefits like insurance programs, clubs, and vacation homes with the expectation that employees refrain from organizing in national unions—the equivalent to today's perks by Google, Facebook, or Amazon. In the early twentieth century, many US companies had taken over social responsibilities for their employees in the face of a weak government and weak unions. Welfare capitalist policies also guarded managerial decisions from negotiating pay and working conditions with union representatives. This essay takes a closer look at IBM to raise questions about tech companies today: Whom do generous company perks benefit? Whose power do they preserve? Why do employees join the family or organize collectively? How are different people—white, Black, men, women, straight, LGBTQ, abled, disabled—integrated in the corporate "family"?[4] And what roles, and rights, do they have within it? What happens when this model is extended globally, as when IBM reorganized its operations in West Germany in the mid-twentieth century?

Sometimes "family" is code for "not under *my* roof." Namely, Mr. Watson insisted on trying to implement IBM's American welfare capitalist culture to its subsidiaries abroad in spite of different local labor practices. IBM's example reveals that organizing a company or industry across different countries and cultures—as is happening in the gaming industry—is fraught because of local differences in labor relations and legislation. West German IBM employees, for example, appreciated the benefits of being part of the "IBM family" while they also organized in national unions, formed works councils, and elected a labor representative to the board of directors— all of which are common forms of labor organization in Germany, and precisely what welfare capitalist companies like IBM sought to avoid. But Mr. Watson's family rhetoric made the American welfare capitalist measures palatable to German workers, who were ensconced in socialist and Christian labor organizations, and it promoted cooperative labor relations. IBM's example shows that tech companies have a long history of deploying gender and class both nationally and internationally, and this history still affects professional culture and labor solidarity in our globalized economy today.

IBM'S PROGRESSIVE LABOR AND GENDER RELATIONS IN THE UNITED STATES

Progressive politics have long served the bottom line. At IBM, Mr. Watson shaped his company's culture like few executives did, and foreign visits and corporate celebrations heightened the personality cult surrounding him. In the United States, he pursued progressive labor and gender policies; for example, in the 1930s he implemented new labor benefits such as air-conditioning, clubs, and insurance programs— the equivalent to today's free cafeterias, bus shuttles, and other perks. IBM was a latecomer to such welfare capitalist measures, which other companies had created in the first three decades of the twentieth century, and which then still counted as progressive compared to the violent suppression of labor unions in other companies. Following the Great Depression, welfare capitalism floundered because many workers began to vote for the Democratic Party in federal elections and to join industrial unions to fight for their rights, bolstered by New Deal legislation easing the organization of workers.[5] But IBM, formed in 1913 through a merger of three fledgling companies, could afford welfare capitalist measures only in the 1930s, thanks partly to profitable contracts for the new Social Security system. By now, the paternalist nature of welfare capitalist labor relations also became clearer as workers in pursuit of industrial democracy—that is, corporate representation through unions—fought to have their voices heard at the negotiating table. By contrast, Mr. Watson, like most welfare capitalists, expected his workers to refrain from organizing in labor unions in return for the benefits granted to them by managerial fiat.

Yet Watson claimed that everyone in IBM was equally important. Such rhetoric has the ring of Silicon Valley's present claims about meritocracy—that everyone can advance equally based on their skills. Both encourage workers to think and act as individuals working together (or competing with each other), rather than as members of the working, middle, or managerial classes. In this volume, Mar Hicks, Sreela Sarkar, and Janet Abbate show that the meritocratic ideal discriminates against those who are less powerful. In Hicks's case, women programmers in postwar Britain lost out despite having more skills than their male replacements, and Sarkar's and Abbate's analyses, respectively, show how the promises of bootstraps meritocracy work against Muslim lower-class women in India and women and minority programmers in the United States today.[6] Like meritocratic ideals, Watson's emphasis on equal importance is uplifting at first sight, but below the surface it relies on assumptions of equality that hide the obstacles employees confront based on who they are, where they come from, and how they are treated in society. Watson's model of equal, individualistic workers meant that workers always confronted management alone. This weakened their position and deprived most employees—men as well as women—of a say in IBM's corporate affairs. Current evidence suggests that the supposedly flat hierarchies in today's tech companies may have similarly disempowering effects.

At IBM, Mr. Watson often emphasized the importance of humans and human relations at seminars or workshops like the one depicted in figure 7.1. He listed different corporate positions from manufacturer and general manager to sales manager and sales man, service man, factory manager and factory man, and office manager and office man, and crossed out syllables until only "man" remained, claiming that IBM employees, regardless of their position, first and foremost counted as humans. However, "man" here came to stand in for all employees, ignoring all women employees—just as Mr. Watson's claim that all employees mattered equally ignored power differences along gender and class lines. It was a first step to making employees forget the impact of the outside world on the shape of their employment at IBM. "The nurturing of human relations is the fundamental task of all humans," Mr. Watson stated, ignoring that our society normally held women responsible for nurturing human relations. "We have different ideas and different areas of work. But when we get to the bottom of it, there is only one thing we deal with in our corporation—and that is the human." The conjuring of equal importance among all employees hid the fact that most employees at IBM—men and women—did not have any say in corporate matters.

Figure 7.1 Thomas J. Watson Sr., IBM's chairman, during a presentation on "The Human as a Task." (*Source*: Combined issue of *IBM World Trade News* and *THINK* [July 1956]: 8. Courtesy of International Business Machines Corporation, © 2019 International Business Machines Corporation.)

Mr. Watson did open management opportunities—albeit limited ones—to women at a time when most companies relegated them to clerical and routine positions. One of the reasons may have been that Anne van Vechten, a college friend of the Watsons' oldest daughter, apparently challenged him to create career paths for young women.[7] The women trained at the company's school in Endicott, New York, where the company's new male employees also prepared for their positions, albeit in separate classes. Also, the women were allowed to prepare only for positions in customer training and human relations—showing that the company imposed strict limits on their careers. Obviously, not everyone was to be equal in IBM's large family.

During their training, the young men and women organized weekly parties, many of which Mr. Watson attended, and more than one couple emerged, likely ending the career of the young women because IBM, like most companies, still implemented a marriage ban and expected married women to resign from salaried work. In 1935, IBM graduated its first class of women systems service professionals. But IBM executives were so reluctant to hire them that Mr. Watson threatened to fire the newly trained men unless the women were hired first. The young women inserted themselves into IBM's male professional culture, which included a strict dress code of dark blue suits, ties, and white shirts. The women typically dressed in professional yet fashionable dresses with checkered fabrics and sported nice hairdos—a far cry from the high-necked white blouses and long dark skirts typically worn by "office girls" only a decade or two earlier.[8] In the following years, Mr. Watson often surrounded himself with young women for luncheons and other occasions, at one time even buying new shoes for a group of them after a rainy morning out and about New York City.[9] The women were family, but IBM's family still had hierarchies.

MR. WATSON'S "IBM FAMILY"

What kind of family is headed by fifteen men and one woman? In 1949, IBM reorganized its international operations in a wholly owned subsidiary, the IBM World Trade Corporation. An image of the new board of trustees shows a group of mostly older, white men around a table, including Mr. Watson at the head, his wife Jeanette K. Watson by his side, and his younger son, Arthur K. Watson, at the table (fig. 7.2). Arthur, who was to chair IBM's international operations, seamlessly blended into the picture. But Mrs. Watson stood out. While all of the men looked straight into the camera, she bent her head toward her husband in a seemingly deferential pose. In contrast to IBM's young women professionals, who typically wore light-colored,

Figure 7.2 Board of trustees of IBM's World Trade Corporation, 1949. (*Source: IBM World Trade News* [October 1949]: 2. Courtesy of International Business Machines Corporation, © 2019 International Business Machines Corporation.)

fashionable dresses, Mrs. Watson had chosen a dark jacket and white blouse that blended in with the men's suits; but her hat set her apart, marking her as the only woman in the room. In addition, her posture distinguished her as an outsider within the circle of accomplished men, raising the question whether she had been included in the image just as a family member.[10]

Mrs. Watson was in fact a regular and full member of the board of trustees, although her corporate role had little resemblance to today's female executives. She had traveled many thousands of miles with her husband to visit IBM's foreign subsidiaries, shaken hands with thousands of IBM employees and their spouses, and hosted countless guests at their New York home for dinners and parties that were more business than pleasure. Her presence symbolically turned IBM employees into corporate children, cared for by the Watsons. But while the family model may have seemed benevolent, it could also make legitimate employee demands look like the whining of ungrateful children and deprive them of a say in the corporation. Through her symbolic role, Mrs. Watson may arguably have played a more crucial role in shaping IBM's labor relations abroad than any of the men in the room.

Although Mrs. Watson presented a deferential façade in public, she appears to have pulled the strings in the background, with her appointment to the board acknowledging her informal influence at IBM. She had apparently connected with her teetotaling husband over the fact that neither of them were drinking alcohol at the social function where they met. The couple married in 1913, shortly before Mr.

Watson became IBM's top executive, and had four children within five years. By the 1920s, Mrs. Watson became more active in company affairs, often accompanying her husband on international trips. Company stories speak to her influence over her husband. Once, during one of his rambling speeches, she had the manager whom she was seated next to in the first row deliver a note to her husband at the podium, interrupting him midsentence. The note simply said: "Shut up." Another time, she abruptly stopped him from opening a kitchen cabinet in a young woman employee's home—possibly saving the woman's career in the notoriously abstinent IBM culture since the cabinet held alcohol.[11]

In private, powerful; in public, pretty. In this way, Mrs. Watson had a lot in common with the generation of young women in the US after World War II, although she was of an older generation herself. Those young women married in greater numbers and at earlier ages than their mothers, and they devoted themselves to their roles as homemakers and mothers in their new suburban homes.[12] Although Mrs. Watson had the help of several servants and threatened divorce over her husband's insensitivity of frequently bringing home company guests unannounced, she never took on salaried employment, fitting the image of the homemaking wife and mother. As such, she provided a gendering foil for her husband, a charismatic but also abrasive executive.

With Mrs. Watson visibly at his side, Mr. Watson frequently invoked the gendered rhetoric of the IBM family to attain his corporate goals. In his speeches he described the labor relations he sought to install in the company's West German subsidiary, as well as in subsidiaries elsewhere, by referencing the "IBM family" that relied on Mrs. Watson's visibly feminine, "wifely" role. During a 1953 visit to Germany, for example, he emphasized that it had "always been his objective for all employees to be dealt with like family members." Mr. Watson also expressed his delight in being able to personally greet all IBM employees, because wherever he and his wife "have been, in each country of the IBM-Family, they have always met the same kind of people. People who think right and people who have always let us feel their friendship." He assured German employees that he had always tried to "return this friendship to the full extent" because friendship was above everything else for him and his wife.[13]

Family rhetoric sells more than warm feelings and kickbacks; it promises good press and (job) security. Like in the United States, IBM paid above-average salaries abroad and implemented benefits like vacation homes and a pension insurance. The company also sought to avoid unionization abroad, and it instead preferred dealing with its employees on an individual basis. To do so, it fostered direct corporate communication through employee magazines that often relayed the company's goals and vision; through a suggestion system that tapped into the employees' experience and knowledge in return for individual rewards; and through the so-called open-door policy that encouraged employees to escalate grievances to upper management rather than work through their union representatives. To this end, Mr. Watson often talked about the IBM family, with himself as the father and his wife as the mother at the helm. While she was soft and caring, he could be strict and authoritarian. This family rhetoric undergirded the welfare capitalist labor measures that IBM pursued in West Germany; together, they cared deeply and provided justly for the members of the large IBM family.

IBM'S GENDER AND LABOR RELATIONS IN GERMANY

German employees related to Mr. Watson's rhetoric of the IBM family, giving it its own local meaning. During a 1953 visit from Mr. and Mrs. Watson, for example, the head of IBM's German works council, W. Berger, presented Mrs. Watson with a Mocca coffee set made from Meissen porcelain during a festive evening reception for 3,600 IBM employees and their spouses. Acknowledging her rather than him with a gift, the German employees related to the couple on a family basis, presenting an appropriately feminine gift. At the same time, they selected a set from Meissen porcelain, which could be expected to be known and appreciated by a New York socialite, whereas a set from a local porcelain manufacturer would have expressed local pride in craftsmanship. At other occasions, Berger also carefully embraced Mr. Watson's family rhetoric and, like Mr. Watson, linked it to the company's labor relations.

When the German subsidiary celebrated Mr. Watson's fortieth IBM anniversary in April 1954, for example, Berger sent the "warmest thank you" across the Atlantic, to the "head of our so often praised IBM family in the name of all IBM employees in Germany and their family members." Appreciating the company's welfare capitalist measures, he praised that Mr. Watson showed his outstanding personality in his direct personal relation to his employees, like through the company's open-door policy and in his company's social benefits such as above-tariff salaries and vacation times, Christmas bonuses, insurance contributions, emergency funding, the pension system, and the recreation homes. All of these, in the end, were due to Mr. Watson's goodwill. Berger then acknowledged the wives of IBM employees, explaining that their invitation to the festivities expressed the company's gratitude for their work as "guardians of domesticity and family life and source of all happiness and creative power" who, in this way, contributed to IBM's undisturbed operation.[14] The women's reproductive work, taking care of the household and kids, allowed IBM's male employees to devote themselves to their corporate jobs, free from outside worries.

The editors of IBM's German employee magazine, *IBM Deutschland*, unabashedly embraced and even heightened the family rhetoric. Reporting on the Watsons' summer 1953 visit, they stated that IBM was more than a mere place to work; employment with IBM was "a chair at the table of a large and faithfully providing family." At IBM, every employee "knows that a large organization stood behind him. He [sic!] knows that IBM will provide him with support, a secure place in a unique community, he knows that he will find help and consolation as well as work and bread."[15] The magazine then reported on a young man who, after reading Mr. Watson's biography, stated that he "no longer had the feeling to just arrive at his workplace when he arrived at IBM in the morning. He now knew that IBM was more for him, thanks to the ingenuity of only a single man"—Mr. Watson.[16] The employee magazine brazenly promoted Mr. Watson's gendered self-presentation as the compassionate patriarch providing for the members of his corporate family.

Still, not everyone at IBM was convinced. German executives copied the Watsons' gendered presentation at company events but employed the family rhetoric more sparingly. At the fortieth-anniversary celebration for Mr. Watson, for example, the German subsidiary invited all employees and their spouses to a festive evening on April 30, 1954—the eve of the international labor movement day, a bank holiday in West Germany to celebrate international labor solidarity and promote workers' rights in memory of the Chicago Haymarket riots. Similar to the festive events marking the Watsons' visit to Germany the previous summer, now the German executives Johannes H. Borsdorf and Oskar E. Hörrmann and their wives personally greeted every guest. They copied the Watsons' habit of shaking the hands of hundreds of employees and their spouses—a habit that Mr. Watson often described as an expression of IBM's family spirit, despite the fact that Mrs. Watson wore long gloves on the greeting line that she dispensed with later in the evening, which might suggest that she had second thoughts about the close physical contact. As they did for the Watsons' visit the previous summer, the company again invited employees with their spouses outside of their usual work time, marking IBM's reach into family time and the private sphere. And again, the attending ladies received a properly gendered gift, this time a bottle of eau de cologne, acknowledging their roles as trim caretakers of their homes.

Sometimes progress follows accidentally from attempts to secure the status quo. In their speeches, Borsdorf and Hörrmann reminded IBM employees of Mr. Watson's habit of thanking his wife for her involvement with the company, a habit that most audience members would have witnessed the previous summer. They continued that Mr. Watson was aware that his wife always influenced him positively, and that the wives of IBM employees likewise exerted positive influence on their spouses. Therefore, they explained, Mr. Watson had wished that the wives of IBM employees be invited to his anniversary celebrations.[17] Assuming that IBM employees were men, this gendered rhetoric rendered the company's women employees invisible.[18] Nevertheless, through its hospitality, the company extended its reach into IBM families, seeking to enlist spouses to support the company's bidding; Borsdorf and Hörrmann encouraged all audience members to follow the Watsons' example.

Mr. Watson's conservative family rhetoric, based on clearly divided gender roles between husband and wife, found easy acceptance in West Germany's conservative political climate. During the 1950s, many West Germans longed to return to the supposed "normalcy" of the nuclear family, with its wage-earning father, homemaking mother, and two to three children.[19] Throughout the decade, half the women of eligible age in Germany worked, with their employment increasing from 47.4 percent in 1950 to 54.1 percent in 1960, and their higher visibility in the service sector fueling growing public concern about women's work.[20] Representatives across the political spectrum passed legislation that promoted women's family roles. For example, the 1949 German Basic Law—the German constitution—granted special protection of marriage, motherhood, and families as well as equal rights to women for the first time. But guided by conservative family value, the law also stated that housework constituted women's primary contribution to their families and that women could hold gainful employment only when it did not interfere with their duties as housewives and mothers.[21] The conservative family policies captured the dominant social values that provided a fruitful ground for the notion of the IBM family.

The strong emphasis on the nuclear family in West German legislation must be seen in the larger Cold War context. The East German government gave women full equal rights in 1949, and the 1950 Law for the Protection of Mothers and Children created social services for working mothers, expanded women's opportunities to work, and mandated equal wages and affirmative action programs. In West Germany, by contrast, the notion of the "family as a refuge" from outside influences (*Fluchtburg Familie*) promoted the idea that families provided security against the intrusions of (Communist or National Socialist) totalitarian governments as well as against the degrading forces of individualism, materialism, and secularism.[22]

LABOR ORGANIZATION IN IBM'S GERMAN SUBSIDIARY

Of course, there was no such thing as an IBM family. But Mr. Watson's gendered family rhetoric shaped IBM's German labor relations, even if not precisely in the way he intended. In particular, Mr. Watson proved unable to achieve the major goal of welfare capitalism: to prevent union organization. The way that unionization and employee relations played out differently in IBM's German subsidiary reminds us that labor relations are local affairs, different from country to country and sometimes even region to region—an important consideration in an international unionization drive. In Germany, the corporate constitution law of 1952 regulated a form of corporate governance called codetermination.[23] Like all large companies—with over 1,000 employees in the iron and steel industry and over 2,000 employees in all other sectors—IBM Germany therefore had an elected works council that represented employees in grievance procedures and advocated for social benefits such as a pension system in the mid-1950s. IBM's German board of directors also included five labor representatives, including one directly elected by IBM employees, in addition to the five employer representatives and the head of the board.

If IBM was a family, what was the local labor union—the divorce lawyer or the marriage counselor? At the quarterly company assemblies, the head of IBM's works council, Berger, reported on employee numbers as well as on the rate of union organization: of 2,141 employees in Böblingen and Sindelfingen in 1953, for example, 1,611 were union members, a total of 75 percent of all employees, with 93.8 percent of the 1,425 blue-collar employees and 38.4 percent of the 716 white-collar employees being organized in national unions.[24] Berger also explained the reasons for organizing in national unions and even urged IBM employees to join their unions and achieve 100 percent union organization.[25] German IBM employees thus continued their German tradition of organizing in national unions.

At the same time, local traditions shaped IBM's labor relations. The majority of the labor movement in Germany was socialist, organized in socialist-leaning unions associated with the Social Democratic Party, and dominated by the umbrella organization Deutscher Gewerkschaftsbund (DGB) after World War II. Yet labor unions in Germany associated with political parties across the spectrum, including Christian unions close to the Christian Democratic Union (CDU) and its predecessor, the

Catholic Center Party. Founded in the 1890s as counterweight to atheist socialist unions, they built on the principles of Catholic ethics. Outlined in the 1891 papal encyclical *Rerum Novarum,* these principles included human dignity (*Personalität*) and subsidiarity. Christian unions dominated in Catholic regions, including Swabia, the region in southwest Germany where IBM's main manufacturing facilities were located after World War II.

While little is known about the background and union affiliation of the head of IBM's works council, Berger, he and his works council related Mr. Watson's family rhetoric to Catholic social ethics. In a 1954 editorial in the IBM employee magazine, for example, Berger asked whether workers were "proletarians" or "personalities" (*Persönlichkeiten*). He explained that the Marxist labor movement initially saw workers as proletarians who, in their class fight, had to eliminate other social classes, especially the capitalist class. By the mid-twentieth century, according to Berger, the labor movement shaped societal relations to allow everyone to work and live according to his or her abilities. In traditionally Catholic Swabia, a Marxist proletarian movement had never taken hold, in Berger's view. Instead, the local precision industry promoted workers who were "smart in their heads and clever with their hands, and worked in a precise, conscientious, reliable and loyal way," or who, in Berger's words, became proud craftsmen and citizens—or personalities—rather than proletarians.[26]

Berger easily associated this argument with the Catholic principle of human dignity and Mr. Watson's frequent focus on the "human." In German, the linguistic similarity helped link personality and human dignity (*Persönlichkeit* and *Personalität*). Catholic social ethics states that humans are created in the image of God, and therefore each person should be treated with respect, and social institutions, including economic relations, ought to enable individuals to live in human dignity. At IBM, Watson declared the (hu)man as one of the company's main tasks, as the above-described workshop presentation indicated. One of the German executives remembered that at IBM, "it was always about the human," promoting people rather than projects.[27] IBM's human relations easily fit into Catholic social principles. Likewise, Mr. Watson's emphasis on the family translated into the principle of subsidiarity, the idea of responsibility and self-help in small social groups, starting with the family. Thus, Mr. Watson's welfare capitalist family rhetoric easily adapted to the Christian union thinking prevalent in southern Germany. It also served to create a cooperative workforce for IBM.

Nothing clarifies the "human as a task" idea like a fight over who will do the task. This became clear in the summer of 1954, when labor relations at IBM came to a head because IG Metall, the largest and one of the more belligerent unions in Germany, demanded contract negotiations. IG Metall organized workers, including many IBM employees, in the metalworking and steel industries. Following a period of economic growth, IG Metall demanded wage increases to guarantee workers their fair share. In Germany, unions conducted contract negotiations with employer associations at the state level, and IG Metall forged negotiations in the West German states of Bavaria, Berlin, and Baden-Württemberg, where IBM was located.[28] The ongoing negotiations forced IBM employees to take a stance on whether they would support a strike threat in solidarity with employees at other metalworking and steel companies.

In the context of these contract negotiations, Berger elaborated the IBM family rhetoric in an editorial that also explained his view of labor relations. He noted that employees often talked about the IBM family, appealing to each other to do the best for a company that was like a family where all members belonged and felt responsible for each other. At the same time, he cautioned that differences existed between the corporate family and one's own family, and that the notion of family needed to be used carefully. Yet he asserted that after a century of Marxist class conflict, by the mid-twentieth century, "partnership" had grown between employers and employees; "what used to be a fight against each other has turned into a struggle about cooperation."[29] In other words, Berger acknowledged the forming of IBM's welfare capitalist family culture as a way of overcoming labor conflict.

When contract negotiations between IG Metall and the employer association came to a head in 1954, Berger informed the local union office that the impending strike threat negatively affected labor relations at IBM and signaled that IBM employees would, if at all possible, seek to avoid a strike—a controversial step that required extensive explanation to his fellow workers because it violated the socialist principle of worker solidarity. Once the strike was averted, however, IBM's executives offered voluntary salary increases above the newly negotiated tariff as appreciation for the cooperative behavior of IBM employees during the contract negotiations.[30] Although IBM's welfare capitalism did not stop German IBM employees from organizing in national unions, the company did achieve cooperative relations with its employees. Welfare capitalist measures had thus created a corporate climate that rendered employees strike-averse although they organized in national unions.

CONCLUSION

Mr. Watson used gendered family rhetoric to shape his company's labor relations at home and abroad. Mrs. Watson's silent presence allowed her husband to present himself as family father at the helm of the company, providing for IBM employees the same way he provided for his large family of four kids (two of whom he had already installed as top executives within the company), their spouses, and over a dozen grandchildren. However, Mr. Watson's family rhetoric revealed unintended insights into class and gender at IBM. Family relations are not egalitarian along age and gender lines, and neither was Watson's welfare capitalist approach. Even though he often emphasized that everyone in IBM mattered, what he said determined the shape of the IBM family's future. Rather than allowing employees and their representatives to decide what was important to them and meeting them as partners at the negotiating table, Mr. Watson and IBM's management single-handedly determined the company's benefits and declared them by managerial fiat. In so doing, Mr. Watson materially affected the lives of thousands of employee families, determining their futures according to what was best for the company.

Today, the lavish perks and idea of meritocratic equality in many tech companies still hide the fact that, ultimately, employees are not equal to those who manage them, and they have no official say about the direction in which their CEO intends to take the company. Like IBM, tech companies today seek to create docile workforces by convincing workers they do not need unions, rather than through overt union-busting. Despite such pressures, IBM employees in Germany organized in national labor unions; but they also founded their social and corporate identity on their professional skills, building on Catholic social ethic ideas about human dignity and personality rather than Marxist ideas about class solidarity. They believed in personal ability and meritocratic values, unlike the women trainees in Seelampur whom Sreela Sarkar discusses in this volume, who questioned and subverted the promise that professional skills would set them free. But by building on meritocracy rather than unions, professionals remain isolated in the face of corporate power. For example, in 2015, almost 65,000 IT professionals in Silicon Valley joined a class action lawsuit against an illicit anti-poaching scheme among Bay Area tech companies. They achieved a meager settlement of under $6,000 apiece—for many, probably less than their monthly salary—rather than organize collectively in labor unions built on solidarity among workers to change policy and company practices.[31] Silicon Valley companies quietly cooperated to maintain a tractable labor force, just as Mr. Watson used IBM's "family" to keep workers docile. Though the tactics change, the objective remains the same.

IBM's labor relations were highly gendered, and they affected men and women differently. Sure, when corporate "sons" intonated songs celebrating their corporate patriarch, they marked clear lines of power that silenced men—like women—in the corporate hierarchy. Yet, despite Mr. Watson's progressive efforts at opening executive careers to young women, the women still had to overcome many hurdles in IBM's corporate culture that did not affect men, from the marriage ban to being barred from careers outside of customer training and human relations and managers resisting to hire them. Women and other gender minorities continue to face many obstacles in today's tech companies—some blatant and some more subtle. They range from a lack of women in leadership positions and ineffective responses to sexual harassment allegations, to the effects of whiteboard interviews that Janet Abbate discusses in this volume. When tech entrepreneurs publicly bask in their startup culture of late nights at the office, drinking Coke or beer and playing jokes on each other while also getting some programming done, they coincidentally celebrate a culture of masculine bonding and competition that excludes those men and women who seek private time in the evening to spend with their family or significant others; those who don't find sexist jokes funny; and those who seek to relate to their colleagues in cooperative rather than competitive ways. As labor and gender relations intermingle, they define how employees relate to each other, and who is included or excluded. Corporate cultures mold and produce labor forces that will best reproduce their needs and hierarchies. In this context, gender becomes an important corporate tool in the service of disciplining and creating tractable, easy-to-control white-collar labor forces. Gender, it turns out, is just as powerful in business as it is in people's everyday lives, and often people overlook how it is used to discipline employees into particular performances of class and labor—not just to divide workers along gendered lines.

Finally, labor relations remain local. Even at IBM, where Mr. Watson strove for a homogeneous corporate culture across the company's global operations, employees in different countries adapted the Watsons' welfare capitalism to their own labor and gender norms, which were influenced by their local traditions. In Germany, the socialist traditions of the labor movement guided even IBM employees to join labor unions. But traditions of the Catholic labor movement, which was prevalent in Swabia, where IBM's main manufacturing facilities were located, also made them receptive to Watson's family rhetoric and helped create goodwill even in situations of labor strife. An international union organization drive like the one forming in the gaming industry will have to accommodate different labor traditions and legislations. For example, in the United States, unions typically engage in plant-level collective bargaining of wage levels and work conditions such as seniority rights and grievance procedures. By contrast, in Germany, collective bargaining occurs between employer associations and labor union representatives in a given industry at the state level, and larger companies have works councils and labor directors with a say in decisions concerning labor, such as the introduction of new work processes and technologies. While an international labor movement will benefit from the solidarity of employees across national borders, differences in labor legislation and culture across countries pose a challenge.

Local labor traditions also influence corporate labor relations. Thus, it matters for a company's labor and gender relations whether it is, like IBM, located in Swabia with its more skill-conscious workforce influenced by the teachings of Catholic social ethics, or in Berlin, the emerging hub for IT startups in Germany today with a more individualized professional workforce.[32] Likewise, the United States has different local labor cultures in the more unionized north and the union-free south, and between bureaucratized companies of the East Coast like Xerox and IBM, and the venture-capital-driven startups of Silicon Valley. It is no coincidence that foreign automobile manufacturers such as Mercedes, Volkswagen, and Toyota have not located their US manufacturing plants in the Detroit area, where they would have had access to local talent and suppliers; rather, they opted for the union-free south. The same is true for tech companies. Silicon Valley provides them with a professional workforce that believes in meritocratic advancement based on individual skills rather collective improvement through solidarity. If Silicon Valley is to have a more diverse, collectively oriented workforce, employees will need to challenge corporate rhetoric that suggest differences do not exist, whether they come in the form of an egalitarian "family" ideal or believing in professional meritocracy. They can turn up the heat on their employers by insisting on labor relations that serve the needs of all workers—men and women of any ethnicity, ability, and sexual orientation—more than the needs of the corporation.

ACKNOWLEDGMENTS

I would like to acknowledge insightful comments by the editors, the conference participants at Shift CTRL: New Perspectives on Computing and New Media (Stanford University, May 6–7, 2016, Jen Light, and my RIT colleagues Heidi Nickisher and Amit Ray. Research for this essay was supported through NSF Scholar's Award 1457134.

NOTES

1. This often-repeated quote appears to have originated in a 1982 interview with Noyce. Everett M. Rogers and Judith K. Larson, *Silicon Valley Fever: Growth of High-Tech Culture* (New York: Basic Books, 1984), 191. The authors conclude that Silicon Valley firms treat their employees with "a high degree of paternalistic care" to avoid unionization, and they describe union-busting efforts against manual workers. Decades after Noyce's statement, Amazon and Tesla both fight unionization of their large manual-labor forces. Michael Sainato, "'We Are No Robots': Amazon Warehouse Employees Push to Unionize," *Guardian* (January 1, 2019); Verne Kopytoff, "How Amazon Crushed the Union Movement," *Time* (January 16, 2014); and David R. Baker, "Tesla, Labor Officials Spar over Fremont Factory Union Drive in Hearing," *San Francisco Chronicle* (June 11, 2018).

2. IBM's corporate songbook was first published in 1927 and reissued in 1937. "IBM Rally Song, Ever Onward Transcript," https://www-03.ibm.com/ibm/history/multimedia/everonward_trans .html, and Lee Hutchison, "Tripping Through IBM's Astonishingly Insane 1937 Corporate Songbook," *Ars Technica* (August 29, 2014), accessed February 25, 2017, https://arstechnica.com/ business/2014/08/tripping-through-ibms-astonishingly-insane-1937-corporate-songbook/.

3. In this essay, I refer to the Watsons as "Mr." and "Mrs." in order to use the same respectful appellation for both, rather than, for example, asymmetrically referring to him by his last name and her by her first name. Mrs. Watson was born Jeanette M. Kittredge, and she turned her maiden name into her middle name when she took her husband's last name upon marriage. Of course, the naming convention of referring to Mrs. Watson by her husband's last name also denotes her social dependence on him as a married woman. Contemporary sources, such as IBM magazines, commonly referred to Mrs. Watson by her husband's first *and* last name as "Mrs. Thomas Watson"—an unusual appellation by today's common practices that highlights her dependence even more.

4. How the corporate family model affects minorities, people of color, and persons of different abilities remains a subject for further research.

5. Sanford M. Jacoby, *Modern Manors: Welfare Capitalism since the New Deal* (Princeton: Princeton University Press, 1997); and Lizabeth Cohen, *Making a New Deal. Industrial Workers in Chicago, 1919–1939* (New York: Cambridge University Press, 2008 [1990]). For IBM, see David L. Stebenne, "IBM's 'New Deal': Employment Policies of the International Business Machines Corporation, 1933–1956," *Journal of the Historical Society* 5, no. 1 (2005): 47–77.

6. Mar Hicks, "Sexism Is a Feature, Not a Bug"; Sreela Sarkar, "Skills Will Not Set You Free"; and Janet Abbate, "Coding Is Not Empowerment," all in this volume.

7. Insights into IBM's treatment of women and gender relies on anecdotal evidence from Mr. Watson's biography; no systematic study of the subject has been attempted to date. Kevin Maney, *The Maverick and His Machine* (Hoboken, NJ: John Wiley and Sons, 2003), 144, 163–168.

8. An image of the first IBM training class for women systems service professionals in 1935 is available at http://www-03.ibm.com/ibm/history/ibm100/us/en/icons/employeeedu/. While the earlier clothing had been derived from middle-class daytime attire, by the 1920s, women office workers developed an office "uniform" that resembled their leisure wear with elaborate hairstyles, jewelry, light sandals, and soft, colorful fabric. Angel Kwolek-Folland, *Engendering Business: Men and Women in the Corporate Office, 1870–1930* (Baltimore: The Johns Hopkins University Press, 1994), 175.

9. Maney, *Maverick*, 165. Maney suggests that the young women allowed Mr. Watson to live out a more feminine side of his personality, with interests in beauty, fashion, and art, which he couldn't share with Jeannette. Yet there are no suggestions of any infidelity. For Mr. Watson's mentorship and support for Dorothy Shaver, see Stephanie Amerian, *Fashioning a Female*

Executive: Dorothy Shaver and the Business of American Style, 1893–1959 (Los Angeles: University of California Press, 2011).

10. For a more extensive analysis of IBM's gender and labor relations, including the board of trustees image, see Corinna Schlombs, *Productivity Machines: German Appropriations of American Technologies from Mass Production to Computer Automation* (Cambridge, MA: MIT Press, 2019), 195–218; see also Corinna Schlombs, "The 'IBM Family': American Welfare Capitalism, Labor and Gender in Postwar Germany," *IEEE Annals of the History of Computing* 39, no. 4 (2017): 12–26.

11. Maney, *Maverick*, 131–132, 134.

12. For a history of women and families in the United States, see Elaine Tyler May, *Homeward Bound: American Families in the Cold War Era* (New York: Basic Books, 2008 [1988]); Jessica Weiss, *To Have and to Hold: Marriage, the Baby Boom, and Social Change* (Chicago: University of Chicago Press, 2000); Alice Kessler-Harris, *In Pursuit of Equity: Women, Men, and the Quest for Economic Citizenship in the Twenty-First Century* (Oxford: Oxford University Press, 2001); and Colin Gordon, *New Deals: Business, Labor and Politics in America, 1920–1935* (New York: Cambridge University Press, 1994).

13. "Die Rede von Thomas J. Watson," *IBM Deutschland* (August 1953): 4. At the same time, the family rhetoric emasculated employees, particularly male office workers, who often sought to reassert their position in relation to their women colleagues. Kwolek-Folland, *Engendering Business*, 169–170.

14. "Unsere Feier auf dem Killesberg," *IBM Deutschland* (June 1954): 5.

15. "Gedanken zum Besuch von Thos. J. Watson. Grosse Tage für die IBM Deutschlanad," *IBM Deutschland* (August 1953): 3.

16. "Gedanken zum Besuch," 3.

17. "Unsere Feier auf dem Killesberg," 4.

18. While no demographic information on IBM's workforce in Germany is available, numerous images in the employee magazines document that significant numbers of women worked in clerical as well as manufacturing positions.

19. Historians coined the term "normalization" (*Normalisierung*) for this prevalent social sentiment of the first postwar decade in Germany. Of course, the "normalcy" of the nuclear family had been an ideal rather than a reality even in prewar times.

20. Adelheid zu Castell, "Die demographischen Konsequenzen des Ersten und Zweiten Weltkrieges für das Deutsche Reich, die Deutsche Demokratische Republik und die Bundesrepublik Deutschland," in *Zweiter Weltkrieg und sozialer Wandel. Achsenmächte und bestetzte Länder*, ed. Waclaw Dlugoborski (Göttingen: Vandenhoeck, 1981).

21. In 1957, legislation encoded the so-called "housewife marriage" (*Hausfrauenehe*). Now women no longer needed their husbands' approval to work, giving them at least in principle the right to engage in gainful employment; but at the same time, women remained tied to their primary duties as housewives and mothers. For German family legislation, see Robert G. Moeller, *Protecting Motherhood: Women and the Family in the Politics of Postwar West Germany* (Berkeley: University of California Press, 1993).

22. Elizabeth D. Heineman, *What Difference Does a Husband Make? Women and Marital Status in Nazi and Postwar Germany* (Berkeley: University of California Press, 1999), 146–147; Ute Frevert, *Women in German History: From Bourgeois Emancipation to Sexual Liberation* (Oxford: Berg, 1990), 265; and Moeller, *Protecting Motherhood*, 71.

23. Gloria Müller, *Mitbestimmung in der Nachkriegszeit: Britische Besatzungsmacht—Unternehmer—dGewerkschaften* (Düsseldorf: Schwann, 1987); and Gloria Müller, *Strukturwandel und Arbeitnehmerrechte. Die wirtschaftliche Mitbestimmung in der Eisen-und Stahlindustrie 1945–1975* (Essen: Klartext-Verlag, 1991).

24. "Betriebsversammlungen in Berlin und Sindelfingen," *IBM Deutschland* (July 1953): 1.

25. "Unsere Betriebsversammlungen am 18. Dezember 1953," *IBM Deutschland* (January–March 1953): 12; and "Wir sprechen miteinander über unsere betrieblichen Tagesfragen," *IBM Deutschland* (September 1954): 2. In later employee meetings, Berger continued to report on employee statistics but no longer mentioned the rate of union organization.

26. W. B., "Proletarier oder Persönlichkeiten?" *IBM Deutschland* (August 1954): 2. The debate continued six months later: "Entproletarisieren! Ein Diskussionsbeitrag des Betriebsrates Sindelfingen/Böblingen," *IBM Deutschland* (February 1955): 3. In this concluding statement, backing up Berger's earlier editorial, the works council built its position on the work of the Catholic philosopher Josef Pieper.

27. Karl Ganzhorn, personal interview with the author (April 27, 2007).

28. Wolfgang Krüger, "Lohnpolitik noch immer ohne Massstab. Streiks werden uns der sozialen Gerechtigkeit nicht näher bringen," *ZEIT* (July 22, 1954). IBM employees usually organized in IG Metall or a suitable white-collar union such as the Deutsche Angestellten Gewerkschaft (DAG). Contract negotiations in Germany are typically held at the state level between the regional union representatives and the regional representatives of the employer association.

29. "Das Haus und das Werk," *IBM Deutschland* (December 1954): 2.

30. "Wir sprechen miteinander," 1–2; and "Herr Hörrmann über Arbeits-, Lohn-, Produktionsprobleme," *IBM Deutschland* (September 1954): 3–4.

31. Jeff J. Roberts, "Tech Workers Will Get Average of $5,770 under Final Anti-Poaching Settlement," *Fortune* (September 3, 2015), http://fortune.com/2015/09/03/koh-anti-poach-order/.

32. For further reading, see Sareeta Amrute, *Encoding Race, Encoding Class: Indian IT Workers in Berlin* (Durham, NC: Duke University Press, 2016).

8

SIRI DISCIPLINES

Halcyon M. Lawrence

VOICE TECHNOLOGIES ARE NOT REVOLUTIONARY

Voice technologies are routinely described as revolutionary. Aside from the technology's ability to recognize and replicate human speech and to provide a hands-free environment for users, these revolutionary claims, by tech writers especially, emerge from a number of trends: the growing numbers of people who use these technologies,[1] the increasing sales volume of personal assistants like Amazon's Alexa or Google Home,[2] and the expanding number of domestic applications that use voice.[3] If you're a regular user (or designer) of voice technology, then the aforementioned claim may resonate with you, since it is quite possible that your life has been made easier because of it. However, for speakers with a nonstandard accent (for example, African-American vernacular or Cockney), virtual assistants like Siri and Alexa are unresponsive and frustrating—there are numerous YouTube videos that demonstrate and even parody these cases. For me, a speaker of Caribbean English, there is "silence" when I speak to Siri; this means that there are many services, products, and even information that I am not able to access using voice commands. And while I have other ways of accessing these services, products, and information, what is the experience of accented speakers for whom speech is the primary or singular mode of communication? This so-called "revolution" has left them behind. In fact, Mar Hicks pushes us to consider that any technology that reinforces or reinscribes bias is not, in fact, revolutionary but oppressive. The fact that voice technologies do nothing to change existing "social biases and hierarchies," but instead reinforce them, means that these technologies, while useful to some, are in no way revolutionary.[4]

One might argue that these technologies are nascent, and that more accents will be supported over time. While this might be true, the current trends aren't compelling. Here are some questions to consider: first, why have accents been primarily developed for Standard English in Western cultures (such as American, Canadian, and British English)? Second, for non-Western cultures for which nonstandard accent support has been developed (such as Singaporean and Hinglish), what is driving these initiatives? Third, why hasn't there been any nonstandard accent support for minority speakers of English? Finally, what adjustments—and at what cost—must standard and foreign-accented speakers of English make to engage with existing voice technologies?

In this essay, I argue that the design of speech technology is innately biased. It is important to understand the socioeconomic context in which speech technologies are developed and the long history of assimilation—rooted in imperialist, dominant-class ideologies—that nonnative and nonstandard speakers of English have had to practice in order to participate in global economic and social systems. Voice technologies do nothing to upset this applecart. I hypothesize that any significant change in these software development practices will come from developers in the periphery[5] invested in disrupting the developing world's practice of unequal consumption of technology vis-à-vis creation and innovation. This deimperializing wave of change will come from independent developers because they embrace the notion that "support for other cultures is not a feature of software design, but a core principle."[6] In proposing a way forward, I examine the results from my study that investigated the acceptability of foreign-accented speech in speech technology as well as contexts in which accented speech can be used in speech technology interactions. I conclude with guidelines for a linguistically inclusive approach to speech interaction design.

SIRI DISCIPLINES

In his slave biography, Olaudah Equiano said, "I have often taken up a book, and have talked to it, and then put my ears to it, when alone, in hopes it would answer me; and I have been very much concerned when I found it remained silent."[7] Equiano's experience with the traditional interface of a book mirrors the silence that nonstandard and foreign speakers of English often encounter when they try to interact with speech technologies like Apple's Siri, Amazon's Alexa, or Google Home.

Premised on the promise of natural language use for speakers, these technologies encourage their users not to alter their language patterns in any way for successful interactions. If you possess a foreign accent or speak in a dialect, speech technologies practice a form of "othering" that is biased and disciplinary, demanding a form of postcolonial assimilation to standard accents that "silences" the speaker's sociohistorical reality.

Because these technologies have not been fundamentally designed to process non-standard and foreign-accented speech, speakers often have to make adjustments to their speech—that is, change their accents—to reduce recognition errors. The result is the sustained marginalization and delegitimization of nonstandard and foreign-accented speakers of the English language. This forced assimilation is particularly egregious given that the number of second-language speakers of English has already exceeded the number of native English-language speakers worldwide.[8] The number of English as a Second Language (ESL) speakers will continue to increase as English is used globally as a *lingua franca* to facilitate commercial, academic, recreational, and technological activities. One implication of this trend is that, over time, native English speakers may exert less influence over the lexical, syntactic, and semantic structures that govern the English language. We are beginning to witness the emergence of hybridized languages like Spanglish, Konglish, and Hinglish, to name a few. Yet despite this trend and the obvious implications, foreign-accented and nonstandard-accented speech is marginally recognized by speech-mediated devices.

YES, ACCENT BIAS IS A THING

Gluszek and Dovidio define an accent as a "manner of pronunciation with other linguistic levels of analysis (grammatical, syntactical, morphological, and lexical), more or less comparable with the standard language."[9] Accents are particular to an individual, location, or nation, identifying where we live (through geographical or regional accents, like Southern American, Black American, or British Cockney, for example), our socioeconomic status, our ethnicity, our cast, our social class, or our first language. The preference for one's accent is well-documented. Individuals view people having similar accents to their own more favorably than people having different accents to their own. Research has demonstrated that even babies and children show a preference for their native accent.[10] This is consistent with the theory that similarity in attitudes and features affects both the communication processes and the perceptions that people form about each other.[11]

However, with accents, similarity attraction is not always the case. Researchers have been challenging the similarity-attraction principle, suggesting that it is rather context-specific and that cultural and psychological biases can often lead to positive perceptions of nonsimilar accents.[12] Dissimilar accents sometimes carry positive stereotypes, which lead to positive perceptions of the speech or speaker. Studies also show that even as listeners are exposed to dissimilar accents, they show a preference for standard accents, like standard British English as opposed to nonstandard varieties like Cockney or Scottish accents (see table 8.1 for a summary of findings of accent studies).[13]

Table 8.1 Summary of Findings of Accented-Speech Perception Studies

Area of research	Finding	Authors
Negative perception of nonnative accented speech	People who speak nonnatively accented speech are perceived more negatively than speakers with native accents	Bradac 1990; Fuertes et al. 2009; Lindemann 2003, 2005
Strength of accent	The stronger the accent, the more negatively the accented individual is evaluated	Nesdale and Rooney 1996; Ryan et al. 1977
Similarity attraction	Babies and children show a preference for their native accent	Kinzler et al. 2007, 2009
Nonnative speech evoking negative stereotypes	Nonnative speakers seen as less intelligent	Bradac 1990; Lindemann 2003; Rubin et al. 1997
	Nonnative speakers seen as less loyal	Edwards 1982
	Nonnative speakers seen as less competent	Boyd 2003; Bresnahan et al. 2002
Language competency	Nonnative speakers perceived as speaking the language poorly.	Hosoda et al. 2007; Lindemann 2003
Positive stereotypes	Certain accents are seen as prestigious standard UK and Australian accents	Lippi-Green 1994; Giles 1994
Nonnative speech evoking discriminatory practices	Discrimination in housing	Zhao et al. 2006
	Discrimination in employment	Kalin 1978; Matsuda 1991; Nguyen 1993
	Discrimination in the courts	Frumkin 2007; Lippi-Green 1994
	Lower-status job positions	Bradac et al. 1984; de la Zerda 1979; Kalin 1978

On the other hand, nonsimilar accents are not always perceived positively, and foreign-accented speakers face many challenges. For example, Flege[14] notes that speaking with a foreign accent entails a variety of possible consequences for second-language (L2) learners, including accent detection, diminished acceptability, diminished intelligibility, and negative evaluation. Perhaps one of the biggest consequences of having a foreign accent is that L2 users oftentimes have difficulty making themselves understood because of pronunciation errors. Even accented native speakers (speakers of variants of British English, like myself, for example) experience similar difficulty because of the differences of pronunciation.

Lambert et al. produced one of the earliest studies on language attitudes that demonstrated language bias.[15] Since then, research has consistently demonstrated negative perceptions about speech produced by nonnative speakers. As speech moves closer to unaccented, listener perceptions become more favorable, and as speech becomes less similar, listener perceptions become less favorable; said another way, the stronger the foreign accent, the less favorable the speech.[16]

Nonnative speech evokes negative stereotypes such that speakers are perceived as less intelligent, less loyal, less competent, poor speakers of the language, and as having weak political skill.[17] But the bias doesn't stop at perception, as discriminatory practices associated with accents have been documented in housing, employment, court rulings, lower-status job positions, and, for students, the denial of equal opportunities in education.[18]

Despite the documented ways in which persons who speak with an accent routinely experience discriminatory treatment, there is still very little mainstream conversation about accent bias and discrimination. In fall 2017, I received the following student evaluation from one of my students, who was a nonnative speaker of English and a future computer programmer:

> I'm gonna be very harsh here but please don't be offended—your accent is **horrible**. As a non-native speaker of English I had a very hard time understanding what you are saying. An example that sticks the most is you say goal but I hear ghoul. While it was **funny** at first it got **annoying** as the semester progressed. I was left with the impression that you are very proud of your accent, but I think that just like movie starts [sic] acting in movies and changing their accent, when you profess you should try you speak clearly in **US accent** so that non-native students can understand you better.

While I was taken aback, I shouldn't have been. David Crystal, a respected and renowned British linguist who is a regular guest on a British radio program, said that people would write in to the show to complain about pronunciations they didn't like. He states, "It was the extreme nature of the language that always struck me. Listeners didn't just say they 'disliked' something. They used the most emotive words they could think of. They were 'horrified,' 'appalled,' 'dumbfounded,' 'aghast,' 'outraged,' when they heard something they didn't like."[19] Crystal goes on to suggest that reactions are so strong because one's pronunciation (or accent) is fundamentally about identity. It is about race. It is about class. It is about one's ethnicity, education, and occupation. When a listener attends to another's pronunciation, they are ultimately attending to the speaker's identity.

As I reflected on my student's "evaluation" of my accent, it struck me that this comment would have incited outrage had it been made about the immutable characteristics of one's race, ethnicity, or gender; yet when it comes to accents, there is an acceptability about the practice of accent bias, in part because accents are seen as a mutable characteristic of a speaker, changeable at will. As my student noted, after all, movie stars in Hollywood do it all the time, so why couldn't I? Although individuals have demonstrated the ability to adopt and switch between accents (called code switching), to do so should be a matter of personal choice, as accent is inextricable to one's identity. To put upon another an expectation of accent change is oppressive; to create conditions where accent choice is not negotiable by the speaker is hostile; to impose an accent upon another is violent.

One domain where accent bias is prevalent is in seemingly benign devices such as public address systems and banking and airline menu systems, to name a few; but the lack of diversity in accents is particularly striking in personal assistants like Apple's Siri, Amazon's Alexa, and Google Home. For example, while devices like PA systems only require listeners to comprehend standard accents, personal assistants, on the other hand, require not only comprehension but the performance of standard accents by users. Therefore, these devices demand that the user assimilate to standard Englishes—a practice that, in turn, alienates nonnative and nonstandard English speakers.

FRESHWATER YANKEE: FROM ASSIMILATION TO ALIENATION

According to the *Dictionary of the English/Creole of Trinidad and Tobago*, the term "freshwater Yankee" is used to describe someone who "speaks with an American accent or adopts other American characteristics, esp. without ever having been to the US."[20] During the First World War, the presence of American servicemen at air bases in Trinidad was contentious, and "for those who lived in wartime Trinidad, the close encounter with the US empire meant more than anything else endless internecine friction and confrontation."[21] Hence when "freshwater Yankee" began to be used as a Trinidadian Creole term, it was not complimentary. As Browne states, "Caribbean people who immigrated to America prior to 1965 have traditionally been viewed as assimilationists and frowned upon by those in their native islands. For example, the image of the freshwater Yankee has made its way into the cultural consciousness as a powerful critique of assimilators and an equally powerful deterrent to what was considered 'pretending' to be American."[22]

In December 2016, as I neared the completion of my annual pilgrimage to Trinidad, I spent time with a friend who is an iPhone user. I watched in fascination as she interacted with Siri, first in her native Trinidadian accent and then, when Siri did not understand her, repeating her question in an American accent. Siri sprang to life. It occurred to me that I had just witnessed yet another context in which "freshwater Yankee" could be used. This personal anecdote demonstrates the very practical ways in which nonstandard accented speakers of English must assimilate to "participate" in emerging technologies.

The assimilation by accented speakers using speech technology can also be alienating. In human–human interaction, both the speaker and the listener have an opportunity to negotiate the communicative process. For example, over the last ten years of living in the US and speaking a British West Indian variant of English, there have been daily instances where I have not been understood; but I have almost always had the freedom to negotiate the communicative process. Since changing my accent is nonnegotiable, I have deployed clear speech strategies[23] to become more intelligible (slowing my rate of speech, hyperarticulating,[24] and making lexical changes, to name a few). None of these choices have in anyway made me feel inauthentic or alienated, as these strategies do not involve me taking on another's accent to be understood. Yet in speech technologies, because there is no opportunity to negotiate with the device, a change in accent is almost always required for accented speakers who are not understood but wish to engage with the technology. What evolves over time is a sense of alienation, so poignantly described by Nieto:

> Having learnt that the existence and dialect of the dark-skinned is the incarnation of the bad, and that one can only hate it, the colonized then has to face the fact that "I am dark-skinned, I have an accent," at this crossroads there seems to be only one possible solution, namely, becoming part of the superior, being one of them, speak their language. Nevertheless that solution is hopeless, therefore the oppressed faces alienation for the first time, the sense that one has lost one's place in the world, our words are meaningless, our spirits are powerless.[25]

The assimilation and subsequent alienation described above in the digital sphere is by no means a new phenomenon. Assimilation and alienation have always been the impact of imperialist ideology, and language has always been a tool of imperialist practices.

LANGUAGE AS A TOOL OF IMPERIALISM

The practice of accent bias is far more complex than just an individual's prejudice for or against another's accent. In much the same way that institutionalized practices of racism and sexism have a long and complex history, accent bias is better understood against the broader ideology of imperialism, facilitated by acts of conquest and colonialism. For millennia, civilizations have effectively leveraged language to subjugate, even erase, the culture of other civilizations.

One problem with the discussion about the history of English as a global language is that the discourse is often couched in organic language, using linguistic terms like "spread," "expansion," "movement," and "death," signaling the inevitability of language change over time through agentless, natural forces.[26] But language change isn't always an organic process; sometimes it comes through the violent subjugation of a people and their culture. This is not to be mistaken for some haphazard, unintentional by-product of colonialism; as Phillipson argues, the rise of English "was not accidental, but [was] carefully engineered."[27]

The physical, then cultural, subjugation of peoples also characterized Europe's colonization of the New World. In the case of the Americas and the Caribbean, when Native Americans were decimated by disease, Europeans—in search of a new source of labor on their plantations—captured, enslaved, and transported millions of West Africans to the New World. However, physical subjugation was only part of the expansion of European control. Many scholars[28] have documented the ways in which language was used as a tool of imperialist expansion in the New World, as the colonizer's language was exclusively used in her newly captured territories. Migge and Léglise note that "language policies became an essential way in which the colonization of peoples was made complete . . . the colonizers' language . . . became a necessity for all those who wished to advance socially and to participate in the colony's public sphere."[29]

Therefore, Like Latin before it, the emergence of English and its status as *lingua franca* is fundamentally a story of the exercise of British and American power facilitated through military force, imperialist ideology, and cultural imposition over the last 300 years. As Crystal notes:

Why a language becomes a global language has little to do with the number of people who speak it. It has much more to do with who those speakers are. Latin became an international language throughout the Roman Empire, but this was not because the Romans were more numerous than the peoples they subjugated. They were simply more powerful.[30]

It is important to note that not all scholars agree that the dominance of English as *lingua franca* is problematic; for example, Quirk and Widdowson argue that English has become globalized for practical and historical reasons and that it can help the development of poor countries without necessarily endangering their cultures.[31] This ideology is still very much present in modern-day imperialist ideologies, as internet connectivity and access to digital technologies are not only seen as a good thing for developing nations but purported as a human right.[32] However, as astutely noted by Bill Wasik, "in old-fashioned 19th-century imperialism, the Christian evangelists made a pretense of traveling separately from the conquering colonial forces. But in digital imperialism, everything travels as one, in the form of the splendid technology itself: salvation and empire, missionary and magistrate, Bible and gun."[33]

Consequently, digital media—and by extension, the language of digital media—is arguably one of the most powerful tools of neo-imperialism. For evidence of this claim one need look no further than the dominance of English as the language of the internet.

IN THE DIGITAL WORLD, NOT ALL LANGUAGES ARE CREATED EQUAL

Phillipson, perhaps one of the most recognized and contentious voices on the topic of linguistic imperialism, defines the term as a "theoretical construct, devised to account for linguistic hierarchization, to address issues of why some languages come to be used more and others less, [and] what structures and ideologies facilitate such processes."[34] He argues that linguistic imperialism is a subtype of linguicism, akin to other hierarchization based on race, ethnicity, gender, and so on. While much of Phillipson's work focuses on education's role as a "vital site" for processes of linguistic hierarchization, I propose that the digital economy provides a less formal, but perhaps just as powerful, ubiquitous site for linguistic hierarchization to emerge. As stated by Watters, "empire is not simply an endeavor of the nation-state—we have empire through technology (that's not new) and now, the technology industry as empire."[35]

After almost thirty years of the World Wide Web being used commercially, English is still its dominant language. Consider the following: as of November 2017, 51 percent of web pages are written in English, with Russian having the next largest share, at 6.7 percent; over twenty-five languages have 0.1 percent representation, and over 100 languages have fewer than 0.1 percent representation online.[36] Software is still written in Latin alphabets because computational technology was never conceived with the ideal of being inclusive of language. For example, the ASCII character code only supports 128 different characters, so as a result, complex script languages like Chinese, which need more characters, are pushed to the margins.

A similar language divide also exists in speech technologies, as software is primarily developed for the English-speaking market (standard American, British, and Australian). Error rates across languages differ significantly; for example, for Google's voice recognition system, the word error rate for US English is 8 percent, and for tier two languages (i.e., languages spoken in emerging markets, like India), the error rate is above or around 20 percent, which renders it functionally unusable.[37]

Within the English-speaking market, development of technologies for nonstandard (creole) and nonnative (indigenized) accents are even more telling of the language divide and of bias. To describe the patterns of language development of speech technologies, we can look to the theory of concentric circles of English, developed by Kachru, who suggests that the world's Englishes could be classified by three circles.[38] First, the Inner Circle represents "traditional bases of English," which include the United Kingdom, the United States, Australia, New Zealand, Ireland, Anglophone Canada, and some of the Caribbean territories. The Outer Circle is characterized by countries where English is not the native language but is designated as an official language, most often because of the legacy of colonialism. This includes countries such as India, Nigeria, the Philippines, Bangladesh, Pakistan, Malaysia, Tanzania, Kenya, and non-Anglophone South Africa and Canada. Finally, the Expanding Circle of English includes countries where the language has no historical or governmental role but is nevertheless used widely as a foreign language to conduct business. This circle represents the rest of the world's population: China, Russia, Japan, most of Europe, Korea, Egypt, Indonesia, etc.

Mufwene presents a more scathing classification of the world's Englishes and instead designates two categories: legitimate and illegitimate offspring.[39] The legitimate offspring of English is typically spoken by descents of Europeans, while

Table 8.2 Patterns of Language Support of Major Speech Technologies

Mufwene's legitimate and illegitimate offspring of English		Kachru's Concentric Circle Model of English	English variety	Englishes supported in major speech technologies
Legitimate offspring	English typically spoken by Europeans	Inner Circle: Traditional bases of English	United Kingdom, the United States, Australia, New Zealand, Ireland, Anglophone Canada	United Kingdom, the United States, Australia, New Zealand, Ireland, Anglophone Canada
Illegitimate offspring of English	English pidgins and creoles	Outer Circle: English is not the native language but is the designated official language	India, Nigeria, the Philippines, Bangladesh, Pakistan, Malaysia, Tanzania, Kenya, non-Anglophone South Africa and Canada	Hinglish, South African English ONLY
	Nonnative or indigenized varieties of English	Expanding Circle: English has no historical or governmental role, but is used widely as *lingua franca*	China, Russia, Japan, most of Europe, Korea, Egypt, Indonesia, Caribbean Territories, South America	Singaporean English

illegitimate offspring of English are pidgins, creoles (aligned with Kachru's Outer Circle of English) and nonnative or indigenized varieties (aligned to Kachru's Expanding Circle of English). Table 8.2 illustrates how these two classification systems are aligned. It is clear as we look at tables 8.2 and 8.3[40] (with the exception of Caribbean territories, whose markets are far too small to be considered lucrative) that the locus of development in speech technology for personal assistants (Siri, Alexa, and Google Home) is happening in Kachru's Inner Circle/Mufwene's Legitimate Offspring of English. Any development of other Englishes represents technology companies' pursuit of emerging and lucrative markets like Singapore and India. Consider the example of the development of Hinglish for speech technology.

In 2017, Amazon's Alexa rolled out support for Hinglish. There were also reports that Apple was planning similar support for the hybridized language.[41] But who are the speakers of Hinglish? Where do they come from? And how is the hybridized language perceived and used in India, a nation of 1.3 billion people? Hinglish is defined as a "colloquial umbrella-term spanning isolated borrowings of indigenized

Table 8.3 Accents/Dialects supported by Amazon's, Google's and Apple's Personal Assistants

English accents/dialects	Amazon's Alexa	Google Home	Apple's Siri
American	✔	✔	✔
British	✔	✔	✔
Australian	✔	✔	✔
Canadian	✔	✔	✔
Indian	✔		✔
Singaporean			✔
South African			✔
Irish			✔

Indian English forms within otherwise Monolingual Hindi or English, to rich code-switching practices unintelligible to Monolingual Hindi or English speakers."[42] It is estimated that approximately 350 million people speak Hinglish (more than even the number of native speakers of English worldwide). Additionally, researchers indicate that the emergent language is being used by the elite and is perceived as being more prestigious and modern than even English itself.[43] Given this fact, nothing about Apple's or Amazon's initiative is surprising. The decision to break into the Hinglish market is not being driven by inclusionary ideals, it is about accessing an emerging profit center. Given that market forces will continue to drive the commercial decisions about which accents are represented in speech technologies, it is more than likely that some accents will never be represented in speech technologies, because no lucrative market for these accents exists.

A RALLYING CALL FOR CHANGE

Despite over fifty years of development, the inability of speech technology to adequately recognize a diverse range of accented speakers is a reflection of the lack of diversity among employees of technology firms. This lack of diversity means that primarily dominant-class cultural norms are represented in the design choices of technology devices. As Alper observes, "the values, desires, and ideals of dominant cultural groups are systematically privileged over others in societies, and these biases are built into the structures of organizations, institutions, and networks."[44] As a result, our current speech technologies represent the values, ideals and desires of the Global North and as tools of digital imperialism, they bear "cultural freight as they cross borders."[45]

Speech technology is trained using speech corpora—a collection of thousands of voices speaking on a wide range of topics. Many of the early speech technologies used the now thirty-year-old corpus Switchboard, from the University of Pennsylvania's Linguistic Consortium, to train on; the accents represented there are largely American Midwestern. The result is that speech recognition error rates for marginalized voices are higher than others; as Paul notes, "a typical database of American voices, for example, would lack poor, uneducated, rural, non-white, non-native English voices. The more of those categories you fall into, the worse speech recognition is for you."[46] Work by Caliksan-Islam, Bryson, and Narayanan has empirically demonstrated what we have already suspected: that beyond just marginalization of people, "human-like semantic biases result from the application of standard machine learning to ordinary language."[47] Simply put, speech technologies replicate existing language biases.

Collecting data for a corpus is an expensive endeavor, and quite often existing corpora like Switchboard are all that smaller firms have as a resource. Larger firms, like Apple and Amazon, are building their own corpora, which they collect in different ways, including from the training data we freely provide when we set up our voice assistants or when we use dictation features on our software. Naturally, not only is this data not publicly shared but very little is known about the user demographics: which accents are captured, who participates in the transcription, verification process, and so on. The entire process is shrouded in mystery; market competitiveness demands that it be so.

Given the exclusive practices of big tech companies, we must look to independent developers—and free and open-source software (FOSS) initiatives that support independent developers—to make greater strides in disrupting the system. Ramsey Nasser,[48] for example, comes to mind as an independent developer working on an Arabic coding language. Ramsey understands and embraces the importance of having one's culture and language represented in the technologies we use.[49] Another exciting project called Project Common Voice was launched by Mozilla in 2017. To date the project has collected over 300 hours of speech contributed by nearly 20,000 people globally. According to Mozilla, "individual developers interested in innovating speech technologies for marginalized accents have a corpora that in time will provide them with diverse representations."[50]

Part of the challenge of democratizing the process and the technology is the prevailing misconception that accents are not only undesirable but unintelligible for use in speech technologies. Deeply unsettled by this notion, I conducted a study in 2013 to answer the following:

1. What is the relationship between user perception of accented speech compared to user performance in response to accented speech?
2. How do participants perform on different types of accented-speech tasks?

This research was motivated by the fact that many studies about accented speech examine how participants perceive accented speech and accented speakers.[51] Bolstered by the finding that sometimes nonsimilar accents carry positive stereotypes, I was interested in participants' performance in response to accented speech, not just their perception of that speech.[52] For example, even if a listener held a particular perception about an accent, could that listener successfully complete a task that was presented by an accented speaker? And if so, what types of tasks? Were some tasks more challenging if the listener were presented with accented speech? I theorized that if listeners could indeed successfully complete certain tasks, then there may be room for the deployment of accented speech in contexts where standard accents are currently used. There were three relevant findings from the study:

1. The talkers' accent did not have a significant effect on the listeners' performance.
2. There was a significant interaction between listeners' accent and talkers' accent on time taken.
3. Accent had a significant effect on the listeners' ability to recall information.

Taken together, this is good news. The findings suggest that the use of nonnative speech can be deployed in technology but needs additional scaffolding, such as visual support, longer processing times, repetition, reinforcement, and training. However, it should not be used if recall of data is necessary or if processing time is limited. So while emergency systems, for example, should continue to be designed using standard and native accents, perhaps there is room for us to hear accented speakers in other contexts, like information systems that provide banking information, flight, and weather updates.

The benefit here is that listeners begin to hear accented speech in everyday contexts. Rubin indicates that listeners benefit from the practice of listening to foreign-accented speech, both in terms of their improved comprehension and in changed attitudes toward accented speakers.[53] The entertainment industry is one place where we are beginning to see slow but high-profile change in this regard—Ronnie Chiang of *The Daily Show* and Diego Luna of *Rogue One: A Star Wars Story* are two actors who perform their roles in their nonnative English accents, for example.

CONCLUSION

My goal in writing this chapter was to dispel the notion that existing speech technologies provide a linguistic level playing field for all users. Speech technologies are not revolutionary, and it is erroneous to claim that they are. They are biased, and they discipline. As more devices embed speech as a mode of communication using standard accents, speakers who cannot or will not assimilate will continue to be marginalized. However, given the growing services across a range of devices and contexts, speech technologies are an ideal platform for the performance and representation of linguistic diversity. Given this potential, Meryl Alper challenges us to ask the critical question with regards to the design and development of speech technologies: "who is allowed to speak and who is deemed to have a voice to speak in the first place?"[54] The question is an important one as it goes to the heart of representation and inclusion. Connectivity and access to technology cannot be ethically purported as a human right, if all people don't have an opportunity to hear themselves represented in the technology that they use.

Looking back at the last fifty years of speech technology development, the fact that we have figured out how to teach and train our technology to speak and recognize human speech is in and of itself an amazing technological accomplishment. However, it is not enough that our technologies *just* speak, in the same way that it is not enough that our children *just* have a school to attend or that adults *just* have access to health care. In the same way we expect our physical spaces—where we work, play, and learn—to support and reflect our identity, culture, and values, so too we should expect and demand that our technology—as *tools* that support our work, play, and learning—also reflect our identity, culture, and values. I, for one, look forward to the day when Siri doesn't discipline my speech but instead recognizes and responds to my Trinidadian accent.

ACKNOWLEDGMENTS

I would like to thank Dr. Karen Eccles, Maud Marie Sisnette, and Keeno Gonzales, who directed me to the Lise Winer Collection housed at the Alma Jordan Library, the University of the West Indies, St. Augustine, Trinidad and Tobago.

NOTES

1. Jayson DeMers, "Why You Need to Prepare for a Voice Search Revolution," *Forbes* (January 9, 2018), https://www.forbes.com/sites/jaysondemers/2018/01/09/why-you-need-to-prepare-for-a -voice-search-revolution/#66d65a2434af.

2. Herbert Sim, "Voice Assistants: This Is What the Future of Technology Looks Like," *Forbes* (November 2, 2017), https://www.forbes.com/sites/herbertrsim/2017/11/01/voice-assistants-this -is-what-the-future-of-technology-looks-like/#2a09e5ce523a.

3. Nick Ismail, "2018: The Revolutionary Year of Technology," *Information Age* (May 15, 2018), https://www.information-age.com/2018-revolutionary-year-technology-123470064/.

4. Mar Hicks, "Computer Love: Replicating Social Order through Early Computer Dating Systems," *Ada New Media* (November 13, 2016), https://adanewmedia.org/2016/10/issue10-hicks/.

5. In the narrow sense, I refer to independent computer scientists not affiliated with large software firms, but also in the broader sense, I refer to those developers who come from or are affiliated with developing economies.

6. Ramsey Nasser, "Command Lines: Performing Identity and Embedding Bias," panel at Computer History Museum (April 26, 2017), accessed January 03, 2018, http://bit.ly/2r1QsuD.

7. Olaudah Equiano, *The Life of Olaudah Equiano, or Gustavus Vassa, the African* (North Chelmsford, MA: Courier Corporation, 1814), 107.

8. David Crystal, *English as a Global Language* (Cambridge: Cambridge University Press, 2003).

9. Agata Gluszek and John F. Dovidio, "The Way They Speak: A Social Psychological Perspective on the Stigma of Nonnative Accents in Communication," *Personality and Social Psychology Review* 14, no. 2 (2010): 215.

10. John R. Edwards, "Language Attitudes and Their Implications Among English Speakers," in *Attitudes Towards Language Variation: Social and Applied Contexts* (1982): 20–33; H. Thomas Hurt and Carl H. Weaver, "Negro Dialect, Ethnocentrism, and the Distortion of Information in the Communicative Process," *Communication Studies* 23, no. 2 (1972): 118–125; Anthony Mulac, "Evaluation of the Speech Dialect Attitudinal Scale," *Communications Monographs* 42, no. 3 (1975): 184–189; Ellen Bouchard-Ryan and Richard J. Sebastian, "The Effects of Speech Style and Social Class Background on Social Judgements of Speakers," *British Journal of Clinical Psychology* 19, no. 3 (1980): 229–233; Ellen Bouchard-Ryan, Miguel A. Carranza, and Robert W. Moffie, "Reactions Toward Varying Degrees of Accentedness in the Speech of Spanish-English Bilinguals," *Language and Speech* 20, no. 3 (1977): 267–273; Katherine D.Kinzler, Emmanuel Dupoux, and Elizabeth S. Spelke, "The Native Language of Social Cognition," *Proceedings of the National Academy of Sciences* 104, no. 30 (2007): 12577–12580; and Katherine D. Kinzler, Kristin Shutts, Jasmine Dejesus, and Elizabeth S. Spelke, "Accent Trumps Race in Guiding Children's Social Preferences," *Social Cognition* 27, no. 4 (2009): 623–634.

11. Donald L. Rubin, "Nonlanguage factors Affecting Undergraduates' Judgments of Nonnative English-Speaking Teaching Assistants," *Research in Higher Education* 33, no. 4 (1992): 511–531; and Donn Byrne, William Griffitt, and Daniel Stefaniak, "Attraction and Similarity of Personality Characteristics," *Journal of Personality and Social Psychology* 5, no. 1 (1967): 82.

12. Andreea Niculescu, George M. White, See Swee Lan, Ratna Utari Waloejo, and Yoko Kawaguchi, "Impact of English Regional Accents on User Acceptance of Voice User Interfaces," in *Proceedings of the 5th Nordic Conference on Human-Computer Interaction: Building Bridges* (ACM, 2008), 523–526.

13. Howard Giles, "Evaluative Reactions to Accents," *Educational Review* 22, no. 3 (1970): 211–227. All bibliographic information has been provided in the body of this essay. Table 8.1 just provides a summary of what has been already discussed.

14. James Emil Flege, "Factors Affecting Degree of Perceived Foreign Accent in English Sentences," *Journal of the Acoustical Society of America* 84, no. 1 (1988): 70–79.

15. Wallace E. Lambert, Richard C. Hodgson, Robert C. Gardner, and Samuel Fillenbaum, "Evaluational Reactions to Spoken Languages," *Journal of Abnormal and Social Psychology* 60, no. 1 (1960): 44.

16. James J. Bradac and Randall Wisegarver, "Ascribed Status, Lexical Diversity, and Accent: Determinants of Perceived Status, Solidarity, and Control of Speech Style," *Journal of Language and Social Psychology* 3, no. 4 (1984): 239–255; Jairo N. Fuertes, William H. Gottdiener, Helena Martin, Tracey C. Gilbert, and Howard Giles, "A Meta-Analysis of the Effects of Speakers' Accents on Interpersonal Evaluations," *European Journal of Social Psychology* 42, no. 1 (2012): 120–133; Stephanie Lindemann, "Koreans, Chinese or Indians? Attitudes and Ideologies About Non-Native English Speakers in the United States," *Journal of Sociolinguistics* 7, no. 3 (2003): 348–364; Stephanie Lindemann, "Who Speaks "Broken English"? US Undergraduates' Perceptions of Non-Native English," *International Journal of Applied Linguistics* 15, no. 2 (2005): 187–212; and Drew Nesdale and Rosanna Rooney, "Evaluations and Stereotyping of Accented Speakers by Pre-Adolescent Children," *Journal of Language and Social Psychology* 15, no. 2 (1996): 133–154; Bouchard-Ryan, Carranza, and Moffie, "Reactions Toward Varying Degrees of Accentedness," 267–273.

17. Bradac and Wisegarver, "Ascribed Status, Lexical Diversity, and Accent," 239–255; Donald L. Rubin, Pamela Healy, T. Clifford Gardiner, Richard C. Zath, and Cynthia Partain Moore, "Nonnative Physicians as Message Sources: Effects of Accent and Ethnicity on Patients' Responses to AIDS Prevention Counseling," *Health Communication* 9, no. 4 (1997): 351–368; Edwards, "Language Attitudes and Their Implications among English Speakers," 20–23; Sally Boyd, "Foreign-Born Teachers in the Multilingual Classroom in Sweden: The Role of Attitudes to Foreign Accent," *International Journal of Bilingual Education and Bilingualism* 6, no. 3–4 (2003): 283–295; Mary Jiang Bresnahan, Rie Ohashi, Reiko Nebashi, Wen Ying Liu, and Sachiyo Morinaga Shearman, "Attitudinal and Affective Response toward Accented English," *Language & Communication* 22, no. 2 (2002): 171–185; Megumi Hosoda, Eugene F. Stone-Romero, and Jennifer N. Walter, "Listeners' Cognitive and Affective Reactions to English Speakers with Standard American English and Asian Accents," *Perceptual and Motor Skills* 104, no. 1 (2007): 307–326; Stephanie Lindemann, "Who Speaks 'Broken English'?," 187–212; and Laura Huang, Marcia Frideger, and Jone L. Pearce, "Political Skill: Explaining the Effects of Nonnative Accent on Managerial Hiring and Entrepreneurial Investment Decisions," *Journal of Applied Psychology* 98, no. 6 (2013): 1005.

18. Bo Zhao, Jan Ondrich, and John Yinger, "Why Do Real Estate Brokers Continue to Discriminate? Evidence from the 2000 Housing Discrimination Study," *Journal of Urban Economics* 59, no. 3 (2006): 394–419; Rudolf Kalin and Donald S. Rayko, "Discrimination in Evaluative Judgments against Foreign-Accented Job Candidates," *Psychological Reports* 43, no. 3, suppl. (1978): 1203–1209; Mari J. Matsuda, "Voices of America: Accent, Antidiscrimination Law, and a Jurisprudence for the Last Reconstruction," *Yale Law Journal* (1991): 1329–1407; Beatrice Bich-Dao Nguyen, "Accent Discrimination and the Test of Spoken English: A Call for an Objective Assessment of the Comprehensibility of Nonnative Speakers," *California Law Review* 81 (1993): 1325; Lara Frumkin, "Influences of Accent and Ethnic Background on Perceptions of Eyewitness Testimony," *Psychology, Crime & Law* 13, no. 3 (2007): 317–331; Rosina Lippi-Green, "Accent, Standard Language Ideology, and Discriminatory Pretext in the Courts," *Language in Society* 23, no. 2 (1994): 163–198; James J. Bradac, "Language Attitudes and Impression Formation," in *Handbook of Language and Social Psychology*, ed. H. Giles and W. P. Robinson (New York: John Wiley & Sons, 1990), 387–412; Nancy De La Zerda and Robert Hopper, "Employment Interviewers' Reactions to Mexican American Speech," *Communications Monographs* 46, no. 2 (1979): 126–134; Kalin and Rayko, "Discrimination in Evaluative Judgments," 1203–1209; and William Y. Chin, "Linguistic Profiling in Education: How Accent Bias Denies Equal Educational Opportunities to Students of Color," *Scholar* 12 (2009): 355.

19. David Crystal, "Sound and Fury: How Pronunciation Provokes Passionate Reactions," *Guardian* (January 13, 2018), https://www.theguardian.com/books/2018/jan/13/pronunciation -complaints-phonetics-sounds-appealing-david-crystal.

20. Lise Winer, *Dictionary of the English/Creole of Trinidad and Tobago: On Historical Principles* (Montreal: McGill-Queen's Press-MQUP, 2009), 365.

21. Harvey R. Neptune, *Caliban and the Yankees: Trinidad and the United States Occupation* (Chapel Hill: University of North Carolina Press, 2009), 2.

22. Kevin Adonis Browne, *Mas Movement: Toward a Theory of Caribbean Rhetoric* (Philadelphia: Pennsylvania State University, 2009), 4.

23. Michael A. Picheny, Nathaniel I. Durlach, and Louis D. Braida, "Speaking Clearly for the Hard of Hearing II: Acoustic Characteristics of Clear and Conversational Speech," *Journal of Speech, Language, and Hearing Research* 29, no. 4 (1986): 434–446.

24. Interestingly, many of the clear speech strategies that work in human–human interactions can be misunderstood as a display of angry speech in human–computer interactions. Some systems are now designed to recognize angry speech (marked by a slower rate of speech and hyperarticulation of vowel sounds), and transfer the speaker to a live agent.

25. David Gonzalez Nieto, "The Emperor's New Words: Language and Colonization," *Human Architecture* 5 (2007): 232.

26. Robert Phillipson, "Realities and Myths of Linguistic Imperialism," *Journal of Multilingual and Multicultural Development* 18, no. 3 (1997): 238–248.

27. Bettina Migge and Isabelle Léglise, "Language and Colonialism," *Handbook of Language and Communication: Diversity and Change* 9 (2007): 299.

28. Aimé Césaire, *Discours sur le colonialisme* (Paris: Edition Réclame, 1950); Frantz Fanon, *Peau noire masques blancs* (Paris: Le seuil, 1952); Ayo Bamgbose, "Introduction: The Changing Role of Mother Tongues in Education," in *Mother Tongue Education: The West African Experience* (Hodder and Staughton, 1976); Ayo Bamgbose, *Language and the Nation* (Edinburgh: Edinburgh University Press, 1991); Ayo Bamgbose, *Language and Exclusion: The Consequences of Language Policies in Africa* (Münster: Lit Verlag, 2000).

29. Migge and Léglise, "Language and Colonialism," 6.

30. Crystal, *English as a Global Language.*

31. Randolph Quirk and Henry G. Widdowson, "English in the World," *Teaching and Learning the Language and Literatures* (1985): 1–34.

32. Maeve Shearlaw, "Mark Zuckerberg Says Connectivity Is a Basic Human Right—Do You Agree?," *Guardian* (January 3, 2014), https://www.theguardian.com/global-development/poverty -matters/2014/jan/03/mark-zuckerberg-connectivity-basic-human-right.

33. Bill Wasik, "Welcome to the Age of Digital Imperialism," *New York Times* (June 4, 2015), https:// www.nytimes.com/2015/06/07/magazine/welcome-to-the-age-of-digital-imperialism.html.

34. Phillipson, "Realities and Myths of Linguistic Imperialism," 238.

35. Audrey Watters, "Technology Imperialism, the Californian Ideology, and the Future of Higher Education," *Hack Education* (October 15, 2015), accessed May 9, 2018, http://hackeducation .com/2015/10/15/technoimperialism.

36. "Usage of Content Languages for Websites," W3Techs, accessed March 31, 2018, W3techs .com/technologies/overview/content_language/all.

37. Daniela Hernandez, "How Voice Recognition Systems Discriminate against People with Accents: When Will There be Speech Recognition for the Rest of Us?" *Fusion* (August 21, 2015), http://fusion.net/story/181498/speech-recognition-ai-equality/.

38. Braj B. Kachru, "The English Language in the Outer Circle," *World Englishes* 3 (2006): 241–255.

39. Salikoko Mufwene, "The Legitimate and Illegitimate Offspring of English," in *World Englishes 2000*, ed. Larry E. Smith and Michael L. Forman (Honolulu: College of Languages, Linguistics, and Literature, University of Hawai'i and the East-West Center, 1997), 182–203.

40. Source of table 8.3: "Language Support in Voice Assistants Compared," *Globalme Language and Technology* (April 13, 2018), https://www.globalme.net/blog/language-support-voice-assistants -compared.

41. Saritha Rai, "Amazon Teaches Alexa to Speak Hinglish. Apple's Siri Is Next," Bloomberg .com (October 30, 2017), https://www.bloomberg.com/news/articles/2017-10-30/amazon-teaches -alexa-to-speak-hinglish-apple-s-siri-is-next.

42. Rana D. Parshad et al., "What Is India Speaking? Exploring the "Hinglish" Invasion," *Physica A: Statistical Mechanics and Its Applications* 449 (2016): 377.

43. Parshad et al., "What Is India Speaking?," 375–389.

44. Meryl Alper, *Giving Voice: Mobile Communication, Disability, and Inequality* (Cambridge, MA: MIT Press, 2017), 2.

45. Wasik, "Welcome to the Age of Digital Imperialism."

46. Sonia Paul, "Voice Is the Next Big Platform, Unless You Have an Accent," *Wired* (October 17, 2017), www.wired.com/2017/03/voice-is-the-next-big-platform-unless-you-have-an-accent/.

47. Aylin Caliskan-Islam, Joanna J. Bryson, and Arvind Narayanan, "Semantics Derived Automatically from Language Corpora Necessarily Contain Human Biases," *arXiv preprint arXiv:1608.07187* (2016): 1–14.

48. Nasser's work can be followed at http://nas.sr/.

49. Ramsey Nasser, "Command Lines: Performing Identity and Embedding Bias," panel at Computer History Museum (April 26, 2017), accessed June 1, 2017, http://bit.ly/2r1QsuD.

50. https://voice.mozilla.org/en.

51. Halcyon M. Lawrence, "Speech Intelligibility and Accents in Speech-Mediated Interfaces: Results and Recommendations" (PhD diss., Illinois Institute of Technology, 2013).

52. Niculescu et al., "Impact of English Regional Accents," 523–526.

53. Rubin et al., "Nonnative Physicians as Message Sources," 351–368.

54. Alper, *Giving Voice*, 2.

9

YOUR ROBOT ISN'T NEUTRAL

Safiya Umoja Noble

INTRODUCTION

In my long-term research agenda, I have been mostly concerned with who benefits, who loses, and what the consequences and affordances are of emergent technologies, and now I am pointing these lenses from the fields of Black studies, gender studies, and critical information studies to the field of robotics.[1] Recently, I've been thinking about the algorithmically driven software embedded in anthropomorphized computers—or humanlike robots—that will enter the market soon. My aim with this chapter is to offer a series of provocations and suggest that we continue to gather interdisciplinary scholars to engage in research that asks questions about the reinscribing of race and gender in both software and hardware.

One of the most fiery debates in contemporary society, particularly in the United States, is concerned with the treacherous implementation of racist ideologies and legal and social practices that structurally displace, demean, and deny rights to indigenous people, African Americans, Latinx communities, and immigrants from around the world, particularly Asian Americans, Southeast Asians, and Pacific Islanders. In fact, as this book goes to press, hate crimes and violence against the aforementioned groups have seen a significant uptick, fueled by viral vitriol disseminated with intense speed and in great volume by digital platforms and computing networks. How then will robotics be introduced into these fiery times, and with what kinds of consequences? These are the questions that lie before us, despite the notions of robotics existing somewhere between a new form of assistive and nonthreatening labor and autonomous droids that dominate humans. Unfortunately, as we grapple with competing cultural ideas about whether robots will be helpful or harmful, robotic devices are already being inscribed with troublesome ideas about race and gender. In fact, these concerns are intertwined with the ideas Halcyon Lawrence has raised in this book about language, dialects, and virtual assistants and only underscore how much is at stake when we fail to think more humanistically about computing.

YOUR ROBOT IS A STEREOTYPE

In January of 2018, I had a novel experience of meeting with some of the leading women in the field of robotics in the United States, gathered by Professor Londa Schiebinger and the Gendered Innovations initiative at Stanford University to share research about the ways that gender is embedded in robotics research and development. One key takeaway from our meeting is that the field of robotics rarely engages with social science and humanities research on gender and race, and that social scientists have much to offer in conceptualizing and studying the impact of robotics on society. An important point of conversation I raised at our gathering was that there is a need to look closely at the ways that race and gender are used as hierarchical models of power in the design of digital technologies that exacerbate and obfuscate oppression.

At that Stanford gathering of women roboticists, the agenda was focused explicitly on the consequences and outcomes of gendering robots. Social scientists who study the intersection of technology and gender understand the broader import of how human beings are influenced by gender stereotypes, for example, and the negative implications of such. Donna Haraway helped us understand the increasingly blurred lines between humans and technology vis-à-vis the metaphor of the cyborg.[2] More recently, Sherry Turkle's research argued that the line between humans and machines is increasingly becoming indistinguishable.[3] Jennifer Robertson, a leading voice in the gendering of robots, helps contextualize the symbiotic dynamic between mechanized technologies like robots and the assignment of gender:

> Roboticists assign gender based on their common-sense assumptions about female and male sex and gender roles. . . . Humanoid robots are the vanguard of posthuman sexism, and are being developed within a reactionary rhetorical climate.[4]

Indeed, the ways that we understand human difference, whether racialized or gendered, is often actualized in our projections upon machines. The gender and robotics meeting left me considering the ways that robots and various forms of hardware will stand in as human proxy, running on scripts and code largely underexamined in their adverse effects upon women and people of color. Certainly, there was a robust understanding about the various forms of robots and the implications of basic forms of gendering, such as making pink versus blue robot toys for children. But the deeper, more consequential ways of thinking about how gender binaries are constructed and replicated in technologies that attempt to read, capture, and differentiate among "male" versus "female" faces, emotions, and expressions signal we are engaging territories that are rife with the problems of gendering.

The complexity of robots gendering us is only made more complicated when we engage in gendering robots. In the field of human–robot interaction (HRI), ideas about genderless or user-assigned gender are being studied, particularly as it concerns the acceptability of various kinds of tasks that can be programmed into robots who will interact with human beings across the globe. For example, questions as to whether assistive robots used in health care, who may touch various parts of the body, are studied to see if these kinds of robots are more easily accepted by humans when the gender of the robot is interpreted as female. Søraa has carefully traced the most popular commercial robots and the gendering of such technologies, and further detailed how the narratives we hold about robots have a direct bearing on how they are adopted, perceived, and engaged with. This is an important reason why gender stereotyping of robots plays a meaningful role in keeping said stereotypes in play in society. For example, Søraa deftly describes the landscape of words and signifiers that are used to differentiate the major commercial robotics scene, stemming largely from Japan, and how male and female androids (or "gynoids," to use a phrase coined by Robertson) are expressions of societal notions about roles and attributes that "belong" or are native to binary, naturalized notions of biological sex difference.[5]

Robots run on algorithmically driven software that is embedded with social relations—there is always a design and decision-making logic commanding their functionality. If there were any basic, most obvious dimensions of how robots are increasingly embedded with social values, it would only take our noticing the choices made available in the commercial robotics marketplace between humanoid robotics and other seemingly benign devices, all of which encode racial and gender signifiers. To complicate matters, even the nonhumanlike (nonanthropomorphic) robotic devices, sold by global digital media platforms like Google, Apple, and Amazon, are showing up in homes across the United States with feminine names like

Figure 9.1   Image of Microsoft's Ms. Dewey, as researched by Miriam Sweeney.

Amazon's Alexa or Apple's Siri, all of which make "feminine" voices available to users' commands.

The important work of Miriam Sweeney at the University of Alabama shows how Microsoft's Ms. Dewey was a highly sexualized, racialized, and gendered avatar for the company (see fig. 9.1). Her *Bitch* interview about fembots is a perfect entry point for those who don't understand the politics of gendered and anthropomorphized robots.[6] It's no surprise that we see a host of emergent robotic designs that are pointed toward women's labor: from doing the work of being sexy and having sex, to robots that clean or provide emotional companionship. Robots are the dreams of their designers, catering to the imaginaries we hold about who should do what in our societies.

But what does it mean to create hardware in the image of women, and what are the narratives and realities about women's labor that industry, and its hardware designers, seek to create, replace, and market? Sweeney says:

> Now more than ever it is crucial to interrogate the premise of anthropomorphization as a design strategy as one that relies on gender and race as foundational, infrastructural components. The ways in which gender and race are operationalized in the interface continue to reinforce the binaries and hierarchies that maintain power and privilege. While customization may offer some individual relief to problematic representations in the interface, particularly for marginalized users, sexism and racism persist at structural levels and, as such, demand a shifted industry approach to design on a broad level.[7]

To Sweeney's broader point, it's imperative we go beyond a liberal feminist understanding of "women's roles" and work. Instead, we need to think about how robots fit into structural inequality and oppression, to what degree capital will benefit from

Figure 9.2   Google image search on "robots." (*Source*: Google image, February 8, 2018.)

the displacement of women through automation, and how the reconstruction of stereotypical notions of gender will be encoded in gender-assigned tasks, free from other dimensions of women's intellectual and creative contributions. Indeed, we will continue to see robots function as expressions of power.

WHY ARE ALL THE ROBOTS WHITE?

In my previous work, I've studied the images and discourses about various people and concepts that are represented in dominant symbol systems like Google search (see fig. 9.2).[8] Robots have a peculiar imaginary in that they're often depicted as servile and benevolent. Often encased in hard white plastics, with illuminated blue eyes, they exist within the broader societal conceptions of what whiteness means as metaphor for what is good, trustworthy, and universal. White robots are not unlike other depictions where white or light-colored skin serves as a proxy for notions of goodness and the idealized human.

Racialized archetypes of good versus evil are often deployed through film and mass media, such as the film *I, Robot* (2004), starring Will Smith, which explores the ethics of robots while mapping the terrain and tropes of (dis)trust and morality

Figure 9.3 Still image from the film *I, Robot*. (*Source*: https://www.rogerebert.com/reviews/ i-robot-2004.)

in machines (see fig. 9.3).[9] Drawing down on the racist logics of white = good and Black = bad, what is the robotics industry (and Hollywood, for that matter) attempting to take up in discussions about the morality of machines? Indeed, an entire new area of study, better known among roboticists as "machine morality," must still contend with histories and cultural logics at play in design and deployment, and these seem to be well-studied and understood best in partnership with humanists and social scientists, for whom these ideas are not novel but extend for hundreds of years if not millennia.[10] Indeed, despite computer scientists' rejection of the humanities, these egregious notions of conceptually introducing machine morality into systems of racial and gender oppression are quite obviously prone to disaster to those humanists and social scientists who study the intersections of technology and the human condition. To be blunt, humans have yet to figure out and proactively resolve the immorality of racism, sexism, homophobia, occupation, ableism, and economic exploitation. It's quite a wonder that roboticists can somehow design robots to contend with and resolve these issues through a framework of machine morality; indeed, we could certainly use these problem-solving logics in addressing extant human misery with, or without, robots.

YOUR ROBOT'S "BRAIN" IS RUNNING ON DIRTY DATA

The digital is one powerful place where gendered and racialized power systems are obscured, because when artificial intelligence or automated decision-making systems discriminate against you, it is difficult to even ascertain what has happened.

Automated discrimination is difficult to parse, even though there is clearly distinguished literature that has carefully documented how it works and how prevalent it is becoming. As Cathy O'Neil warns, it's difficult to take AI to court or exert our rights over computer code and hardware.[11] Of concern are the lack of clearly defined linkages between the making of data sets and the historical and social practices that inform their construction. For example, when data is developed from a set of discriminatory social processes, such as the creation of statistics on policing in a municipality, it is often impossible to recognize that these statistics may also reflect procedures such as overpolicing and disproportionate arrest rates in African American, Latinx, and low-income neighborhoods.[12]

Early work of educational theory scholar Peter Smagorinsky carefully articulates the dominant metaphors that are used with our conceptions of "data" that infer a kind of careful truth construction around purity and contamination, not unlike new frameworks of "biased" versus "objective" data:

> The operative metaphors that have characterized researchers' implication in the data collection process have often stressed the notion of the purity of data. Researchers "intrude" through their media and procedures, or worse, they "contaminate" the data by introducing some foreign body into an otherwise sterile field. The assumption behind these metaphors of purity is that the researcher must not adulterate the social world in which the data exist. . . . The assumption that data are pure implies that researchers must endeavor strongly to observe and capture the activity in a research site without disrupting the "natural" course of human development taking place therein.[13]

We must continue to push for a better, common-sense understanding about the processes involved in the construction of data, as evidenced by the emergent and more recent critiques of the infallibility of big data, because the notion that *more* data must therefore lead to *better* data analysis is incorrect. Larger flawed or distorted data sets can contribute to the making of worse predictions and even bigger problems. This is where social scientists and humanists bring tremendous value to conceptualizing the paradigms that inform the making and ascribing of values to data that is often used in abstract and unknown or unknowable ways. We know that data is a social construction as much as anything else humans create, and yet there tends to be a lack of clarity about such in the broad conceptualization of "data" that informs artificial decision-making systems. Often, the fact that data—which is the output on some level of human activity and thought—is not typically seen as a social construct by a whole host of data makers makes intervening upon the dirty or flawed data even more difficult.[14]

In fact, I would say this is a major point of contention when humanists and social scientists come together with colleagues from other domains to try to make sense of the output of these products and processes. Even social scientists currently engaging in data science projects, like anthropologists developing predictive policing technologies, are using logics and frameworks that are widely disputed as discriminatory.[15] The concepts of the purity and neutrality of data are so deeply embedded in the training and discourses about what data *is* that there is great difficulty moving away from the reductionist argument that "math can't discriminate because it's math," which patently avoids the issue of *application* of predictive mathematical modeling to the social dimensions of human experience. It is even more dangerous when social categories about people are created as if they are fixed and natural, without addressing the historical and structural dimensions of how social, political, and economic activities are shaped.

This has been particularly concerning in the use of historical police data in the development of said predictive policing technologies, which we will increasingly see deployed in a host of different robotics—from policing drones to robotic prison guards—where the purity and efficacy of historical police data is built into such technology design assumptions.

Typically, such products use crime statistics to inform processes of algorithmic modeling. The resulting output is a "prediction" of precise locations of future crime (see fig. 9.5).[16] This software, already used widely by police agencies across the United States, has the potential to grow exponentially in its power and impact when coupled with robots, drones, and other mechanical devices. It's important, dare I say *imperative*, that policy makers think through the implications of what it will mean when this kind of predictive software is embedded in decision-making robotics, like artificial police officers or military personnel that will be programmed to make potentially life-or-death decisions based on statistical modeling and a recognition of certain patterns or behaviors in targeted populations.

Crawford and Shultz warn that the use of predictive modeling through gathering data on the public also poses a serious threat to privacy; they argue for new frameworks of "data due process" that would allow individuals a right to appeal the use of their data profiles.[17] This could include where a person moves about, how one is captured in modeling technologies, and use of surveillance data for use in big data projects for behavioral predictive modeling such as in predictive policing software:

> Moreover, the predictions that these policing algorithms make—that particular geographic areas are more likely to have crime—will surely produce more arrests in those

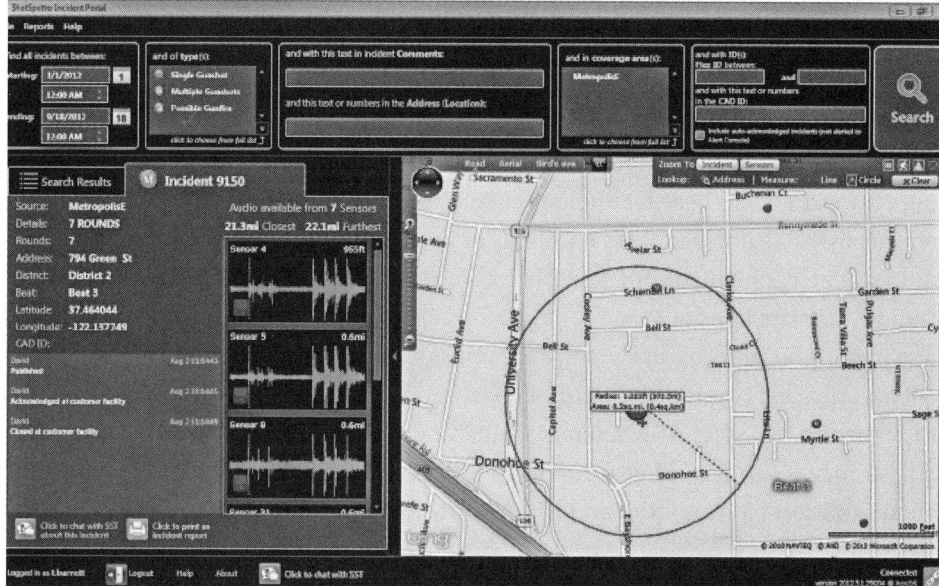

ר example of a gunfire detection technology incident map. Photo courtesy of Shot Spotter Inc.

Figure 9.4 Predictive policing software depicting gunfire detection technology. (*Source:* Shot Spotter.)

areas by directing police to patrol them. This, in turn, will generate more "historical crime data" for those areas and increase the likelihood of patrols. For those who live there, these "hot spots" may well become as much PII [personally identifiable information] as other demographic information.[18]

In the case of predicting who the likely perpetrators of "terrorist crime" could be, recent reports on the Immigration and Customs Enforcement (ICE) agency's predictive decision-making through its "Extreme Vetting Initiative" have been found to be faulty, at best.[19] In fact, many scholars and concerned scientists and researchers signed a petition to Microsoft CEO Satya Nadella and other key executives, expressing concern and signaling widespread condemnation among those who understand the potency and power of using such technologies in discriminatory ways.[20] Their petition was in direct solidarity with Microsoft employees who objected to the use of Microsoft's technology in the Azure Government platform that would be used to accelerate abuses of civil and human rights under policies of the Trump Administration, and it also amplified the work of Google employees who had rebuffed Google executives' efforts to engage in Project Maven, a contract with the US Department of Defense to process drone footage used in military operations, another example of the melding of robotic technology (drones) with AI software meant to increase the potency of the product.[21]

Facial-recognition software is but one of the many data inputs that go into building the artificial backbone that bolsters predictive technologies. Yet there is alarming concern about these technologies' lack of reliability, and the consequences of erroneously flagging people as criminals or as suspects of crimes they have not committed. The deployment of these AI technologies has been brought to greater scrutiny by scholars like computer scientists Joy Buolamwini and Timnit Gebru, whose research on facial-recognition software's misidentification of women and people of color found that commercial technologies—from IBM to Microsoft or Face++—failed to recognize the faces of women of color, and have a statistically significant error rate in the recognition of brown skin tones.[22] Buolamwini even raised her concerns about Amazon's Rekognition and its inaccuracy in detecting women of color to CEO Jeff Bezos.[23]

Drones and other robotic devices are all embedded with these politics that are anything but neutral and objective. Their deployment in a host of social contexts where they are charged with predicting and reporting on human behavior raises important new questions about the degree to which we surrender aspects of our social organization to such devices. In the face of calls for abolition of the carceral state (prisons, jails, and surveillance systems like biometric monitors), robots are entering the marketplace of mass incarceration with much excitement and little trepidation. Indeed, the robotic prison guard is estimated to cost about $879,000, or 1 billion Korean won, apiece.[24]

> Steep price tag notwithstanding, prison authorities are optimistic that, if effective, the robots will eventually result in a cutting of labor costs. With over 10.1 million people incarcerated worldwide, they see the implementation of robotic guards as the future of penal institution security. . . . For their part, the designers say that the next step would be to incorporate functionality capable of conducting body searches, though they admit that this is still a ways off—presumably to sighs of relief from prisoners.[25]

Predictive technologies are with us, particularly in decisions that affect policing, or insurance risk, or whether we are at risk for committing future crime, or whether we should be deported or allowed to emigrate. They are also part of the logic that can drive some types of robots, particularly those trained on vast data sets, in order to make decisions or engage in various behaviors. New robotics projects are being developed to outsource the labor of human decision-making, and they will open up a new dimension of how we will legally and socially interpret the decisions of robots that use all kinds of unregulated software applications, with little to no policy oversight, to animate their programmed behavior.

BUT MY ROBOT IS FOR SOCIAL GOOD!

The prevailing thinking about robots is far less nefarious than drones using faulty facial-recognition systems before a military or police strike on various global publics. On a subsequent visit to Stanford University some months after the gathering of women roboticists and social scientists, I encountered a small robot on the street not unlike the newest food delivery robot, Steve, launched in August 2019 in San Francisco (see fig. 9.5).

The Postmates robot, Steve, requires a human operator—a reminder from the work of Sarah Roberts in this volume that "your AI is a human," and in this case your robot is basically managed and manipulated by a human standing about thirty

Postmates received a permit to start testing its autonomous delivery robot, Serve, in San Francisco. The permit is limited to an industrial area of the city and allows the firm to test only three devices at a time.
Photo: Postmates

Figure 9.5 "Steve" is Postmates' autonomous food delivery robot. (*Source*: Postmates.)

feet away.[26] Steve and the media coverage about the glory of human partnerships with robots to reduce traffic, parking hassles, and time spent to get food a few feet out of the restaurant to the delivery driver are typical of the way robots are framed as helpful to our local economies. But the darker stories of how the entire enterprise of food delivery companies like Postmates, Grubhub, Uber Eats, and DoorDash are putting our favorite local restaurants out of business for the exorbitant fees they charge for their services are often hidden in favor of from mainstream media stories like the launch of Steve. Meanwhile, community activists are increasingly calling on the public to abandon these services at a time when we are being persuaded through cute little robots like Steve to look past the exploitive dimensions of the business practices of their tech company progenitors.[27]

CONCLUSION

The future of robotics, as it currently stands, may be the continuation of using socially constructed data to develop blunt artificial intelligence decision-making systems that are poised to eclipse nuanced human decision-making, particularly as it's deployed in robotics. This means we must look much more closely at privatized projects in industry-funded robotics research and development labs, which are deployed on various publics. We have very little regulation about human–robotics interaction, as our legal and political regimes are woefully out of touch with the ways in which social relations will be transformed with the introduction of robotics we engage with at home and at work.

In 2014, Professor Ryan Calo issued a clarion call for the regulation of robots through a Federal Robotics Commission that would function like every other US consumer safety and protection commission.[28] Four years later, on June 7, 2018, the *MIT Technology Review* hosted the EmTech Next conference, a gathering of people interested in the future of robotics. Most of the attendees hailed from the finance, education, government, and corporate sectors. According to reporting by *Robotics Business Review*, five key themes emerged from the conference: (1) to see "humans and robots working together, not replacing humans," (2) to deal with the notion that "robots are needed in some markets to solve labor shortages," (3) that "education will need to transform as humans move to learn new skills," (4) that "lower skilled workers will face difficulty in retraining without help," and (5) "teaching robots and AI to learn is still very difficult."[29] Not unlike most of the coverage for *Robotics Business Review*, the story represented the leading thinking within the commercial robotics industry: robots are part of the next chapter of humanity, and they are here to stay. Yet Calo's concerns have largely gone unaddressed in the call for coordinated and well-developed public policy or oversight.

We have to ask what is lost, who is harmed, and what should be forgotten with the embrace of artificial intelligence and robotics in decision-making. We have a significant opportunity to transform the consciousness embedded in artificial intelligence and robotics, since it is in fact a product of our own collective creation.

NOTES

1. This chapter is adapted from a short-form blog post at FotoMuseum. See Safiya Umoja Noble, "Robots, Race, and Gender," Fotomuseum.com (January 30, 2018), https://www.fotomuseum.ch/en/explore/still-searching/articles/154485_robots_race_and_gender.

2. Donna J. Haraway, "A Manifesto for Cyborgs: Science, Technology, and Socialist Feminism in the 1980s," *Australian Feminist Studies* 2 (1987): 1–42.

3. Sherry Turkle, *Life on the Screen* (New York: Simon & Schuster, 2011).

4. J. Robertson, "Gendering Humanoid Robots: Robo-Sexism in Japan," *Body & Society* 16 (2010): 1–36.

5. Roger Andre Søraa, "Mechanical Genders: How Do Humans Gender Robots?," *Gender, Technology and Development* 21 (2017): 1–2, 99–115, https://doi.org/10.1080/09718524.2017.1385320; Roger Andre Søraa, "Mecha-Media: How Are Androids, Cyborgs, and Robots Presented and Received through the Media?," in *Androids, Cyborgs, and Robots in Contemporary Culture and Society* (Hershey, PA: IGI Global, 2017).

6. See Sarah Mirk, "Popaganda: Fembots," *Bitch* (December 7, 2017), https://www.bitchmedia.org/article/popaganda-fembots-westworld-female-robots-siri-ex-machina-her-metropolis-film.

7. Miriam Sweeney, "The Intersectional Interface," in *The Intersectional Internet: Race, Gender, Class & Culture*, ed. Safiya Umoja Noble and Brendesha M. Tynes (New York: Peter Lang, 2016).

8. Safiya Umoja Noble, *Algorithms of Oppression: How Search Engines Reinforce Racism* (New York: NYU Press, 2018).

9. Christopher Grau, "There Is No 'I' in 'Robot': Robots and Utilitarianism," in *Machine Ethics*, ed. M. Anderson and S. L. Anderson (Cambridge: Cambridge University Press, 2011), 451–463.

10. For a definition of the concept of "machine morality," see John P. Sullins, "Introduction: Open Questions in Roboethics," *Philosophy & Technology* 24 (2011): 233, http://doi.org/10.1007/s13347-011-0043-6.

11. Cathy O'Neil, *Weapons of Math Destruction: How Big Data Increases Inequality and Threatens Democracy* (New York: Crown Publishing, 2016).

12. Kate Crawford and Jason Schultz, "Big Data and Due Process: Toward a Framework to Redress Predictive Privacy Harms," *Boston College Law Review* 55, no. 93 (2014), https://ssrn.com/abstract=2325784.

13. Peter Smagorinsky, "The Social Construction of Data: Methodological Problems of Investigating Learning in the Zone of Proximal Development," *Review of Educational Research* 65, no. 3 (1995): 191–212, http://dx.doi.org/10.2307/1170682.

14. Imanol Arrieta Ibarra, Leonard Goff, Diego Jiménez Hernánd, Jaron Lani, and E. Glen Weyl, "Should We Treat Data as Labor? Moving Beyond 'Free,'" *American Economic Association Papers & Proceedings* 1, no. 1 (December 27, 2017), https://ssrn.com/abstract=3093683.

15. See letter from concerned faculty to LAPD, posted by Stop LAPD Spying on *Medium* (April 4, 2019), https://medium.com/@stoplapdspying/on-tuesday-april-2nd-2019-twenty-eight-professors -and-forty-graduate-students-of-university-of-8ed7da1a8655.

16. Andrew G. Ferguson, "Predictive Policing and Reasonable Suspicion," *Emory Law Journal* 62, no. 2 (2012), http://law.emory.edu/elj/content/volume-62/issue-2/articles/predicting-policing-and -reasonable-suspicion.html.

17. Crawford and Shultz cite the following: Ferguson, "Predictive Policing" (explaining predictive policing models).

18. Crawford and Schultz, "Big Data and Due Process."

19. Drew Harwell and Nick Miroff, "ICE Just Abandoned Its Dream of 'Extreme Vetting' Software That Could Predict Whether a Foreign Visitor Would Become a Terrorist," *Washington Post* (May 17, 2018), https://www.washingtonpost.com/news/the-switch/wp/2018/05/17/ice-just-abandoned -its-dream-of-extreme-vetting-software-that-could-predict-whether-a-foreign-visitor-would -become-a-terrorist/?utm_term=.83f20265add1.

20. "An Open Letter to Microsoft: Drop your $19.4 million ICE tech contract," petition, accessed July 29, 2018, https://actionnetwork.org/petitions/an-open-letter-to-microsoft-drop-your-194 -million-ice-tech-contract.

21. See Kate Conger, "Google Plans Not to Renew Its Contract for Project Maven, a Controversial Pentagon Drone AI Imaging Program," *Gizmodo* (June 1, 2018), https://gizmodo.com/google -plans-not-to-renew-its-contract-for-project-mave-1826488620.

22. Joy Buolamwini and Timnit Gebru, "Gender Shades: Intersectional Accuracy Disparities in Commercial Gender Classification," Proceedings of the 1st Conference on Fairness, Accountability and Transparency, *Proceedings of Machine Learning Research* 81 (2018): 77–91.

23. See Ali Breland, "MIT Researcher Warned Amazon of Bias in Facial Recognition Software," *The Hill* (July 26, 2018), http://thehill.com/policy/technology/399085-mit-researcher-finds-bias -against-women-minorities-in-amazons-facial.

24. Lena Kim reported the robotic prison guards are developed by the Asian Forum for Corrections, the Electronics and Telecommunications Research Institute, and manufacturer SMEC. See Lena Kim, "Meet South Korea's New Robotic Prison Guards," *Digital Trends* (April 21, 2012), https://www.digitaltrends.com/cool-tech/meet-south-koreas-new-robotic-prison-guards/.

25. Kim, "Robotic Prison Guards."

26. See story by Sophia Kunthara and Melia Russell, "Postmates Gets OK to Test Robot Deliveries in San Francisco," *San Francisco Chronicle* (August 15, 2019), https://www.sfchronicle.com/ business/article/Postmates-gets-OK-to-test-robot-deliveries-in-San-14305096.php.

27. See Joe Kukura, "How Meal Delivery Apps Are Killing Your Favorite Restaurants," *Broke-Ass Stuart* (February 14, 2019), https://brokeassstuart.com/2019/02/14/how-meal-delivery-apps-are -killing-your-favorite-restaurants/.

28. Ryan Calo, "The Case for a Federal Robotics Commission," *Brookings Institution Center for Technology Innovation* (September 2014), https://ssrn.com/abstract=2529151.

29. Keith Shaw, "Five Robotics and AI Takeaways from EmTechNext 2018," *Robotics Business Review* (June 7, 2018), https://www.roboticsbusinessreview.com/news/5-robotics-and-ai-takeaways -from-emtechnext/.

10

BROKEN IS WORD

Andrea Stanton

Typewriters and computers were designed with English and other left-to-right-scripts in mind, while Arabic is written right to left. As a result, technicians have described adapting these technologies to Arabic script as a particularly thorny challenge. But behind the technical problems of typing in Arabic (or Persian or Urdu) stands a world on fire—and not the world of the Middle East. Our assumptions about technology and its relationship to the Roman alphabet, and to the unidirectional scripts it supports, have led us to ask the wrong questions—and to seek the wrong solutions. Why has typing in Arabic been considered such a technical challenge, and why are the solutions—past and present—so inadequate? Why do users who work in multiple languages and scripts find it so difficult to produce legible word-processed text in Arabic and English, or any other combination of Arabic-script and Roman-script languages? Why are the solutions so technical? Arabic is not the problem—thinking of Arabic script as a problem is.

At heart, typing in Arabic—especially in a document that also has Roman-script text—is not a technical issue and needs more than a technical solution. Thinking of Arabic-script typing as a technical problem is the result of societal assumptions made long before the computer and word processing were developed—assumptions that shaped the development of Arabic-script typewriters. Those assumptions are, first, that Arabic script was a particularly "challenging" one that required modifications in order to be used on a typewriter, and, second, that typists and word processors would work in one script at a time. Asking the wrong questions has led us to seek partial, technical solutions. Asking better questions can help us address the culturally laden assumptions about unidirectional script as a typing norm and embrace the reality of a global world in which many people live and work across multiple script forms.

Both the typewriter and the word-processing component of the computer first developed in the United States. They were conceived, designed, prototyped, manufactured, marketed, and distributed in English, and with English in mind. The standards that developed with them reflected that fact: Roman script, unidirectional left-to-right, no accents, and no cursive. Yet from the typewriter's mass production in the 1870s, the companies that produced it, the consular officials who promoted American products overseas, and entrepreneurial inventors around the world also sought to make word-processing tools compatible with other languages and other scripts. Remington and other typewriter manufacturers knew that, after Roman script, Arabic script was the next most widely used script in the world, used for Arabic, Persian, Urdu, and—until the 1920s—Turkish.

Given these numbers, it is revealing how much of the history of Arabic word processing involves technicians faulting Arabic, and—even into the 1970s—calling for users to reform or change the script. Did they never wonder whether the technology, and its designers, might be the main issue? This chapter surveys twentieth-century efforts to redesign Arabic-script typewriter layouts to fit English-language typist norms. It then chronicles some of the attempts to frame the challenges of letter form and width, as well as key placement and type bar alignment, as problems that led to delays in rolling out Arabic-script word-processing programs, and as bugs that prevent toggling between Arabic and Roman scripts. This chapter argues that talk of such technical glitches and bugs obscures underlying hegemonic norms. The history of Arabic word-processing talk rehearses much older European-origin assumptions about the incompatibility not only of scripts but also of the social worlds that support them. As Mar Hicks has argued, these assumptions have the effect of making technological advances regressive rather than revolutionary, entrenching bias in new forms.[1] As Halcyon Lawrence notes in this volume, similar assumptions about normative accents have reinforced voice technologies' exclusion of nonnative and nonstandard speakers.[2] Arabic is no more a conceptual problem for word-processing technologies than accented English is for voice technologies, and neither are bidirectional scripts—and nor are the people who daily operate in multiple languages and scripts.

Twitter's ability to allow users to include multiple scripts in one tweet offers a technical fix—but the overall problem will not be resolved until we recognize that

Euro-American cultural assumptions about unidirectional scripts and single-script fixity lie at its root. Once we address these assumptions, we can conceptualize technologies that fuel a global world in which many people live and work across multiple script forms.

"HERE INDEED WAS A MEDLEY OF PROBLEMS": ARABIC AND THE TYPEWRITER

In September 1873, New York State–based Remington Works—best known for its firearms—released its first commercial typewriter. Remington's new product was a simple device that lacked a shift key and so printed every word in capital letters—although with a QWERTY keyboard that would remain the standard for any word-processing machine. It was promoted as a time-saving device for letter writing, with the added advantage of increased readability. However, because would-be typists first had to learn the QWERTY system, it was not until the emergence of typing schools and training courses in the late 1870s that the machine began to catch on in the United States—first as a business tool, and only decades later as a personal or leisure tool.[3] Decades later, it finally delivered on its promise of "legibility, rapidity, ease, convenience, and economy," in the words of the first typewriter catalog—effectively bringing the printing press revolution in reach of the typist (or at least the person or company she worked for).[4]

Accents were layered in by using "dead" keys, which type over the previous letter rather than moving to the next space. Non-Roman scripts—including Cyrillic, Greek, Hindi, and Arabic—presented more serious challenges: the typewriter's keyboard layout, carriage return orientation, and kerning (spacing). "Of all the languages now written on the typewriter," a 1923 industry history suggested, "the Arabic group presented the gravest mechanical difficulties":

> The Arabic character, as written, is not subject to any of the usual rules. It has in its own complete alphabet over one hundred individual characters; it writes backwards, i.e. from right to left; the characters are written on the line, above the line and below the line, and they are of various widths, requiring full spacing, half spacing, and no spacing at all. Here indeed was a medley of problems well calculated to tax ingenuity to the limit, and the Arabic typewriter is a crowning triumph of mechanical skill.[5]

With twenty-eight letters but multiple forms of each, the Arabic script, it argued, posed the largest and most difficult problem in typewriter design. Solving it was heralded as a singular triumph.[6]

That English became the norm was a historical accident, but it stuck. American typewriter designers privileged English language, not just Roman scripts, even when celebrating typewriter designs for other languages. For example, a 1903 *New York Times* article on the many type shuttle languages noted that an Arabic keyboard design had recently been patented and that a Japanese version was in process. Those designs, the article stated, featured the versatility of typewriters to tackle even the most challenging languages. But, it added, even Romance languages posed their own difficulties: "French, Spanish, and Scandinavian machines are like ours except that the keyboards contain certain accents that are not needed by us," the article stated.[7] Anything other than English, even if typed in Roman script, was a deviation from the norm.

The Ottoman citizen Selim Haddad obtained a United States patent for what is considered the first Arabic-ready typeface in 1899. A graduate of Beirut's Syrian Protestant College, later known as the American University of Beirut, he had taught at a New York City–based night school for Syrian (Levantine) immigrants, run by his brother Ameen, that taught English to Arabic speakers.[8] Starting in the late 1890s, United States consular officials around the Middle East had begun suggesting a need for Arabic typewriters, noting that "millions of people" in the region and elsewhere might appreciate the American import. In 1902, Gabriel Bie Ravndal wrote from Beirut: "There will be immediately a great demand for the Arabic typewriter, and before long, I take it for granted, several [companies] in the United States will be manufacturing and exporting Arabic machines."[9]

In November 1899, Haddad filed Patent Number 637,109 with the United States Patent Office, containing a set of Arabic, Persian, and Ottoman Turkish types that could be installed on "any of the well-known typewriting machines, whether it be of the bar or lever description or of the wheel or segment or plate description." His innovations started a lasting shift in Arabic type. "By the present alphabet the successive letters of a word are often connected in an oblique direction above or below the writing line," he noted. "My letters are constructed so that all connect on a horizontal line and all bear a certain fixed relation to a base or writing line." These changes economized space, permitting three typed lines in the vertical space usually taken by two handwritten lines. Haddad's typeface also regularized the width of the letters, with a standard and double width, rather than individual variance among letters. His design thus fit Arabic within what Mullaney identifies as the key features of the QWERTY keyboard, with letters that all fell around a single baseline.[10]

But Haddad was unable to make Arabic letters isomorphic, with the same shape regardless of position, or to fit them all to the same amount of horizontal space. What he was able to do was to discipline the letters to minimize and regularize their variations in shape and position. He claimed that his type required only fifty-eight characters, thanks to his streamlining the positional variation of each letter.[11] This remarkably simplified the 114 individual characters. In the Arabic script, as with other "cursive" or connected scripts, each letter has approximately four forms: a freestanding form (used when writing the alphabet), an initial form (used when the letter starts a word), a medial form (used when the letter is in the middle of a word), and a final form (used when the letter ends a word). There is no distinction between cursive and block print Arabic-script writing as there is, for example, in Roman-script writing. By Haddad's count, the Arabic alphabet had twenty-nine letters or characters (including a connected lam/alif for "la"). His "new alphabet" streamlined these four forms into only two. Many critics would miss this feature, complaining that Arabic has too many letters, rather than having too many letter forms. Major United States newspapers covered Haddad's invention as noteworthy news, including a 1902 article in the *Washington Post*.[12] In 1904, a *New York Times* article noted that Haddad had prototype machines built as gifts for the Ottoman sultan and Egyptian khedive (ruler), but that commercial manufacture was still under development.[13]

Designing the type was only the first step: the machine itself, with Arabic-script type and a reversed carriage return, still needed to be built. Haddad's typewriter layout addressed in theory the *complexity* of the Arabic script: the multiple forms of each letter, based on its position within a word. However, it was unable to address the bidirectionality of Arabic (and other right-to-left script languages like Hebrew): while Arabic words are written and read right to left, cardinal numbers are written and read left to right. Someone using this typewriter design would have needed to type out cardinal numbers in reverse order, starting with the ones or smallest digit rather than the largest.[14] Haddad's approach addressed both the typeface and how to lay it out on a standard typewriter, while other patent filers in the same period, like Arthur Guest and Ernest Richmond, addressed only the typeface. Guest and Richmond requested a patent for Arabic characters used in typewriters or printing, in which the characters had the same width at their base and all operated from one common point of alignment. "All the types made in accordance with this invention are on blocks of equal width [and] the character on each type commences from a point at the right-hand side, which point is the same in all the types and is herein referred to as the "point of [align]ment," the patent specification explained. No effort was made to lay this typeface out on an actual typewriter; the inventors noted that "we do not limit ourselves to the precise forms of the characters illustrated in the examples shown, because they may be somewhat varied." Instead, they sought only to standardize all Arabic letter base heights and widths and to fit a "standard typewriter." But to actually type, one would have to replace the shuttle, wheel, and bar.[15]

Haddad's innovations sparked early-twentieth-century efforts to restandardize the Arabic script, focusing primarily on the multiple forms of each letter and their varying widths. These script reforms did not explicitly address the font, but *naskh* became the standard font for Arabic typing on the typewriter, and later the computer, and *nastaliq* for Persian and Urdu—the equivalents of Courier and the similar-looking Remington fonts.[16] For example, in 1922 Seyed Khalil was granted a patent for the keyboard layout that he had designed for the Underwood Typewriter Company. Initially submitted in 1917, this layout assumed that one form of each letter would be enough. "I have found that if I use forms of letters which are nearly perfect representations of the letters standing alone," Khalil explained, "it becomes unnecessary to provide types for writing the modified connecting forms of these letters."[17] His layout provided terminal forms for those letters whose "terminal forms are very different from the medial forms," but regularized their location so that terminal letters were all in the "uppercase" position and the nonterminal letters were all in the "lowercase" position of the keyboard, an arrangement he called "absolutely logical." What logic justified reducing the script to a fourth of its original self? His design, he claimed, avoided changing existing typewriters, and also sped the typing process. His "cursive Arabic" script would bear only limited resemblance to Arabic, Persian, or Turkish handwritten texts. But it would produce a standardized script—and all it required was for hundreds of millions of people to relearn how to read. As with twenty-first-century voice technologies and non–native English speakers, typewriting technology expected users to adjust themselves to its limitations. In Lawrence's words, these technologies are biased, and they discipline users for speaking with the "wrong" accent or using a script whose letters had too many forms.[18]

Kemal Ataturk ended Turkish's script experimentation by romanizing the alphabet and script in 1928. The Latin alphabet was praised as supporting mass literacy, by making it easier for Turks to learn to read and write. It separated them from their rich Ottoman textual heritage, part of Ataturk's efforts to give Turkey a national history that minimized the Ottoman role. In 1929, Remington shipped three thousand typewriters to fill an order from the Turkish government, for which it "had to construct new dies [the individual letters used in the typewriter], bringing the total cost to $400,000"; Remington's foreign director explained the government wanted to "speed up its work by the use of these typewriters," rather than writing documents and filling forms by hand.[19] The *New York Times* described the typewriters as "aiding" Ataturk in his Westernization effort, and the "rat-a-tat-tat" of their keys represented the sound of modernity. (It also echoed the larger rat-a-tat-tat of the machine guns that contributed to the Ottoman Empire's defeat in World War I and allowed France and Great Britain to crush the independence movements of the Ottoman Levant (Iraq, Lebanon, Palestine, and Syria).

Efforts to streamline the Arabic script continued into the mid-twentieth century. Mohammed Lakhdar, a Moroccan official who worked in the Ministry of Education, devised a script system incorporated into a typewriter in 1957, which the Ministry promoted as an aid to literacy. Arabic-script typewriters only included long vowels in their keys; they did not include short vowels, which are represented by diacritical marks, because designers felt that they would require too many keystroke combinations and complications. Yet because typed Arabic did not include vowelization, Lakhdar and others argued, only those with high literacy could read printed texts. Lakhdar's system incorporated diacritical vowel markings but could do so on a standard typewriter layout, with no need for additional keys. "Machines with Latin characters can be easily utilized for Arabic with the Lakhdar system at the same cost," supporters argued. The Lakhdar system promised to speed typing and reading further by requiring only one form for each Arabic letter.[20]

As an invention, typewriters are what historian Uri Kupferschmidt calls "small technologies" that overlap with consumer products or durables. "They were adopted, acquired, or appropriated as commodities on the [Middle Eastern] market, initially by elites, but gradually trickled down" toward mass consumption—like watches, sewing machines, gramophones, radios, umbrellas, or safety razors.[21] It often took some time: these products had been developed elsewhere, and were only brought to Middle Eastern markets after having achieved mass-market success in the United States or Western Europe.[22] Yet these small technologies found differing receptions. Unlike the sewing machine, which spread rapidly, typewriters penetrated the Middle Eastern business market slowly and shallowly. A lower adoption rate for some "small technologies" may have been due to the region's smaller middle class, with less disposable income[23] than in the United States. But in the case of the typewriter, Kupferschmidt suggests that lower literacy rates, the privileging of male scribal jobs, and the lag to a "pink collar" female stenographical job force, as well as the minimal private or leisure market, all negatively contributed to typewriter sales.[24] Historical data regarding Middle East literacy rates is limited, but rates in the 1950s were significantly lower than those in Western Europe and North America.

But small technological inventions can have significant consequences. The typewriter shaped nontyped writing structure and script form: it set the standards, even though relatively few people owned or used a typewriter. This is most visible at the keyboard level: the typewriter layout envisioned first by Haddad has become the standard layout for Arabic computer keyboards, with a few small modifications (for example, the placement of the letter dal—Arabic's "d" sound)—much as the QWERTY Latin-script keyboard transferred from typewriter to computer. The regularized letter widths and streamlined letter positions that characterized Khalil's typewriter design have become standard for computer fonts—including Microsoft's Traditional Arabic, which first appeared with Windows 2000.[25]

Arabic-script typing continued to develop into the later twentieth century. Seyed Khalil obtained another patent for Arabic-script typing in 1960, for example, which blamed "the many idiosyncrasies of the language" for inhibiting the development of "a practical Arabic typewriter . . . capable of operating at speeds and efficiencies of English typewriters." He argued that the large number of Arabic characters in their various forms had forced the production of special typewriters with additional keys and space bars. He described Arabic's connected script—"as in English handwriting"—as posing additional challenges, and repeated earlier assertions that the varying width of Arabic letters made typewritten interconnection even more difficult. As a result, he stated, aligning the type bar or bars required much more precision for Arabic-script machines. He critiqued existing typewriter designs—even those that streamlined Arabic letter forms—as failing to address the hyper-importance of type bar alignment in order to produce smoothly interconnected letters. Khalil's new design still limited the Arabic letter forms to two per letter, while better aligning the type bar. Unlike the early-twentieth-century patents, Khalil's described in meticulous detail each step in his typewriter mechanism design. Finally, the patent suggested that the new design could be applied equally well to a portable, standard, or electric typewriter—indicating that it would set the standard for future typewriter technology.[26]

As the century wore on, efforts to improve Arabic-script typewriting borrowed from the advances in early computing. For example: "A device developed by a Pakistani-born Canadian scientist promises to speed up typing Arabic script while preserving its aesthetic qualities," noted a 1973 *New Scientist* article. "The pilot model . . . contains a logic circuit that permits an operator to use a standard 32-character western-style typewriter keyboard to print the 140 characters used on Arabic typewriters." The machine, not the typist, calculated which letter form to use based on the letters immediately before and after it. As a result, it was supposed to make typing in Arabic faster. Typing a document in Arabic was described as taking "up to four times as long" as typing the same document in Roman script.[27] Machines like this one were bridging the gap between traditional typewriters—manual or electric—and the emergent arena of computer-based word processing.

"SOMETHING OF A PATHOLOGICAL CASE": ARABIC SCRIPT IN COMPUTER SOFTWARE

By the late twentieth century and early twenty-first century, the Arabic script was the third most commonly used, after Roman script and Chinese, and included those using Arabic script for Persian, Urdu, and other languages. Euro-American tech efforts shifted to developing electric typewriter and word-processing capacities for Arabic script, continuing the script modification and regularization approaches, but with less urgency and with no greater success.[28] The electric typewriter offered only an stepwise improvement on the typewriter, rather than a wholesale innovation, although the 1973 machine's "script processor" did point toward word processing and its emphasis on speed.

The personal computer, by contrast, rooted itself into the business and personal practices of the Middle East. Kupferschmidt describes the case as an instance of "technological leapfrogging," defined as "the process whereby some developing countries can jump over several [developmental] stages to move rapidly from standard-modern to highly-modern technologies."[29] Historically, one key example is that of nineteenth-century France and Germany adopting British industrial technology like steam engines and railroad networks, which sped industrialization in those countries.[30] European and North American businesses and private consumers were quick to adopt the personal computer, which facilitated its adoption in the Arabic-script world. The typewriter's relatively low adoption also appears to have helped. Neither typewriters nor typing, or stenography as a profession, had developed deep roots in the region, as attested by the series of patents and reports on technological advances, which all describe Arabic typewriters as inefficient and difficult to use. Hence businesses and private consumers had less investment in typewriters and typewriting, making it possible for computers to play a positive rather than negative disruptive role. Few typists lost their livelihood due to the shift toward computers, for example.

Computers promised both powerful computational capabilities and potent word-processing capacities (at least in Roman-script languages). Word processing offered multifaceted corporate appeal, particularly with its on-screen editing capabilities, which streamlined document revisions. Like the typewriter, computer word-processing programs assumed a fixed writing orientation (left to right, at least initially), one default language, and one script form (Roman, at least initially). These assumptions of fixity all came from the typewriter, as well as United States–based developers' experience working in monolingual professional environments—more specifically, in English and Roman script. Typewriter designers and computer programmers would share these assumptions with internet developers as well. Internet developers "were generally American, and were implicitly thinking only about how to facilitate communication in English." ASCII, the standard and purportedly universal text-transmission protocol introduced in the 1960s, initially offered nothing for languages with accents and non-Roman scripts.[31] Unlike the typewriter, computer word-processing programs promised instead many fonts and font sizes with a click of the mouse. (Some typewriters permitted formatting or emphasis shifts—italics and underlining, for example—but required the user to remove one "daisy wheel" and insert another when switching between italicized and regular text.)

An American University of Beirut research group described their efforts in the early 2000s to collaborate with MIT colleagues in developing a multilingual sustainability web platform that, in part, required the use of Lotus Notes, an IBM software platform for business that included email, calendar, contacts, discussion forums, file sharing, instant messaging, and database capabilities. The team reported that "the Arabic language version of this software was not fully developed" and had multiple flaws, stemming from the fact that Windows did not yet support Arabic fonts.[32] In order to design fairly simple features like website buttons that Arabic-speaking users could click on for more information about a particular aspect of the initiative, the team laboriously pasted buttons and Arabic text into Word documents, imitating an Arabic graphical software interface. Even in the mid-2000s, Microsoft noted that its operations management suite "officially [did] not support . . . bidirectional, complex script languages, such as Arabic or Hebrew."[33] The same issues from a century before—a script running right to left, with numerals running left to right, and letters with multiple forms—remained stumbling blocks even for Microsoft's robust server management suite.

The technical challenge, as described in a conceptual 1987 article on Arabic word processing, was twofold: doing the technical development work to create joinable letters and, as Becker put it, "intermixing scripts that run in opposite directions," which he described as "a design puzzle."[34] Development efforts appear to have focused more on the first challenge than the second. For example, by 2001 Microsoft had developed a list of standard features for the Arabic script, derived from its OpenType font, Arabic Typesetting.[35] The script features included not only the basic isolated, initial, medial, and final forms of each letter but also diacritical mark and alternative ligature (connecting) positions—suggesting a sophistication previously unseen. These efforts—to enable the use of Arabic-script typing rather than mixed-script typing, or the ability to move between multiple scripts in the same document—had mixed results. Users continued to report challenges using Arabic script in Office programs, and the Office approach to Arabic script was fairly well reflected in the title of its font: typesetting, which suggested its conceptual debt to typewriting and the printed word.

The issues with Arabic script in Microsoft-based word-processing applications have shifted over time but have not disappeared, particularly for Mac users. An English-language Apple discussions thread from January 2012, which began with novice user mka8b4, represents numerous similar discussions. "Just asking how to type Arabic letters in word, in more specific how to type letter or paragraph in Arabic language in Microsoft winword," mka8b4 posted. "Your question refers to the generic problem of right-to-left (RTL) scripts," replied a more advanced user. "Did you activate Arabic keyboard layout in System Preferences . . . ?" the responder continued. "If Word cannot handle RTL, you must choose something else; Nisus or Mellel [two word-processing systems that focus on multiscript and multidirectional functionality] are your best choices." Another advanced user, identified as Tom Gewecke, added: "No version of MS Word for Mac has ever supported Arabic. You must use another app." Picking up the thread again in August 2013, another user suggested that the issue was whether a particular font supported Arabic, but Gewecke argued that "MS itself agrees if you ask them that MS Word for Mac does not support Arabic (or any Indic script)," but that if the user could create an Arabic-script document in Word for Windows, they could open it successfully in Word for Mac.[36] Five years later, a science-writing blog offered tips to manage the "bugs" that remained in Microsoft Word, suggesting that users add two additional toolbars in order to manage left-to-right and right-to-left formatting errors. The fix would "need to be done for each document in which you encounter this issue," the writer cautioned, "but once you've done it, that style version should stay with the document even if you send it to someone else."[37]

Critics of Word for Mac abound. "The issue is not that Apple doesn't support Arabic language (it does)," noted Arabic–English translation firm Industry Arabic in January 2013. "The problem is that Microsoft Office for Mac is not built to work with right-to-left languages." The firm suggested several work-arounds, including importing the document to Google Drive and editing it as a Google Doc, purchasing an Office clone like NeoOffice, or operating Windows through a virtualization program. "It remains frustrating that Mac users are forced to resort to these clunky work-arounds to work professionally in the world's fifth most spoken language," the firm concluded.[38] Microsoft Office's bidirectional capabilities have improved considerably since then, although it was not until March 2016 that Microsoft announced a security update that it said would allow Word, PowerPoint, and Excel to "fully suppor[t] Arabic and Hebrew."[39] Microsoft offers detailed instructions for changing various Word features for right-to-left languages—including column orientation and cursor movement.[40] Yet users continued to experience difficulties, as this August 2017 query indicates:

> When I'm working in English in MS Word for Mac (v 15.32), and switch to another application like a web browser in which I use Arabic, and then go back to Word, the document is no longer in English (I mean the language indicator on the bottom on the window). The keyboard layout is not Arabic, however, I cannot type left-to-right English in the doc. The only solution I've found is to save, close, and reopen the file, and even then I have to delete some text.[41]

What is being described here is a bidirectional language-enabled copy of Office for Mac, in which toggling between Word and another Arabic-enabled application confuses the orientation and encoding standards of the Word document. This in turn suggests that Office continues to have difficulty moving back and forth between uni- and bidirectional languages, and also that the bidirectional modifications remain unstable.

This is an important point. That the world's dominant word-processing software has difficulty accommodating bidirectional and unidirectional scripts in the same document is a roadblock for the many people around the world whose life and work bridge script directions. It has produced notable gaffes in public signage, as the well-known Tumblr "Nope, Not Arabic," which highlights Arabic word-processing errors, attests.[42] The regular misunderstandings that arise between different linguistic cultures multiply when the very people who can work in both languages literally cannot type in both within the same document. This hurdle mandates awkward work-arounds. "Whenever I need to include Arabic in a presentation handout," notes one scholar who gives presentations in English on Arabic and Hebrew materials, "I take a photograph of the text and insert it into a Word document as a jpeg."[43] Despite enormous complexities and advances in other fields, word processing still largely assumes that the word is not only linear—it is unidirectional.

Where you buy your computer matters. As Knut Vikør, professor of Islamic history at the University of Bergen, notes in his The Arabic Mac resource website, "On computers bought in the West, Arabic is normally just an extra language that is added"—just an add-on to "your basically English-European Mac."[44] And if it is not "added on" as part of the capability of each program used, it remains an invisible part of the computer. "Arabic capabilities do not reside in any particular program[;] they are generic to your computer itself." Vikør blames cost-savings logics: "It has to do with saving money . . . [and] many companies cannot be bothered." Those who might purchase a specialized program like Nisus Writer or use the free processors OpenOffice or Pages should recognize them as work-arounds, he notes, since sending a file created in those programs to someone using Word or another Arabic-incompatible word processor will likely strip the original document of its bidirectional functionality.

The 2017 release of new typefaces in Arabic (or, as in the Dubai font launched for Microsoft Office 365, simultaneously in Roman and Arabic script) redresses the early twentieth century framing of the "problems" of Arabic script, and a new belief in the positive power of font design. Sheikh Hamdan bin Mohammed bin Rashid al Maktoum, Crown Prince of Dubai, who spoke at the font's launch, stated that the preciseness and distinctness of the design licensed "each human being" their "right to expression," leading to an increase in "tolerance and happiness."[45] The Dubai font is claimed to be the first font "simultaneously designed in both Arabic and English" for general usage, with the capacity to operate in twenty-one additional languages.[46] In other words, users must only toggle between keyboards in order to switch languages within the same document, rather than switching fonts; the line spacing is wide enough to accommodate both scripts without awkward intervals between Arabic and English text.[47] Microsoft has set the font to install automatically in Office 2016's base programs—Word, Outlook, Excel, and PowerPoint—for Windows and Mac users, and notes that it "works just as any other font in Office. Simply select the font from the font picker in your application to apply it."[48] Its clean, simple lines were intended to be legible and appropriate for multiple kinds of documents, as well as to privilege neither language in terms of aesthetics or design choices.[49] It was intended as a showpiece, highlighting the ability of one font to serve two scripts, in two directions—and to enhance Dubai and the United Arab Emirates' reputations on the global tech stage.[50] And, in a way, it does. In 2017 a "solution" arrived. It is now possible to type bidirectionally in one document—thanks to this innovation from the Gulf.

BEYOND TECHNICAL FIXES: WHAT ARE THE BETTER QUESTIONS?

The Dubai font casts new light on a 2003 special issue of the *Journal of Computer-Mediated Communication*. Editors Danet and Herring asked how people "communicating online in languages with different sounds and different writing systems adapt to ASCII environments." Recognizing the then-dominant position of ASCII, they focused on users of non-English and non-Roman-script languages and their efforts to adapt to ASCII protocol, rather than vice versa. "What problems do they encounter," they continued—a potentially technical question about communication glitches— "and what are the social, political, and economic consequences if they do (or do not) adapt?" ASCII's difficulty with accented letters and with non-Roman scripts, they acknowledged, were technical problems that produced social, political, and economic consequences. Danet and Herring's queries focused on internet communication rather than word processing. However, their final question highlights the limits of developing technical fixes for problems that stem from cultural assumptions. "Is there evidence of 'typographic imperialism'?" they asked, referencing Robert Phillipson's work on linguistic imperialism. Developing born-bidirectional fonts like the Dubai font may address some of the technical issues with bidirectional and unidirectional script switching. They do not address the social or political hegemonies undergirding the English-language, Roman-script dominance of computer programming, the internet, and computing infrastructure.[51]

Are born-multiscript fonts like Dubai the most appropriate or most likely locus of change for the future? Can they successfully address user difficulties in switching between languages and scripts, whether between applications, within applications, or within the same document? Or is it the operational platform that requires closer scrutiny—especially its creators' underlying assumptions? Although the technical issues may appear resolved, the questions remain. Type design is too often considered primarily in aesthetic terms, and not enough for its support of underlying cultural logics of space, direction, orientation, and power. Type norms reflect our cultural assumptions; we need to address them in order to have more than piecemeal solutions to the challenges that people who live and work across multiple script forms face every time they fire up their laptops. As Hicks argues in this volume, powerful technologies often reinforce rather than relieve inequalities—to "preserve existing hierarchies and power structures."[52] American consular officials began promoting the typewriter in the Arab world during the imperial era, while word-processing software began circulating during the late Cold War; both technologies reflect their times and origins. It's time to bring in more self-reflexive, less reductive ideas about script forms and script directions.

The Arabic script today is used in the official languages of twenty-six Arabic-speaking countries, as well as in a modified form in three Persian-speaking countries (Afghanistan, Iran, and Tajikistan) and two Urdu-speaking countries (India and Pakistan). It is additionally used to write other national or nationally recognized languages, including Uighur in China. While spoken and written in various communities elsewhere, recognition as an official language makes it more likely that government offices and businesses need to produce documents, and accept documents, produced in Arabic or modified Arabic script. Hence they are more likely to require word-processing and other document-production programs that support the use of Arabic script, either exclusively or in conjunction with other scripts or character languages. Individual users who wish to use Arabic-script languages for social purposes may be more likely to turn to transliteration or to communicate via social media, suggesting diverging needs for various Arabic-script users in the twenty-first century. They also provide a reminder that just as technological problems can have societal roots, they may also have societal solutions.

Several newer platforms that fall under the broad rubric of social media seem to operate with much greater fluidity in multiple scripts and languages. Twitter, for example, permits users to switch almost seamlessly between scripts within the same tweet.[53] Facebook also supports multiscript communication. These newer platforms have evolved to offer users greater flexibility and higher functionality across multiple languages and scripts. What will it take for word-processing software to do the same? It might be that word processing, with its roots in typewriting and its founding assumptions about English, unidirectional script, and single-script fixity, is a dying fire, while more nimble platforms for human communication are igniting. They may spark a more agile concept of "document"—and the creative people who envision and develop that document may need to come from physical and conceptual locations beyond Western Europe and North America. Kavita Philip argues in this volume that we need "new narratives about the internet, if we are to extend its global infrastructure as well as diversify the knowledge that it hosts." The same is true for the more foundational infrastructure that makes it possible for us to share knowledge by putting words on screens and pages.[54] Asking better questions can help us blaze a trail away from the culturally laden assumptions that produced technology mired in unidirectional and monoscript norms toward the reality in which many people live and work across multiple script forms.

NOTES

1. Mar Hicks, *Programmed Inequality: How Britain Discarded Women Technologists and Lost Its Edge In Computing* (Cambridge, MA: MIT Press, 2017).

2. See Halcyon Lawrence, "Siri Disciplines," this volume.

3. *The Story of the Typewriter, 1873–1923* (Herkimer, NY: Herkimer County Historical Society, 1923), 64–70.

4. *Story of the Typewriter*, 69.

5. *Story of the Typewriter*, 126–128.

6. *The Story of the Typewriter* stated that no functioning typewriters had yet been designed for ideographic languages like Chinese. See Thomas S. Mullaney, *The Chinese Typewriter: A History* (Cambridge, MA: MIT Press, 2017).

7. "Typewriters for Various Nations," *New York Times* (January 18, 1903), 27.

8. "The Syrian Society of New York," *Stenographer* 4, no. 1 (May 1893): 210.

9. Gabriel Bie Ravndal, Consul in Beirut (November 4, 1902), *Consular Reports: Commerce, Manufactures, Etc.* 71, nos. 268–271 (January, February, March, and April) (Washington: Government Printing Office, 1903).

10. Thomas S. Mullaney, "Typing Is Dead," this volume.

11. Selim S. Haddad, Types for Type Writers or Printing Presses. US patent 637,109, filed August 12, 1899, issued November 14, 1899.

12. "Typewriter Prints Turkish," *Washington Post* (November 10, 1908), 2.

13. "An Inventor from Mount Lebanon," *New York Times* (August 21, 1904), SM8.

14. This point is often highlighted in Arabic textbooks. See, e.g., Sebastian Maisel, "Numbers," *Speed Up Your Arabic: Strategies to Avoid Common Errors* (Oxford and New York: Routledge, 2015), 113.

15. Arthur Rhuvon Guest and Ernest Tatham Richmond, Types for Arabic Characters. US patent 639,379, issued December 19, 1899.

16. For a brief discussion of these stylistic or font standards, see Joseph D. Becker, "Arabic Word Processing," *Communications of the ACM* 3, no. 7 (July 1987): 600–610.

17. Seyed Khalil, Typewriting Machine. US Patent No. 1,403,329, filed April 14, 1917, issued January 10, 1922.

18. Lawrence, "Siri Disciplines."

19. "3000 Typewriters with Turkish Alphabet Shipped to Aid Kemal Westernize His Land," *New York Times* (July 31, 1929), 27.

20. "New Typewriter for Arabs Shown," *New York Times* (November 3, 1957), 121.

21. Uri Kupferschmidt, "On the Diffusion of 'Small' Western Technologies and Consumer Goods in the Middle East During the Era of the First Modern Globalization," in *A Global Middle East*, ed. Liat Kozma, Cyrus Schayegh, and Avner Wishnitzer (IB Tauris, 2014), 229–260.

22. Kupferschmidt, "On the Diffusion of 'Small' Western Technologies," 233.

23. See, for example, Relli Shechter, *Smoking, Culture, and Economy in the Middle East* (IB Tauris, 2006).

24. Kupferschmidt, "On the Diffusion of 'Small' Western Technologies," 239.

25. See https://www.microsoft.com/typography/fonts/font.aspx?FMID=877.

26. Seyed Khalil, Typing Machines for Arabic Group Languages. US Patent 2,940,575, filed December 19, 1957, issued June 14, 1960.

27. "Electronics Make Typing Arabic Easier," *New Scientist* 60, no. 827 (November 15, 1973): 483.

28. Regarding the quote in the section title, see Ramsey Nasser, "Unplain Text," *Increment* 5 (April 2018), https://increment.com/programming-languages/unplain-text-primer-on-non-latin/.

29. Kupferschmidt, "On the Diffusion of 'Small' Western Technologies," 239.

30. M. R. Bhagavan, "Technological Leapfrogging by Developing Countries," in *Globalization of Technology*, ed. Prasada Reddy (Oxford: EOLSS Publishers, 2009), 48–49.

31. Brenda Danet and Susan C. Herring, "Introduction: the Multilingual Internet," *Journal of Computer-Mediated Communication* 9, no. 1 (November 2003), https://doi-org.du.idm.oclc.org/10.1111/j.1083-6101.2003.tb00354.x.

32. Toufic Mezher, "GSSD-Arabic," in *Mapping Sustainability: Knowledge e-Networking and the Value Chain*, ed. Nazli Choucri et al. (Dordrecht: Springs, 2007), 123–142.

33. Kerrie Meyler and Cameron Fuller, *Microsoft Operations Manager 2005 Unleashed* (Indianapolis: Pearson Education, 2007), 180.

34. Becker, "Arabic Word Processing," 605–606.

35. See "Features for the Arabic script," https://www.microsoft.com/typography/otfntdev/arabicot/features.htm, last updated 2002.

36. See "Just asking how to type Arabic letters in word," https://discussions.apple.com/thread/3684500?start=0&tstart=0.

37. See Amanda W., "Fixing right-to-left text in Microsoft Word," BioScience Writers (November 7, 2018), https://www.biosciencewriters.com/Fixing-right-to-left-text-in-Microsoft-Word.aspx.

38. See http://www.industryarabic.com/arabic-word-processing-guide-on-mac/, first published January 14, 2013. The page references a Facebook group and petition effort to convince Microsoft to develop an Arabic-language compatible Office for Mac, but as of August 2017, both had been removed from Facebook.

39. "MS 16-029: Description of the security update for Office 016 for Mac: March 16, 2016," accessed August 28, 2017, https://support.microsoft.com/en-us/help/3138327/ms16-029-description-of-the-security-update-for-office-2016-for-mac-ma.

40. See, for example, "Type in a bi-directional language in Office 2016 for Mac," accessed August 28, 2017, https://support.office.com/en-us/article/Type-in-a-bi-directional-language-in-Office-2016-for-Mac-d7bb1d52-4d82-4482-910f-d74b7c3bd468.

41. Private communication, August 28, 2017.

42. See https://nopenotarabic.tumblr.com/. In general, the submissions show disconnected (freestanding form) Arabic letters, often also reading left to right.

43. Private communication, September 11, 2017.

44. See "An Introduction to Writing Arabic on the Mac," Knut Vikør, in *The Arabic Macintosh*, https://org.uib.no/smi/ksv/ArabicMac.html.

45. "Hamdan bin Muhammad Launches 'Dubai Font' into the World" [in Arabic], Al Bayan (May 1, 2017), https://www.albayan.ae/five-senses/mirrors/2017-05-01-1.2932379.

46. Dubai Font, "Hamdan bin Mohammed Introduces Dubai Font to the World and Directs Dubai Government Entities to Adopt It in Their Correspondence," press release (April 30, 2017), accessed September 13, 2017, https://www.dubaifont.com/pressCenter/2-hamdan-bin

-mohammed-introduces-dubai-font--to-the-world-and-directs-dubai-government-entities--to -adopt-it-in-their-correspondence.html. A critical piece by The National noted that the government of Abu Dhabi had in 2010 commissioned the first Arabic-English font, the Zayed; however, that font is proprietary and not available to the general public. See Nick Leech, "The Fine Print behind Dubai's New Font," *The National* (May 11, 2017), https://www.thenational.ae/uae/ the-fine-print-behind-dubai-s-new-font-1.12697.

47. "How do I access the Latin or Arabic letters in the font?," Dubai Font FAQ—Technical, accessed September 13, 2017, https://www.dubaifont.com/download.

48. "Using the Dubai Font in Microsoft Office," Microsoft Office Support, accessed September 13, 2017, https://support.office.com/en-us/article/Using-the-Dubai-Font-in-Microsoft-Office-c862df16 -ae0d-46d9-b117-aa3f41f9706e.

49. See "The Fine Print behind Dubai's New Font."

50. "Hamdan bin Muhammad."

51. Danet and Herring, "Introduction." See Robert Phillipson, *Linguistic Imperialism* (Oxford: Oxford University Press, 1992) and *Linguistic Imperialism Continued* (Hyderabad/New York: Orient Blackswan and Routledge, 2010).

52. See Mar Hicks, "Sexism Is a Feature, Not a Bug," this volume.

53. See https://support.twitter.com/articles/119137.

54. See Kavita Philip, "The Internet Will Be Decolonized," this volume.

11

YOU CAN'T MAKE GAMES ABOUT MUCH

Noah Wardrip-Fruin

It seems like there are games about a lot of things, if you look at the cover art and read the press releases. But when you play games, you may begin to realize that we can't make games about much.

The realization may even start when you're a kid. Your family might have a set of games in the cabinet that proclaim themselves to be about being part of a boy band, or going fishing, or acquiring high-flying internet startups of the dot-com era. And these might all be versions of *Monopoly*.

When you get one out of the cabinet and start setting it up, you see what the game proclaims itself to be about. The members of One Direction are on the front of the box. Or the pieces you move include a tackle box, boat, and leaping fish. Or the names of the properties include Lycos, Ask Jeeves, and iVillage.com.

But then you start to play. And things stop making sense.

When you play the common version of *Monopoly*, the players are clearly competing property developers. Each attempts to control neighborhoods, develop properties within them, and thereby drive up rents.

But who are we supposed to be when playing the One Direction edition? Some of the spaces are labeled with songs the band has recorded. Others with charity benefits they've played. Yet others with the reality show for which they separately auditioned. Are we members of the band, competing with one another? Are we agents and promoters, vying for portions of the band's money? Fans competing to control their legacy? There's no way to think about play that results in the hodgepodge of labels making sense.

Playing the bass fishing edition might, at first, feel a little more sensible. If we imagine ourselves as anglers, we might encounter all the things on the board, which include different kinds of rods, reels, and bait, as well as competitions such as the Bass Masters Classic. But once we, for example, own all the different types of rods or contests, the game instructs us to develop them—building bait shops and marinas "on" things that clearly aren't physical locations.

The dot-com edition asks us to do something similarly nonsensical—I feel ridiculous putting a building down on top of Ask Jeeves as though it were a plot of land. Perhaps we could imagine this as further investment in the company, but Hasbro specifically suggests a less-sensible framing: "The game still has houses and hotels, but these now represent households and offices that are online." (2).[1] Yes, please pay me more for a visit to my search engine, now that it has become an ISP.

Rather, the only sensible way to look at these games is that they are still the common version of *Monopoly*—still, as Hasbro puts it, the same "Property Training Game." To think sensibly about our play, we need to imagine it in real estate terms, not those of music, or sport, or the 1990s World Wide Web.

In other words, these games aren't actually about what they say they're about. Even a child can realize it. And yet they proclaim themselves so loudly—the pictures, the words, the shapes of the pieces—that we are taken in, over and over.

This chapter is about the current limits on what games can be about—limits that come from the available vocabulary of ways that games operate and communicate. It's also about the vast gulf between what can be said in this vocabulary and what games promise to be about. In this way, this chapter has a similar goal to others in this book. Sarah T. Roberts, for example, exposes the gulf between the presumed automation of social media content moderation and its human reality. Similarly, Safiya Umoja Noble discusses the gulf between the imagined perfection and objectivity of many technologies (from facial recognition to predictive policing) and their inaccurate and biased realities. Mitali Thakor uses the complex relationship between human experts and software, in the case of high-stakes image classification, to outline the gulf between the imagined future in which all labor is automated and the complex sociotechnical realities ahead of us. And there are many other examples between these covers—because tracing such gulfs is one of the foundations of the work in *Your Computer Is on Fire*.

These gulfs matter for a variety of reasons. We must consider the gulf between rhetoric and reality in games, at least in part, because media are key to how we understand ourselves and our world. As computer games become an increasingly important form of media, we must understand their capabilities and limitations more deeply. We must understand why the considerations and debates they promise to ignite so often fail to catch fire, and how their affordances can be used and expanded by those who wish to spark experiences such as learning, empathy, and rejection of injustice.

"SKINNING" GAMES

The practice that gives us endless editions of *Monopoly* is called *skinning*. When we skin a game we take its particular game design—its particular way of operating—and, by changing surface-level things such as images and text, give it the appearance of being about something else.

This can work well, if the underlying operations are compatible with the new skin. For example, when Monopoly is reskinned to represent property development in a different city. As of this writing, Wikipedia lists more than forty US cities for which this has been done. It's also been done for US states and other geographic regions. I've played a couple of these, which usually retain exactly the same board layout with different property names, and I can attest that it feels just like playing the common *Monopoly*.

But skinning can also be more ambitious, while still remaining sensible. It is possible to change what a game is about through skinning—or take a game that doesn't appear to be about anything and give it a clear, sensible theme. For example, you can create versions of *Tetris* that are about death, as we see in examples ranging from those imagined by famous game designers Raph Koster and Clint Hocking to that of the comedy group Monty Python.

DEADLY TETRIS

Koster, in his 2004 book *A Theory of Fun for Game Design,* offers his version of *Tetris* as a thought experiment:

> Let's picture a mass murder game wherein there is a gas chamber shaped like a well. You the player are dropping innocent victims down into the gas chamber, and they come in all shapes and sizes. There are old ones and young ones, fat ones and tall ones. As they fall to the bottom, they grab onto each other and try to form human pyramids to get to the top of the well. Should they manage to get out, the game is over and you lose. But if you pack them in tightly enough, the ones on the bottom succumb to the gas and die.[2]

The game functions exactly the same way, but the player's efficiency is now in the service of one of humanity's great crimes, rather than simply toward a higher score, longer play session, or personal feeling of skill or immersion. Of course, for experienced *Tetris* players, the pull of old motivations would still be present, in conflict with awareness of the task it now serves, as Koster found in 2009 when he played a version of the game implemented by a Brazilian game club:

> I actually didn't finish the first game I played, because of the decapitated heads at the bottom, and then cramming an upside-down person into the gap. Curiously, I started out playing it just as Tetris—and in fact, the tug of doing so was remarkably strong. But the "dressing," the art and especially the sound, was so strong that after a while, I couldn't ignore it, and it made me uncomfortable.[3]

On one level, this certainly demonstrates that abstract games like *Tetris* can be skinned to present something specific. At another level, however, it is clear that the traditional feeling of, and approach to, playing *Tetris* and what is presented in Koster's horrific "skinning" are at odds with each other.

From Hocking's point of view, this tension is part of the point. His version of *Tetris,* presented in a talk at the 2011 Game Developers Conference, takes inspiration from Brenda Romero's game *Train*.[4] *Train* positions players as packing humans into boxcars for the Nazi regime. When each line is "cleared," it represents a car that has left the station toward a death camp.

Hocking presents three ways to play the game: You can play as a dutiful servant of the Reich, which means playing in the normal, score-optimizing way (toward which both the game's design and its familiarity pulls). Or you could play in a role like Oskar Schindler's, optimizing the way you lose the game—carefully leaving one empty space in every row, so that the system backs up. Or you could play as a saboteur, stacking pieces as quickly as possible to reach "game over."

Hocking's point is that the latter two ways of playing would make no sense in the normal *Tetris*. By reskinning the game, even though the underlying systems are the same, he's changed how it is meaningful to play—and made how to play into a meaningful choice.

Of course, skinning doesn't have to leave every element of the original game system unchanged, as we see in *Monty Python & the Quest for the Holy Grail*.[5] In this 1996 title's "Drop Dead" minigame, we play a version of *Tetris* skinned as packing plague victims into a mass grave. Most of the shapes act like normal *Tetris* pieces. But the "I" piece, in which four tiles are stacked in a line—the piece around which experienced *Tetris* players organize their play, in order to get the bonus that comes from clearing four lines simultaneously—rotates itself outside the player's control while declaring, "I'm not dead!" This one small change makes traditional strategic *Tetris* play impossible, revealing the situation as ridiculous (and perhaps horrifying) even to players who have fallen under *Tetris*'s thrall.

BUSINESS BEJEWELED

While *The Quest for the Holy Grail* plays its change of gameplay for laughs, the reskinning of another famous tile-matching game shows how it can be used for critical ends. *Layoff* was conceived by Mary Flanagan and Angela Ferraiolo of Tiltfactor Lab (at Dartmouth College), and created by Tiltfactor and the Rochester Institute of Technology games program, as part of multi-institution research on values in games.[6] It was released in March 2009, at the height of the financial crisis, garnering significant media attention and over a million players its first week.[7]

Layoff employs a version of *Bejeweled*'s gameplay, which takes place in a grid of objects. The player may swap the positions of any two adjacent objects, as long as the result is at least one "match" of three or more objects of the same type. Matched objects disappear, the player scores points (and may receive additional bonuses for matching more than the required three objects), and the remaining objects move down the grid (to be replaced by more coming in from the top). *Bejeweled*'s objects are colored jewels, with the colors being arbitrary (there is no reason for players to prefer one color over another), and with no distinction made between individual jewels of the same type (except that some occasionally carry bonuses).

Layoff, in contrast, uses pattern-matching gameplay to present something much more specific. Each object is an individual worker. Mousing over a worker reveals a capsule story about them—their name, age, what they do, and so on. The player is put in the position of trying to match the workers for "workplace efficiency adjustments." These make the matched workers disappear, as well as adding to the group of people milling around the unemployment office shown at the bottom of the screen. In between the grid of workers and the unemployment office is a text ticker displaying messages such as: "The American economy lost 2.6 million jobs in 2008, but the average performance-based bonus for 132 of the largest US companies increased to $265,594."

The unemployment office and text ticker are not among the elements that are usually displayed together with *Bejeweled*'s object grid. But they leave its fundamental gameplay unchanged. However, *Layoff* then goes further, introducing a new kind of object—suit-wearing workers. When these workers are moused over, no biography is revealed. Instead, the player sees a quote that appears to be from the suit-wearing worker's perspective, such as: "Customers of the new merged corporation should feel secure. Your investments are safe in our hands." And most importantly, suit-wearing workers cannot be matched. They can only be moved to make the layoff of another type of worker possible, and they can never be laid off—even when the movements of other workers result in a match. Given this, successfully playing *Layoff* eventually results in a grid dominated by suit-wearing workers, above a very full unemployment office, with no further changes possible.

How this gameplay feels seems to depend on where you sit. Writing on games site *Kotaku*, Michael McWhertor says: "It's not only educational and entertaining, it's also depressing."[8] On the other hand, Cindy Perman, writing on business-oriented site *CNBC*, suggests:

> it's a fun way to pass the time. And, the next time you're in a bookstore or drugstore and there are a gaggle of employees behind the counter but it's still taking a ridiculously long time, just close one eye and with your thumb, line them up and lay them off.[9]

In a sense, this is the same point that Hocking made. Once one of these games is reskinned to be about doing something negative to people, then how you choose to play becomes a window into your views on those people and that situation.

But I want to make a larger point. There is a reason that *Tetris* has been reskinned, repeatedly, as a game about mass death. There is a reason that *Layoff* is about disposing of workers rather than creating a space for them to flourish. While we can make games appear to be about different things through reskinning, we also cannot get away from what is fundamentally present in the underlying systems. *Tetris* is a game about objects, and if we reskin it to be about people, we will only be able to create games about situations in which people are treated like objects.

If we try to do something else, we will end up with the game system and its skin at odds, such as the nonsensical editions of *Monopoly*. And then, in order to play, players will have to increasingly "look past" the skin so that they can effectively engage the game. We won't actually have produced a game about what the skin proclaims.

PLAYABLE MODELS

Of course, skinning is only one way that a game can be based on another. For example, the FPS (first person shooter) game genre came to prominence with the runaway success of 1993's *DOOM*. After this, its creators (id Software) had a successful business selling its underlying technology, and that of its successor titles, as an "engine" on which others could make games.

We could call these games reskins of *DOOM*, but in most cases that would be misleading. Not only are the graphics changed but so are the geography of the levels, the behavior of enemies, the types of weapons, and so on. Rather, these games are building on *DOOM*'s particular implementations of a set of *playable models*. In particular, they build on *DOOM*'s models of spatial movement and combat, which are designed to complement each other.

From descriptions, you might think these games are quite different from each other. *Heretic* is described as a dark fantasy game about the struggle between the magic-wielding Serpent Riders and the Sidhe elves.[10] *Strife* is described as a science-fiction game about a ragtag resistance movement fighting The Order, a totalitarian regime that took power after a catastrophic comet impact.[11] *Chex Quest*, a game distributed in boxes of cereal, is described as presenting the Chex Warrior's heroic rescue of Chex people from the Flemoids, evil slime creatures.[12]

But what makes these games work is that they are all about moving through constrained spaces and shooting at enemies. Whether one plays an elf, a resistance fighter on a distant planet, or a human encased in cereal armor, the skin/theme is carefully chosen to work with *DOOM*'s particular playable models. In short, while it might seem that a well-developed set of playable models are an invitation to make a game about anything, in fact they are an invitation only to make things that fit in with the gameplay those models enable. Building on the combination of models popularized by *DOOM* isn't a route to an effective game about romance, philosophy, prejudice, or most other things we might care about.

Which isn't to say people won't try, or at least say they're trying. Perhaps the most successful example of this is the work of game designer Ken Levine, his colleagues at Irrational Games, and the publicists who have promoted their work for 2K Games.

PHILOSOPHICAL SHOOTERS

Levine came to prominence as the creative lead for *BioShock*, a game widely discussed as a critique of Ayn Rand–style Objectivist philosophy.[13] *BioShock* builds on an approach to telling stories through spatial models often called "environmental storytelling." As players move through the game, they are exploring an underwater city, Rapture, designed as an Objectivist utopia. They can see—through structural collapses, broken-down equipment, graffiti, discarded protest signs, and more—that the experiment failed. They can discover audio diaries, and listen to them as they move through the space, that tell more of the story of the city and its inhabitants. In short, the spatial model of the FPS was successfully repurposed to reveal a story, through exploration, that portrays the failure of a set of economic/philosophical ideas (ones which are quite influential in US politics). As portrayed by the game's publicists, this was catnip for journalists and reviewers, especially those who had been waiting for games to "grow up" as a medium.

But as anyone who has played the game knows, you don't spend much of your time thinking about Objectivism. *BioShock* is not a game about philosophical debate. It's a shooter—with the twist being that you can turn your player character into a weapon (through genetic modification) rather than only using regular weapons. It is the combat model that dominates gameplay. The spatial model serves primarily as its support, and only in the breaks between battles does the environmental storytelling have any prominence. But you wouldn't know this from most cultural commentary on the game.[14]

Leading up to the release of Levine's successor title, *BioShock Infinite*, it seemed likely it would get away with the same trick.[15] Again, the prerelease reviews were stellar, and the game was described as dealing with difficult issues at the heart of the US—this time, racism and xenophobia—as found in its airborne city of Columbia. But after release, when people really started to play it, the reception got a lot rockier. As Joe Bernstein summarized for *BuzzFeed News*: "While much of the enthusiast community has received the game as an unqualified triumph, the more thoughtful people who work in gaming have argued over every inch of it. People have pointed out problems with its politics, its violence, its racial politics, its plot, its gaminess, its metaphysics, the way it made them feel."[16]

While one might argue that problems with *BioShock Infinite* stem from issues such as being philosophically muddied, or placing less emphasis on the environmental storytelling that was key to the prior title's success, or many other things, I believe the core problem was its gameplay. Not that the gameplay was bad, but that FPS gameplay is as poor a fit for a theme of racism and xenophobia as *Monopoly* gameplay is for the theme of being a boy band. Journalist and critic Leigh Alexander experienced this, even though she had been particularly excited for the game after an interview with key studio staff at Irrational (Levine, director of product development Tim Gerritsen, lead artist Shawn Robertson, and art director Nate Wells).[17] Alexander describes her thoughts about the game this way:

When does a supposed essay on the purposelessness of conflict simply become purposeless? Racism, corpses, endless slaughter—all the things that are supposed to remind me of how horrible Columbia is only makes me think of how horrible games are. There are more dead bodies in this world than live people, and ever moreso [*sic*] as the game progresses.[18]

Or as Bernstein, who liked the game much better, puts it, "the rules of the genre are at odds with the very magnificence of Irrational's game . . . the very last thing I wanted to do as I played this game was to sprint around finding cover and chaining headshots."[19] This is not to say that games cannot be made about ideas. Rather, the ideas must be consonant with the playable models that enable the core gameplay. And the fact is that the combat model that dominates FPS gameplay is not compatible with a story about interacting with anything human, unless our role is to play a soldier, assassin, or psychopath.

This is why environmental storytelling often works better without a combat model added to the spatial model, without "finding cover and chaining headshots." Art and indie game studios such as Tale of Tales, The Chinese Room, Fullbright, and Galactic Cafe have explored this in the genre sometimes called the "walking simulator."

But when traveling through a walking simulator, if the goal is to tell a story of a traditional shape, the space must be contorted to fit—as seen in the strange, circuitous route through the house of *Gone Home* and the spaces that wink in and out of existence in *The Beginner's Guide*.[20] If we actually want players to be able to use the spatial model for more open exploration, environmental storytelling is probably better suited to the kinds of experimental story shapes pioneered by the literary hypertext community, as seen in the town geography of *Marble Springs* or the spaces of the body in *Patchwork Girl*.[21]

All that said—though FPS games are ill-suited to human interaction, and walking simulators have more often focused on storytelling than philosophy—it is not impossible to make a successful game focused on philosophical issues. In fact, the common version of *Monopoly* is a rip-off of just such a game.

PLAYING MONOPOLIST

What happens when you play *Monopoly*? You move your piece around the periphery of the board, using a simple model of space (a track) that you move down a randomly determined number of spaces (by rolling a die). Early in the game, the randomness feels harmless, and you have various decisions to make: Do you buy the properties on which you land? Do you bid on the properties other players don't buy (if your family follows the auction rule)? How much money do you save for future rounds, when players who land on owned properties will have to pay rent?

All of these, of course, are decisions about how much of one resource to convert (or plan to convert) into another. And it is these resources that are at the heart of all versions of *Monopoly*. That is the nature of its playable model.

This becomes more apparent in the middle game, as players begin to invest resources in developing their properties, so that landing on them becomes a much bigger resource drain. But to do this development, players have to own all the properties of a particular set (usually represented by color). This is the last of the main elements of *Monopoly*'s model of real estate: pattern matching. Throughout the middle game, players negotiate over properties, trying to balance using resources to acquire complete sets (which are opportunities for development), using resources to develop these sets, and hoarding resources for potential rent payments to other players. The randomness, which previously felt harmless, now seems more consequential: whether you land on another player's property, or they land on yours, will now make or break your plans.

As the game enters its end phase, the resource system reveals the positive feedback that lies within it. One player lands on another player's highly developed property. The result is that they must mortgage some of their properties or trade them away, reducing their opportunities to collect rent. Randomness becomes all—the positive feedback makes it likely that a player in the lead will win, but it plays out slowly and unpredictably as the rolls of the die determine who loses resources and who collects them.

And this is the point. *Monopoly* is based on *The Landlord's Game*, which was created at the turn of the twentieth century as "a practical demonstration of the present system of land-grabbing with all its usual outcomes and consequences," as characterized by its designer, Elizabeth Magie.[22] The best-known version of Magie's story is told by Mary Pilon in her book, *The Monopolists*. Pilon traces Magie's outlook back to the economic philosophy of Henry George, who argued that people should own the products of their labor, but everything natural—including land—should belong to everyone equally. According to Pilon, Magie taught classes about this philosophy after work, and saw in games "a new medium—something more interactive and creative" for expressing these ideas.[23] In order to dramatize the difference between a Georgist approach and the current one, Pilon reports,

> [Magie] created two sets of rules for her game: an anti-monopolist set in which all were rewarded when wealth was created, and a monopolist set in which the goal was

to create monopolies and crush opponents. Her dualistic approach was a teaching tool meant to demonstrate that the first set of rules was morally superior.[24]

In its systems, the common version of *Monopoly* is based on the second of these rule sets. To put it another way, *Monopoly* is a rip-off of a game that was designed to ask a question: Do we really want a handful of wealthy people, chosen by chance, to drive the rest of us into desperation?

It is these systems, and their question, that shine through—which is why so many *Monopoly* games are never completed. It stops being fun when the realities of its resources and randomness confront us, when we experience its economic philosophy. And it doesn't matter what the supposed version is "about." It might be funny if someone is bankrupted by landing on a property bearing the logo of the long-dead AltaVista, but no one around the table thinks this makes it a fundamentally different game from the version with fish.

Instead, everyone around the table is learning (perhaps unconsciously) that one of the things it is possible to make a game about is *resources*. If they're feeling particularly thoughtful, or if they know the game's history, they may realize that it is possible to make a game about unfair resource systems, perhaps one that reflects those in our own society. Perhaps they may relate it to video games that do this, such as *Cart Life*, a game about the precarious position of street vendors.[25] Or perhaps they may relate it to the many, many games—both physical and digital—that present the avaricious accumulation of resources as the ultimate good.

HUMAN RESOURCES

Just as with the playable models of space and combat that are at the heart of the FPS genre, there is a temptation to believe that the resource models that underlie games like *Monopoly* and *Cart Life* could be used to make a simulation of anything. For example, if we look at *The Sims*, one of the most successful game franchises of all time, we appear to see a game about middle-class, suburban life in the US. Sometimes it has an obvious focus on resources, as when building, remodeling, and outfitting the homes that are key to maintaining the happiness for each group of Sims. But at other times it does not. In more recent entries in the series, Sim families grow up together, with children passing into adulthood and perhaps forming families of their own, while older family members enter old age and then pass away.

So one could argue that *The Sims* titles are one of the few places that video games have taken on the deeply human topic of the passage of time. But at their cores, the characters are a collection of resource meters. A child Sim might feel sad after grandma has passed away, but neither the character nor the game knows or can say anything about that history. Instead, the child just goes to find something in the house to refill her needs meters—perhaps watching TV. In doing so, she's not suppressing feelings she's going to have to deal with later. She's fully meeting what the game systems represent as her need to feel happier.

In short, *The Sims* might be read as a game about the passage of time. But if we read it that way, it says pretty awful things. It's probably better read as what it is—a game about building dollhouses, using human-looking tokens to score our current creations and collect upgrade resources for the future. Which is why we don't feel too bad when we set the tokens on fire, or trap them in the pool, or do other things it would be monstrous to do to actual humans or even well-drawn fictional characters.

This is not to say that we can't imagine a version of *The Sims* that actually captured the bittersweet experience of the passage of time in an extended network of family and friends. But this would need to be a game in which the foundations of the playable model were made up of things like histories and emotions, rather than resources. Similarly, if we were to imagine a game about philosophical issues that are less economically focused than those in *The Landlord's Game*, its playable model would need a foundation other than resources. Which brings us to the question of whether it is possible to make such games, and why.

OPERATIONAL LOGICS

The foundation of any playable model is a set of operational logics.[26] For example, the primary foundation of *Monopoly* is its resource logics, though a pattern-matching logic determines which properties can be further developed. Similarly, the spatial models of shooters and walking simulators have logics such as navigation, movement physics, and collision detection as a foundation.

An interesting thing about these logics is that they cut across a divide in how we often talk about games. Many discussions of games, including the earlier sections of this chapter, make a division between what a game purports to communicate (its skin or theme or fiction) and how it operates and enables play (its rules or mechanics or models). But the foundational logics of games are both at the same time. We can only understand a logic such as "collision detection" by seeing it as something that communicates ("virtual objects can touch") and something that operates algorithmically ("when two entities overlap, do something").

From one perspective, operational logics are a site of remarkable flexibility and creativity. On a technical level, a logic such as collision detection can be implemented in many ways—from the 2-D hardware implementation of the Atari VCS to the 3-D quadtrees or raycasting that might be found in current software. And on a communicative level, collision detection might be used to represent a wall we can't pass through, a bullet hitting a body, or something more unexpected. Game designers have found a remarkable range of things to communicate with this logic, from Pac-Man's eating to the deflecting of hate speech found in Anna Anthropy's *Dys4ia*.[27]

But from another perspective, the very fact that logics communicate something in particular (e.g., the virtual touch of collision detection) means that they can't arbitrarily communicate anything. At the furthest, they can communicate things we understand metaphorically, such as the "deflection" of a statement.

And while some would argue that, therefore, metaphor is the key to making games on a wide array of topics, the metaphorical approach actually has severe limitations. This is because of the nature of metaphor, as explored in conceptual metaphor theory (CMT). Take the metaphor "an idea is a building." For this metaphor, it makes sense to ask questions such as, "What is the foundation of this idea?" But it doesn't make sense to ask, "What is the HVAC system of this idea?" This is because the metaphorical mapping is to an idea of a building that is highly abstract and schematic, without specifics such as HVAC systems.[28]

On the other hand, when we are building a game, the metaphorical mapping is made to every specific element of the game. Getting all these specifics to work appropriately for the metaphor is very challenging, and can prove impossible, as recounted in the development stories in Doris C. Rusch's *Making Deep Games*.[29] This is what makes games that manage a complete metaphorical mapping, such as Jason Rohrer's *Passage*—with each collision with a wall, chest, or person readable both as spatial movement and as a larger life event—so widely discussed.[30] And this is why all such games tend to be quite short, whether they are *Dys4ia*'s microgames or *Passage*, with its five-minute play time.

FLEXIBLE LINKS

Looking at the landscape of games, one might find hope in a different direction: linking logics, which create networks of connections between game spaces, texts, and other objects. We see such logics in a wide variety of places: cave entrances in *The Legend of Zelda*, dialogue trees in *Star Wars: The Old Republic*, textual links in

Twine titles such as *Howling Dogs*, and so on.[31] Because the communicative role of linking logics is so abstract—"this is connected to that"—we can imagine using it for a game about nearly anything.

And in fact the literary hypertext community I mentioned earlier has shown how links can be used to tell stories about a vast number of things. But this is not the same as enabling play about these things. Rather, in most hypertext fiction the audience does not think of itself as taking action within the fictional world. Instead, they are developing an understanding of the fiction (just as with a novel or film) with links allowing and shaping their explorations. In addition, some hypertext works attempt an interesting hybrid—primarily telling a story but using hand-created links (rather than a game system) to offer players key choices about what happens.

Going further, there are also more systematic elements of games implemented as links. A common example is movement—from the textual interactive fiction tradition founded by *Adventure*'s cave system to the evocative island of *Myst* or the world-spanning map of *80 Days*.[32] For this to be successful requires a very consistent use of links, both in how they are presented to players and in how they operate. A common beginner mistake is to implement link-based doors that, unintentionally, don't operate in both directions.

Of course, there are also other ways to implement movement in games. And the problem comes when we try to use links to get beyond what other logics offer, such as toward broader agency in the fictional world, of the sort we might imagine when truly playing a character. For example, when we play *Star Wars: The Old Republic*, one character option is to be a Han Solo–style smuggler. We can imagine ourselves cleverly improvising our way through various situations—but the game never lets us use our quick wits, instead taking us down predetermined story paths. And because such paths are time-consuming to hand author, there aren't very many of them.

This works better in a game like *Howling Dogs*, in which the player character is presented as imprisoned and transported into VR fantasies, with a barely hinted route to potential agency. Or in one like *80 Days*, in which we play a character constrained by servitude, etiquette, outsider status, and lack of knowledge of the surroundings. Such games offer opportunities for reflection and critical thinking, much like the literary hypertexts that preceded them, but they don't offer much of the opportunity to experiment and understand through action—through play—that is key to games from *DOOM* to *Layoff*.

Given this, if we want to create games that are played and are about something, we have no choice but to grapple with the limited number of well-developed operational logics. It is this constraint that limits the set of playable models we can construct, and thereby limits the set of things we can meaningfully make games about.

PAST THE LIMITS?

Of course, as with any time we're told about a constraint, we do have other choices if we redefine the problem. For example, we could make games that are primarily conduits for human communication and creativity, such as tabletop and live-action role-playing games (RPGs). These can offer powerful experiences about topics ranging from our understanding of history (*Microscope*) to the workings of colonialism (*Dog Eat Dog*).[33]

It is only when we want a game that can be enjoyed on its own (like most forms of media) or without requiring extraordinary creativity on the part of players (like nearly all games that aren't RPGs) that the limited set of well-developed logics and models presents a problem. Which is, of course, most of the time.

So what do we do?

One option would be to demand more from the big game companies. Yet this is like saying we should demand more from the part of the film industry that gave us the *Transformers* franchise. We certainly should hold their feet to the fire for the racist, sexist, classist, colonialist ideas embedded within their playable models and fictions—as David Leonard does with the construction of race and racism in *Grand Theft Auto: San Andreas*.[34] And we should treat big-budget, popular games as part of culture, as ideologically revealing as the most popular television or music—as Soraya Murray does in *On Video Games*.[35] But we shouldn't expect studios in search of the next big, safe entertainment to also be seeking new playable models, and new logics to enable them. In deference to their owners and existing audiences, they will largely keep using the familiar models, themed with variations on the same handful of fictional situations, and we will need to look elsewhere for alternatives.

Another option is to look toward the burgeoning scene of indie and art game creators. These creators also face challenges in creating new logics and models, but for a different reason: they generally must create games with limited resources, and often limited time.[36] Nevertheless, these creators have consistently demonstrated that there are many important human experiences that can be presented in terms of spaces or resources or pattern matching or links or one of the few other well-developed types of logics—or that can be evoked by play in these terms—that have been little explored in games. We are even seeing new genres, such as the walking simulator, created through alternative uses of mainstream playable models.

But what if we wanted a genre such as the *talking* simulator? Or games meaningfully engaging, through play, with ideology, history, identity, or resistance? What if we wanted games able to grapple with the key experiences of being human—from caring about someone to finding a way of communicating that caring, from believing in something to taking action in a world of complexity and consequence.

For this we would need a different type of institution, one capable of doing the long-term work of attempting to create new logics and models meant to support new play experiences, and developing experimental games to guide and validate them. Perhaps one might imagine this coming from a company such as Microsoft, which has a culture of research and multiple major game studios within it. But in actuality, Microsoft shut down its Studios University Program, which was one of the few real efforts in the industry to connect research and game making.

Given this, our primary hope appears to be the same place from which so many earlier pioneering efforts in games emerged—from the first video game (Christopher Strachey's *MUC Draughts*) to the first spatial combat game (*Spacewar!*), and from the founder of the game/story hybrid (*Adventure*) to the founder of the generative Rogue-like genre (*Rogue*).[37] All of these, of course, came from universities. And on this front, the news is mixed. Games are now increasingly the subject of academic programs, but these programs are often shockingly vocational.[38] We don't need colleges and universities to teach students to do what the big game companies are already doing. They were doing it just fine before there were academic programs. We need higher education to help students learn to make games that aren't already being made. And we need the university to turn to games as an area of research, to reclaim its role from the moment when video games first developed—to be the place from which new logics, models, and game designs founded upon them emerge.

Until then, you can't make games about much.

NOTES

1. Hasbro Games, rule book for Monopoly Brand Property Trading Game from Parker Brothers: The .com Edition, tabletop game (Hasbro, Pawtucket, RI).

2. Raph Koster, *Theory of Fun for Game Design* (Paraglyph, 2004), 168.

3. Raph Koster, "ATOF Tetris Variant Comes True," *Raph's Website* (blog) (February 13, 2009), https://www.raphkoster.com/2009/02/13/atof-tetris-variant-comes-true/.

4. Clint Hocking, "Dynamics: The State of the Art," In GDC Vault, video record of the Game Developers Conference 2011, San Francisco (February 2011), http://www.gdcvault.com/play/1014597/Dynamics-The-State-of-the; Brenda Romero, *Train*, tabletop game (April 2009).

5. 7th Level, *Monty Python: The Quest for the Holy Grail*, Microsoft Windows 95 (1996).

6. Mary Flanagan, Angela Ferraiolo, Greg Kohl, Grace Ching-Yung, Peng, Jennifer Jacobs, and Paul Orbell, *LAYOFF*, Tiltfactor Lab and the Rochester Institute of Technology (RIT) Game Design and Development Program (Tiltfactor, 2009), http://www.tiltfactor.org/game/layoff/; Tiltfactor, "Tiltfactor | LAYOFF," accessed March 29, 2018, http://www.tiltfactor.org/game/layoff/.

7. Tiltfactor, "Tiltfactor | LAYOFFs—1 Million Players in the First Week; at GDC," accessed March 29, 2018, http://www.tiltfactor.org/layoff-1-million-players-in-the-first-week-at-gdc/.

8. Michael McWhertor, "Layoffs: The Video Game Is Depressing Fun," *Kotaku* (February 16, 2009), https://kotaku.com/5170770/layoffs-the-video-game-is-depressing-fun.

9. Cindy Perman, "Layoffs: The Videogame," *CNBC* (February 20, 2009), https://www.cnbc.com/id/29775890.

10. Raven Software, *Heretic*, MS-DOS 5.x. id Software (GT Interactive, 1994).

11. Rogue Entertainment, *Strife*, MS-DOS (Velocity Inc, 1996).

12. Digital Cafe, *Chex Quest*, Microsoft Windows 3.1 (Ralston-Purina, 1996).

13. 2K Boston and 2K Australia, *BioShock*, Microsoft Windows XP (2K Games, Inc., 2007).

14. Even Hocking, who offered a then-rare critique of the fit between its elements—"Bioshock seems to suffer from a powerful dissonance between what it is about as a game, and what it is about as a story"—and coined the term "ludonarrative dissonance" in his critique, doesn't mention how much more it focuses on shooting than storytelling. Likely due to how (sadly) common this is. Clint Hocking, "Ludonarrative Dissonance in Bioshock," *Click Nothing* (blog) (October 7, 2007), http://www.clicknothing.com/click_nothing/2007/10/ludonarrative-d.html.

15. Irrational Games, *BioShock Infinite*, Microsoft Windows 7 (2K Games, Inc., 2013).

16. Joseph Bernstein, "Why We Can't Talk about 'BioShock Infinite,'" *BuzzFeed News* (April 15, 2013), https://www.buzzfeed.com/josephbernstein/why-we-cant-talk-about-bioshock-infinite.

17. Leigh Alexander, "Irrational Games, Journalism, and Airing Dirty Laundry," *Gamasutra* (February 19, 2014), https://www.gamasutra.com/view/news/211139/Irrational_Games_journalism_and_airing_dirty_laundry.php.

18. Leigh Alexander, "BioShock Infinite: Now Is the Best Time," *Leigh Alexander: On the Art, Culture & Business of Interactive Entertainment, Social Media and Stuff* (blog) (April 11, 2013), https://leighalexander.net/bioshock-infinite-now-is-the-best-time/.

19. Joseph Bernstein, "Why Is 'Bioshock Infinite' a First-Person Shooter?" *BuzzFeed News* (April 1, 2013), https://www.buzzfeed.com/josephbernstein/why-is-bioshock-infinite-a-first-person-shooter.

20. The Fullbright Company, *Gone Home*, Microsoft Windows XP (2013); Everything Unlimited Ltd., *The Beginner's Guide*, Microsoft Windows Vista (2015).

21. Deena Larsen, *Marble Springs*, Apple System 6 (Eastgate Systems Inc., 1993); Shelley Jackson, *Patchwork Girl*, Apple System 6 (Eastgate Systems Inc., 1995).

22. Mary Pilon, *The Monopolists: Obsession, Fury, and the Scandal behind the World's Favorite Board Game* (London: Bloomsbury, 2015), 30.

23. Pilon, *The Monopolists*, 30.

24. Mary Pilon, "Monopoly's Inventor: The Progressive Who Didn't Pass 'Go,'" *New York Times* (February 13, 2015), https://www.nytimes.com/2015/02/15/business/behind-monopoly-an-inventor-who-didnt-pass-go.html.

25. Richard Hofmeier, *Cart Life*, Microsoft Windows 95 (2011).

26. While others may use similar phrasing (for example, Sarah T. Roberts's discussion of the "operating logic of opacity" in "Digital Detritus"), I use the term "operational logics" to refer to a particular idea that I have been developing since 2005, largely in collaboration with colleagues at UC Santa Cruz. Some key publications in this development are: Noah Wardrip-Fruin, "Playable Media and Textual Instruments," *Dichtung Digital* (2005), http://www.dichtung-digital.de/ 2005/1/Wardrip-Fruin/index.htm; Michael Mateas and Noah Wardrip-Fruin, "Defining Operational Logics," in *Proceedings of the Digital Games Research Association* (Brunel University, London, 2009); Mike Treanor, Michael Mateas, and Noah Wardrip-Fruin, "Kaboom! Is a Many-Splendored Thing: An Interpretation and Design Methodology for Message-Driven Games Using Graphical Logics," in *Proceedings of the Fifth International Conference on the Foundations of Digital Games* (Asilomar, CA, 2010); Joseph C. Osborn, Dylan Lederle-Ensign, Noah Wardrip-Fruin, and Michael Mateas, "Combat in Games," in *Proceedings of the 10th International Conference on the Foundations of Digital Games* (Asilomar, CA, 2015); Joseph C. Osborn, Noah Wardrip-Fruin, and Michael Mateas, "Refining Operational Logics," in *Proceedings of the 12th International Conference on the Foundations of Digital Games* (Cape Cod, MA, 2017); Noah Wardrip-Fruin, "Beyond Shooting and Eating: Passage, Dys4ia, and the Meanings of Collision," *Critical Inquiry 45,* no. 1 (September 1, 2018): 137–167. For more in Roberts's concept, see: Sarah T. Roberts, "Digital Detritus: 'Error' and the Logic of Opacity in Social Media Content Moderation," *First Monday* 23, no. 3 (2018), http:// dx.doi.org/10.5210/fm.v23i3.8283.

27. Anna Anthropy (a.k.a. Auntie Pixelante) and Liz Ryerson, *Dys4ia*, Adobe Flash (March 9, 2012).

28. Raymond W. Gibbs Jr., "Evaluating Conceptual Metaphor Theory," *Discourse Processes* 48, no. 8 (October 27, 2011): 529–562, https://doi.org/10.1080/0163853X.2011.606103.

29. For example, Rusch tells the story of the group that would eventually make the game Akrasia (Singapore-MIT GAMBIT Game Lab, Microsoft Windows XP, 2008). They had a set of concepts that most attracted them: "identity, memory, inner demons, and love." But the group didn't end up making a game about any of them, because they couldn't figure out where to start with the metaphorical mapping. So they decided to make a spatial game about addiction. Their first design made addiction the space of a racetrack, and the goal to get off the track. But they had not mapped dependency into the spatial model in any way (and had made the interesting spatial activity the one players should reject). Their second design made addiction the space of a maze, adding a dragon to be chased (for a drug-induced high) and rewards when the dragon was caught. This was too successful at getting players to chase the dragon, so they added a resource system representing health, which lowered when the dragon was caught. But then it was easy to ignore the dragon, so they added a demon that chases the player, representing withdrawal, and were tempted to make the demon also cause health damage. But they realized this didn't make sense to them metaphorically, so instead the demon changes the navigation controls when it catches the player. The team stopped there, but even this final iteration is problematic, which Rusch admits freely. Doris C. Rusch, *Making Deep Games: Designing Games with Meaning and Purpose* (Boca Raton, FL: Focal Press, 2016).

30. Jason Rohrer, *Passage*, Microsoft Windows XP (2007), http://hcsoftware.sourceforge.net/ passage/.

31. Nintendo Co., Ltd., *The Legend of Zelda*, Nintendo Entertainment System (Nintendo of America Inc., 1987); BioWare Austin, LLC, *Star Wars: The Old Republic*, Microsoft Windows 7 (Electronic Arts, Inc., 2011); Porpentine, *Howling Dogs* (2012), http://aliendovecote.com/uploads/ twine/howling%20dogs.html.

32. William Crowther and Donald Woods, *Adventure*, Digital Equipment Corporation PDP-10 (Bolt, Beranek and Newman, Stanford University, 1975); Cyan, Inc., *Myst*, Apple System 7 (Brøderbund Software, Inc., 1993); inkle, *80 Days*, Apple iOS 8 (2014).

33. Ben Robbins, *Microscope: A Fractal Role-Playing Game of Epic Histories*, tabletop role playing game (Lame Mage Productions, 2011); Liam Burke, *Dog Eat Dog*, tabletop role playing game (Liwanag Press, 2012).

34. David Leonard, "Young, Black (& Brown) and Don't Give a Fuck: Virtual Gangstas in the Era of State Violence," *Cultural Studies ↔ Critical Methodologies* 9, no. 2 (April 1, 2009): 248–272, https://doi.org/10.1177/1532708608325938.

35. Soraya Murray, *On Video Games: The Visual Politics of Race, Gender and Space* (London: I. B. Tauris, 2017).

36. However, it should be noted that some indie games have the character of ambitious ongoing research, of the sort from which new logics could emerge. A particular hotbed is the "roguelike" scene of generative games, as exemplified by efforts such as *Dwarf Fortress* and *Caves of Qud* (Tarn and Zach Adams, *Slaves to Armok: God of Blood Chapter II: Dwarf Fortress*, Microsoft Windows XP [Bay 12 Games, 2006]; Freehold Games, LLC, *Caves of Qud*, Apple Mac OS X V10.10 [2015]).

37. Christopher Strachey, *MUC Draughts*, Manchester Mark I (Manchester University, 1951); Stephen Russell, Peter Samson, Dan Edwards, Martin Graetz, Alan Kotok, Steve Piner, and Robert A. Saunders, *Spacewar!*, Digital Equipment Corporation PDP-1 (Massachusetts Institute of Technology, 1962); Michael Toy, Glenn Wichman, and Ken Arnold, *Rogue*, BSD Unix 4.x (University of California Santa Cruz & Berkeley, 1980).

38. This vocational focus is reinforced by the only widely reported ranking of academic programs in games, from The Princeton Review, which appears to assume the role of the academy is to train people for the jobs of today, rather than create new knowledge for the future. For example, the survey from which the rankings are derived asks how many students are placed in the current game industry, but doesn't ask about research output—so the kinds of fundamental advances in game concepts and technology that might be reported in books, journal articles, and conference papers have no role. As a result, the "best" game graduate program in the 2018 rankings is Southern Methodist University's Guildhall, which undertakes no research and describes itself as "Built by the industry, for the industry." See TPR Education IP Holdings, LLC, "Top Game Design Press Release | Public Relations | The Princeton Review | The Princeton Review," accessed March 30, 2018, https://www.princetonreview.com/press/game-design-press-release; Southern Methodist University, "About | SMU Guildhall," accessed March 30, 2018, https://www.smu.edu/Guildhall/About.

III

WHERE WILL THE FIRE SPREAD?

12

CODING IS NOT EMPOWERMENT

Janet Abbate

"Learn to code and change the world!"[1] Does this slogan sound familiar? In the 2010s, the tech industry and nonprofits such as Code.org, Girls Who Code, and #YesWeCode popularized the notion of coding as a way to empower women, African Americans, Latinxs, and other groups underrepresented in computing. Even US President Barack Obama participated in an "Hour of Code" in 2014, posing for photos with young Black students and declaring, "Learning these skills isn't just important for your future—it's important for our country's future."[2] Media stories and widespread hands-on activities reinforce the message that teaching kids to code will solve the tech industry's diversity problems and ensure well-paying jobs for all Americans.[3]

Are these realistic claims that should guide education and workforce policy? Or does the focus on technical skill as a solution to social and economic disparities sidestep more awkward—but urgent—discussions about biases and misconceptions in tech? Mastering simple programming commands and seeing an immediate response on the screen can be exciting and fun for kids, and learning to code may even, with time and further training, lead to a well-paying job in computing. But the popular rhetoric of coding as empowerment glosses over some serious obstacles on the road to employment or entrepreneurship. First, the focus on introductory training hides the industry's failure to hire, retain, and promote already-trained women and minorities. Blaming the educational pipeline for not supplying underrepresented workers deflects attention from the discrimination still faced by those who emerge from it. Second, technical training by itself does not challenge race- and gender-biased ideas about what technically skilled people look like and what constitutes "merit" in the tech workplace. These biased standards perpetuate hiring and promotion disparities. Third, the coding movement's focus on acquiring skills avoids culturally loaded questions about the purpose of software and the social goals it might serve. The preferences of a privileged few tend to shape what types of software are created and made widely available, so that the needs of people lacking access to social and economic services are overshadowed by apps promising entertainment or faster shopping. If we do not take these issues of power and purpose seriously and acknowledge that real solutions may require dominant groups to make uncomfortable changes, the coding movement may simply reproduce current inequities.

DECODING SUPPLY-SIDE ARGUMENTS: IS THE INTERNET A SERIES OF PIPELINES?

Like California's wildfires, the software industry's lack of diversity has generated a lot of heat but no real solutions. NSF data show that between 1993 and 2013, the percentage of women in computer science occupations actually declined from 31 percent to 24 percent.[4] Scandals rocked Silicon Valley in 2017: former Uber engineer Susan Fowler divulged a pattern of sustained and egregious sexual harassment at that company; leading venture capitalists were ousted for sexual misconduct; and a viral Google memo argued that "biological causes . . . may explain why we don't see equal representation of women in tech and leadership."[5] But data points to society—not biology—as the cause, since the proportions are very different in some other countries: in Malaysia, for example, half of all computer science degrees are earned by women.[6] For years the big tech companies resisted disclosing their embarrassingly low diversity numbers, until civil rights activist Reverend Jesse Jackson appeared at a Google shareholders' meeting in May 2014 and demanded the release of this data. Tellingly, Google's agreement may only have happened because there was an African American at Google with enough power to arrange it: David Drummond, "the company's only black high-level executive" and "an old friend of the reverend."[7] Google's disclosure broke the silence, and further pushing from Jackson and his allies led Apple, Facebook, Amazon, Microsoft, and Twitter to follow suit. While everyone seems to agree that the US tech industry's diversity numbers are abysmally low, there is no consensus on the causes and possible cures.

Could early exposure to coding overcome these racial and gender gaps? The logic of this claim rests on an assumption that the main reason women and minorities

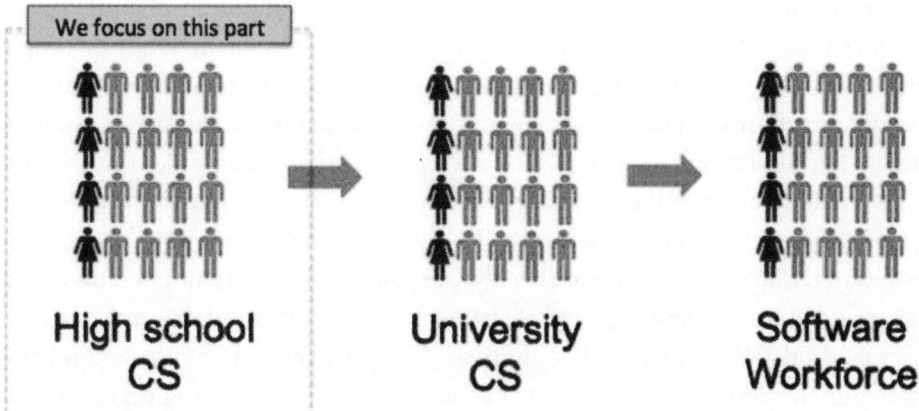

Figure 12.1   The pipeline model. (*Source*: Code.org.)

are underrepresented in computing is that they fail to enter the "pipeline" of technological training at an early enough age, which prevents them from acquiring the
necessary credentials to land a job. Code.org, for example, focuses its efforts on "the
education pipeline," where, it claims, "the diversity problem begins"; its website
displays a diagram with arrows pointing to a simple linear progression from "High
School CS" to "University CS" to "Software Workforce" (fig. 12.1).[8]

But the pipeline argument is contradicted by both data and logic. If underrepresentation were simply a supply problem, women and minorities with CS training
would be snapped up by employers, but they are not. In fact, a 2016 study found
that Blacks and Latinxs were hired by big tech companies at only half the rate at
which they received CS degrees.[9] Girls Who Code founder Reshma Saujani notes
that the "pipeline" loses women and minorities between training and hiring: "If you
were to look at the gender makeup and the racial makeup up of technical interns
versus who gets hired [afterward], there's a huge drop."[10] Rachel Sklar, an activist
for women in tech, calls attention to a double standard whereby women on venture
capital boards are expected to have technical degrees while men are not. "There are
a ton of qualified women out there, women who can measure up to any similarly-
situated man—as long as they're being judged by the same criteria." Sklar adds that
industry support for coding efforts, while valuable, is not a substitute for reform:
"Support for next-gen orgs like Girls Who Code and Yes We Code and Code 2040
and Black Girls Code is fantastic but does not address the very real problems with
this gen."[11]

The pipeline argument implicitly blames women and minorities themselves for not preparing sufficiently. It also reinforces stereotypes about the kinds of people who succeed in tech, equating the masculine pattern of early, obsessive interest in computers with talent and commitment. Influential venture capitalist Paul Graham exemplifies this circular logic in a 2013 interview, saying, "If you go look at the bios of successful founders this is invariably the case, they were all hacking on computers at age 13." Asserting that "if you want to be really good at programming, you have to love it for itself," he assumes that this love for coding will manifest itself at a young age, leading him to dismiss the people who do not learn to program until college— disproportionately women and minorities. This privileged viewpoint ignores the possibility that lack of early involvement could be due to lack of resources or encouragement, rather than lack of interest or talent. Not all thirteen-year-olds have free rein to hack on computers, and young women in particular may be initially turned off by negative stereotypes about computing.[12] While the coding movement will likely help more girls and minority youth spark an interest in coding at a young age, we should beware of assuming that *only* those who show an early interest have the potential to succeed, or that good coders can only be motivated by a love for hacking itself. Like other skilled professionals, such as lawyers or doctors, programmers can be driven to excellence by a desire to serve human needs rather than a fascination with the technical details of their work.

By focusing attention on earlier stages of the pipeline, comments like Graham's shift the responsibility to secondary education: "What that means is the problem is 10 years upstream of us. If we really wanted to fix this problem, what we would have to do is not encourage women to start startups now. It's already too late. What we should be doing is somehow changing the middle school computer science curriculum."[13] A survey of tech company founders by venture capital firm First Round found a similar tendency for male respondents to shift the blame for "tech's diversity problem" from gender bias to early training: "Men are more likely to blame the pipeline into tech; women place greater emphasis on unconscious bias and lack of role models."[14] As Shanley Kane, a tech veteran and cofounder of feminist website *Model View Culture,* argues, "You see venture capitalists talk about the need to get more 10-year-old girls into programming, and that's so far removed from their direct sphere of influence. Meanwhile, there is attrition in every stage of the career path of women once they get into the industry. . . . We are not getting hired, and we are not getting promoted."[15] Sreela Sarkar (ch. 14 in this volume) describes a similar dynamic in India, where training programs that promise to "flatten economic and social barriers for low-income youth" are undermined by the tech industry's bias toward applicants from elite castes and universities. While improving STEM education for girls and African American and Latinx youth is surely desirable, this alone will not result in equality unless the behavior of employers and startup funders is also addressed.

Unconscious bias in hiring is a well-established phenomenon. Numerous studies have found that even employers who advocate for diversity will favor applicants who appear white or male over applicants who have female or "ethnic-sounding" names.[16] In Silicon Valley, race and gender characteristics are combined with behavioral stereotypes to create a model of the ideal coder; this practice is often referred to by the technical term "pattern matching" to give it a gloss of insight and objectivity.[17] Venture capitalists such as John Doerr have openly acknowledged that they favor candidates who fit a particular profile, noting that the founders of big companies such as Amazon, Netscape, Yahoo, or Google "all seem to be white, male, nerds who've dropped out of Harvard or Stanford and they absolutely have no social life. So when I see that pattern coming in—which was true of Google—it was very easy to decide to invest."[18] The tech industry has responded to accusations of exclusion and mistreatment of women and minorities through widespread adoption of implicit bias training programs, yet most of these techniques have no research demonstrating their effectiveness. Such training may even make matters worse: a 2014 study found that when individuals were informed that stereotyping was prevalent, they treated it as normal and felt freer to express biased views and treat others in discriminatory ways.[19]

A more promising alternative is to identify and combat bias in the *practices* involved in recruiting, hiring, promotion, and funding. The problem begins with recruiting. Big tech companies recruit heavily from elite universities like Stanford, Berkeley, or Carnegie Mellon, rather than historically Black colleges and universities or liberal arts colleges where more women and minorities might be found.[20] Job ads use language meant to appeal to men, seeking "rock stars" and "ninjas" who can "crush code." Employers often seek job referrals from their current employees' social networks, which tends to reproduce existing race and gender imbalances. As Catherine Ashcraft of the National Center for Women & Information Technology argues, when employers use the pipeline argument, "It tends to mean that you're not looking in the right places. . . . You haven't done a lot of things to actively expand your networks and look at your job descriptions for bias and be more creative about where you're hiring."[21]

Interview practices often reflect and reward narrow and even counterproductive notions of skill. A standard part of the hiring process is the "whiteboard interview," in which interviewers pose a coding problem and the job candidate must stand at a whiteboard and write code to solve it, using algorithms drawn from memory. Whiteboarding has been criticized as irrelevant to real-world job requirements; as coding instructor Quincy Larson quipped, "The only world where you would actually need to be able to recall an algorithm would be a post-apocalyptic one, where the hard drives of all the computers connected to the internet were fried."[22] In addition, whiteboarding as an artificial, high-stress performance tests the candidate's self-confidence as much as their fitness for the job, and thereby favors white or Asian males who are bombarded with the message that they are naturally better at coding. Diversity consultant Joelle Emerson of Paradigm argues that "whiteboard interviews increase the likelihood of 'stereotype threat,' an anxiety among non-traditional candidates that they'll confirm negative stereotypes about their group."[23] In early 2017 a group of software experts drew attention to the way whiteboarding can unfairly reject good candidates, "confessing" in a series of tweets how they themselves—successful white men—would likely fail such a test. A typical tweet read, "I would fail to write bubble sort on a whiteboard. I look code up on the internet all the time. I don't do riddles." As white men, these coders knew that they would get the benefit of the doubt in interviews, but others would not be so lucky. Indeed, female coders who declined to "confess" pointed out that even lightheartedly admitting inadequacies was a male privilege: "As a woman in tech I don't announce my code shortcomings for fear of consequences"; "If I told you what things I still have to Google, you'd try to use it as 'evidence' that I'm not a 'real' engineer."[24] Potentially less discriminatory alternatives to the whiteboard ritual are available: they include having applicants present a coding project they previously completed; asking them to solve a real coding problem on a computer using the normal development tools; or having the candidate review an existing piece of code.[25]

Retention and promotion are further problems that fall within the employer's part of the pipeline. A 2017 study using a representative nationwide sample found that tech workers who voluntarily left jobs most often did so (42%) because of hostile work environments. The experiences most likely to drive workers to quit—stereotyping, bullying, sexual harassment, and being passed over for promotion—affected women, LBGT, and minority workers roughly twice as much as white males. The report concludes that "diversity numbers may not be changing at least in part because tech companies have become a revolving door for underrepresented groups."[26] The 2015

"Elephant in the Valley" survey found that senior women at tech firms experienced rampant sexist behavior that impaired their advancement and satisfaction at work. Eighty-four percent of women surveyed had been told they were "too aggressive"; 88 percent had had clients or colleagues address questions to male coworkers that should have been addressed to them; 60 percent reported unwanted sexual advances; 75 percent were asked illegal questions about marital status or children in interviews; and 66 percent felt excluded from key networking opportunities because of gender.[27] Combating entrenched discrimination means firms must hold employees accountable for bad behavior—even if they are star performers—and hold managers accountable for diversity results, not just productivity.[28]

The pipeline argument puts the onus on underrepresented groups to solve their own exclusion by learning computer skills at an early age. Coding efforts do have value: underrepresentation is a multipronged problem, and equal access to training is an important part of the solution. But the job-market advantages of white and Asian males go far beyond coding skills: they include fitting the stereotype of what computer expertise "looks like," attending elite schools, having the right social networks, and conforming to the industry's cultural codes. Teaching employers to recruit, hire, promote, retain, and fund minority coders might prove more empowering than teaching minorities to code.

BELIEF IN MERITOCRACY CONSIDERED HARMFUL

"The central issue to the lack of diversity in tech is the tenacious myth that Silicon Valley and tech more broadly are perfect meritocracies."[29] This analysis by Freada Kapor Klein of the Kapor Center for Social Impact critiques the oft-heard message that hiring and promotion are driven solely by talent and performance. While many Americans express faith in economic mobility and a level playing field, this faith seems especially fervent in the tech sector, in part because it rides on a cultural belief that machines are objective and socially neutral.[30] Code.org's Hadi Partovi has argued, "It is one of these skills where the computer doesn't care if you are black or white or a girl or boy or old or young: as long as you know how to speak the language of code, the computer will do what you want."[31] This faith in the fairness of machines is naïve; indeed, when computers are used to select job candidates, the result can be *increased* discrimination. Algorithms are trained using past decisions as models; if employers have historically discriminated against job applicants with "ethnic-sounding" names or degrees from women's colleges, the computer may flag these candidates as less qualified, perpetuating social bias while appearing neutral.[32] Computer experts may also feel that they themselves are strongly rational and objective by virtue of their technical training or scientific outlook. As Jacob Kaplan-Moss, an open-source software pioneer, comments, "Programmers like to think they work in a field that is logical and analytical, but the truth is that there is no way to even talk about programming ability in a systematic way. When humans don't have any data, they make up stories, but those stories are simplistic and stereotyped."[33] As Mar Hicks observes (ch. 6 in this volume), sexist views of merit have become so ingrained that gender disparities seem "natural and unchangeable."

The widely held belief that the tech industry is a meritocracy is both false and harmful. As described above, far from being judged solely on merit, women and minorities face obstacles and double standards at every stage of their educational and employment journey. Laura Weidman—cofounder of Code2040, which promotes Latinxs and African Americans in tech—argues that performance is a function of opportunity: "At Code2040 we believe strongly in the importance of making it on the strength of one's merits. However, the ability to do just that is predicated on having the opportunity to showcase one's merits in the first place." Organizations like Code2040 that base their training and advocacy on the real-life experiences of underrepresented groups can expose inequities that may be invisible to Silicon Valley insiders. Weidman points out how the "old boy network," in which "companies tend to look to their employees' networks for potential hires," reproduces the largely white/Asian male workforce. For minority candidates, "it can be hard to even figure out which companies are hiring or could be a good fit for you without someone to help you navigate the waters."[34]

The dangers of a misguided belief in meritocracy come in many forms. Kapor Klein points out that it has the effect of blaming the victims: "The implication of that is that there's something wrong with African Americans, Latinos, and women from all backgrounds, as opposed to there being something deeply biased about tech."[35] Minorities are judged as "less qualified" for not having had access to educational and work experiences that white male candidates and hiring managers take for granted. When privilege is equated with talent, "diversity" is framed as incompatible with merit, a "lowering of the bar." The racial logic whereby merit is equated with credentials inaccessible to many minority individuals is made painfully clear in a blog post by Leslie Miley, who had been the only Black engineering manager at Twitter until he quit in 2015. Explaining why he had left, Miley recalled:

Hiring Committee meetings that became contentious when I advocated for diverse candidates. Candidates who were dinged for not being fast enough to solve problems, not having internships at "strong" companies and who took too long to finish their degree. . . . A particularly low moment was having my question about what specific steps Twitter engineering was taking to increase diversity answered . . . with, "Diversity is important, but we can't lower the bar."[36]

If the exclusion of minorities is naturalized as reflecting their lack of merit, rather than a moral failing within the industry, then diversity initiatives can only be justified in economic terms—as a strategy to improve products or make companies more competitive. In diversity scholar Ellen Berrey's words, "Rather than a righteous fight for justice or effective anti-discrimination laws, we get a celebration of cultural difference as a competitive advantage. Diversity, we are told, is an end goal with instrumental pay-offs."[37] Programs to teach coding are often framed in those terms—as a business case rather than a fairness issue—and even groups serving minorities seem obliged to repeat that rationale.[38] While the business case for diversity may indeed be compelling, making profitability the main argument for inclusion is problematic. What if a business had no competitive advantage to gain from diversity—would that make discrimination acceptable? Refusing to recognize bias as an ethical problem draws attention away from needed changes in business practices and cultures.

Distorted ideas about meritocracy can foster hostile work environments in a number of ways. Tech companies sometimes interpret "meritocracy" to mean cutthroat competition between employees—exemplified by companies like Uber and Amazon, where zero-sum employee evaluations reward backstabbing coworkers.[39] Do such Darwinian practices really select for the best talent, or do they push out equally capable employees who are unable or unwilling to tolerate these aggressive interactions? Tech startup Zymergen illustrates the opposite approach to pursuing meritocracy. One-third of Zymergen's technical workforce is female—double the industry average—a feat the company managed in part by emphasizing collaborative skills as part of "merit" and valuing "experience and maturity over a hot-shot, one-man show." Zymergen CEO and cofounder Joshua Hoffman argues that his broader view of merit opens up a bigger pool of talent and thereby yields a higher-quality workforce.[40]

The unfounded belief that an organization or industry is meritocratic can lead not just to inaction on diversity but to increased discrimination. This "paradox of meritocracy" was demonstrated in a 2010 study showing that "when an organization is explicitly *presented* as meritocratic, individuals in managerial positions favor a male employee over an equally qualified female employee." The authors suggest that the mantle of meritocracy gives decision makers an unearned sense of "personal objectivity" and "moral credentials" that make them more likely to act on the biases they actually hold.[41] Paradoxically, believing that one's culture is prejudiced (implicit bias) and believing one's culture is *not* prejudiced (meritocracy) both lead to the same result—complacency—unless the desire for fairness is backed up by procedural changes and accountability.

Silicon Valley's self-image as a meritocracy rests on a lopsided view of merit that valorizes narrow technical skills at the expense of social skills that may be equally crucial to the success of a software project.[42] Indeed, social and technical skills are inseparable; communication and cooperation are required both to conceptualize and to successfully implement a software project. The "lone superstar" is wrongly valued over the team player, in part because of a widely held notion that some programmers are ten times as productive as others and therefore hold the key to a company's success. But skeptics have offered a number of reasons to question the existence or value of the "10x" coder. In the first place, coding productivity cannot be measured quantitatively, since what superior programmers produce is not necessarily *more* code but better code. A seemingly brilliant innovation may prove costly in the long run if it is difficult to maintain due to the coder's poor communication skills. And even if a programmer is truly a superstar, the potential benefit of this is lost if the programmer alienates teammates. Kaplan-Moss urges managers to reject "the myth of the brilliant asshole" or "'10x programmer' who is so good at his job that people have to work with him even though his behavior is toxic."[43] As open-source engineer Ryan Scott Brown told a conference audience, "If you have a 10x developer, who is also a 10x jerk and they have driven away nine contributors from your project, he's now a 1x developer, and still a jerk."[44]

The myth of the superstar coder encourages managers to reward men's "heroic" last-minute problem-solving over women's proactive efforts to prevent crises from occurring in the first place. Macho heroics that substitute for collaborative problem-solving have been debunked as dysfunctional since at least 1979, when software engineering pioneer Barry Boehm dubbed such behavior "the Wyatt Earp Syndrome."[45] Yet they are still overvalued as a sign of talent and commitment, as described by Ann Mei Chang, a former senior engineering director at Google. "The people who are most recognized and rewarded are often the guys who pull the all-nighter to fix the bug right before it's supposed to ship or go live. . . . Women are more likely to have tested stuff and worked hard all along the way to make sure there wouldn't be problems at the very end, and that kind of work style is not always as rewarded."[46] In addition to perpetuating counterproductive work styles, favoring heroic last-minute efforts penalizes women (and men) whose family commitments make all-nighters a hardship.

One of the biggest problems with defining merit in terms of heroic solo coding is that it undervalues essential people skills. Referring to the 2017 Google memo controversy, reporter Brian Feldman noted that its sexism reflected Silicon Valley's biased ideas of merit: "The computer industry is built upon the myth of a few anti-social hackers in a garage, creating entire worlds on their own. Qualities like narrowly defined rationality are prized, while qualities like empathy or social aptitude are regarded as irrelevant."[47] Kaplan-Moss argues that "coding is just a small part" of overall programming skill and that communication skills are equally important.[48] A similar belief motivated Zymergen executives to abandon the whiteboard interview: "It doesn't test for things that are important in a workplace. . . . We want to see how well do people perform in a team and perform real work."[49] Zymergen substituted job talks that allow candidates to present past accomplishments on real work projects and describe how they have handled collaborative challenges.

The importance of integrating interpersonal and algorithmic skills is acknowledged in some corners of the coding movement. For example, the curriculum for Girls Who Code Clubs includes typical elements like programming languages and computer science concepts, but also "soft skills" such as leadership, presentation skills, "willingness to ask for help," and "positive attitude."[50] Yet the fact that these vital skills are confined to girls only highlights how they are devalued in masculine computing culture. By and large, computer science classes and coding activities for children teach mastery of tools with little attention to collaborative skills. Focusing on the supposed technical deficiency of women and minorities, while ignoring the well-documented social and cultural deficiencies of the majority, reinforces the assumption that women and minorities need to be "fixed" but the white males who dominate tech are just fine. We have Black Girls Code, but we don't have "White Boys Collaborate" or "White Boys Learn Respect." Why not, if we want to nurture the full set of skills needed in computing?

PURPOSE: IF CODING IS THE ANSWER, WHAT IS THE QUESTION?

What motivates people to learn to code? Many people who love computers from an early age find learning to program intrinsically fascinating and rewarding. Activities aimed at children, such as Hour of Code or coding camps, appeal to that sense of wonder and the thrill of mastery, and for recreational hackers, fun can be enough of a purpose. Silicon Valley caters to the sensibility of workers who are motivated to create great code and make money but may not care which (if any) end users benefit from their efforts. But other potential software developers are much more concerned with the ultimate purpose of their products, and this tends to be especially true for women and minorities. UC Berkeley was able to equalize male and female enrollment in an introductory computer class in part by placing "more emphasis on the impact and relevance of computing in the world."[51] NCWIT has found that young women respond to the message that as an IT professional they can create "products and solutions that save lives, solve health problems, improve the environment, and keep us connected" and that "She can use her skills to help solve pressing social problems in a variety of fields."[52]

One of the most distinctive aspects of coding groups aimed at minorities and women is that they often focus explicitly on how coding skills can be applied to solve social problems that participants care about. Qeyno Group, an organization founded by anti-poverty activist Kalimah Priforce that holds free hackathons for minority youth, describes its events as "teen hackers solving the world's biggest problems."[53] The mission statement of #YesWeCode, founded by former Obama advisor Van Jones with an emphasis on training African American youth, explicitly links learning to code with social justice goals: "Many have been searching for ways to uplift today's urban youth and help them achieve a more promising future. We believe that one solution lies in connecting tech and social justice leaders to spearhead revolutionary tech programs whose benefits extend to the most disadvantaged of society."[54] Jones argues that showcasing minorities as successful coders can also change stereotypes about technical talent: "The minute we have an African-American Mark Zuckerberg or a Latina Sheryl Sandberg, the conversation will shift."[55] The point is not that every female or minority coder sees their work in political terms, or that increasing their representation will automatically make computer culture more democratic—but without groups that are led and defined by underrepresented coders, a shift toward equality is unlikely to happen.

The question for these coding groups is not simply, will this software benefit someone? but also—sometimes pointedly—*whom* will it benefit? Whose problems are prioritized? Whose preferences will be catered to? Which markets are targeted, and which are ignored? Coding enthusiasts talk about "changing the world," but commercial software products and services serve a narrow slice of that world, prioritizing advertisers and consumers with disposable income. Minority coding groups have called out the racial and class bias of the industry's concept of what makes a software product revolutionary. As #YesWeCode's chief innovation officer, Amy Henderson, points out, "A lot of the people who develop apps today are affluent white men, and so they build apps that solve *their* communities' problems."[56] Qeyno's Priforce rejects the tech industry's overemphasis on providing recreational services for a privileged user community while others are ignored: "We are not interested in launching hackathons to build more photo sharing apps or finding pizza at 3am. We want to build and design technology that makes sense to a population that needs it, and who better to lead the development of this technology than the youth in this community."[57] Jones argues, "Silicon Valley is supposedly built on genius, that's their mythology. But you're not including all the geniuses that are out there. How many products and services is Silicon Valley not creating because they don't have people from different backgrounds in the room to come up with the next big idea?"[58] As Freada Kapor Klein points out, "When you don't have representation, you can't possibly come up with the products and services needed for the population, no matter how popular or trending you might be." Her investment firm funds startups run by women and underrepresented people of color, which are more likely to focus on issues such as "diversity, education, nutrition and community outreach."[59] Including more voices can also avoid what diversity activist Y-Vonne Hutchinson calls the "echo chamber" effect, where bias against underrepresented groups can slip unnoticed into algorithms. "When workforces are as homogenous as they commonly are in the Valley, it is inevitable that the products are full of blindspots," she argues. She points to "the Google Photos facial recognition app that compared black people to gorillas, the Amazon Prime delivery-system that did not serve predominantly black neighborhoods, the AI-driven website, Beauty.AI, which only deemed white people as 'beautiful' and Microsoft's AI-bot, Tay, that posted anti-Semitic content."[60] Halcyon Lawrence (ch. 8 in this volume) demonstrates how voice-recognition apps discriminate against people with "nonstandard" accents, meaning that supposedly universal products malfunction for a significant portion of the population. If even *one* developer had had an accent themselves, would this problem have been caught and fixed?

Examples from projects run by African Americans offer a vision of how coding can immediately and concretely help people who are most in need and least served by commercial software offerings. #YesWeCode's first hackathon at the 2014 Essence Festival in New Orleans produced apps focused on health issues affecting the Black community, such as obesity, diabetes, and a lack of fresh food; services for youth in foster care; and help for victims of human trafficking. At a hackathon in Philadelphia the following November, the winning app, called Creating Your Community, allowed young people to take photos of derelict buildings and propose new uses for them while also connecting with local developers, designers, and contractors who could make the vision a reality. #YesWeCode's Kwame Anku described the app as a tool for community self-determination: "We talk about gentrification, but these young people are saying, 'Hey, we can transform our own community, conceptualize what *we* want, and then be able to have the local dollars stay in the community and have the community we want."[61] Perhaps the most striking example of the gap between the needs of underserved communities and the users privileged by Silicon Valley came from Qeyno's 2015 Oakland Hackathon, where the Best Impact winner was an app addressing the life-or-death implications of Black youths' encounters with police in the wake of nationally publicized shootings. Fourteen-year-old George Hofstetter designed "a mobile application for teenagers like myself—African American—to help them figure out how to act . . . towards police officers. So much has happened over the past few years that have resulted in death, that have to do with teenagers around my age. I would like to help them to know what to do."[62] Extending its efforts to the Latinx community, Qeyno's 2018 #DREAMerHack mobilized code to help children of undocumented immigrants, inviting participants "to build 'sanctuary' apps to address the growing insecurities and safety-risks for those affected by forced displacement. . . . In the tradition of the hackers who conducted the Underground Railroad, #DREAMerHack is focused on 'hacking alienation.'"[63]

These young coders and their mentors define empowerment in a much more radical way than getting an entry-level job in Silicon Valley. By making purpose an essential part of the coding conversation, they challenge the goals and culture of the tech industry, rather than simply training underrepresented groups to conform to the status quo. Coding efforts led by women and minorities—and designed specifically with their interests, strengths, and constraints in mind—can provide a more holistic approach that addresses structural and cultural obstacles to participation in computing. The software industry is unlikely to achieve equal representation until computer work is equally *meaningful* for groups who do not necessarily share the values and priorities that currently dominate Silicon Valley.

CONCLUSION: WE CAN AIM HIGHER

Teaching more girls and African American and Latinx youth to code is a positive step toward correcting both Silicon Valley's lack of diversity and the unequal earning power of women and minorities. But superficial claims that learning to code will automatically be empowering can mask a lack of commitment to structural change. To focus only on getting individuals into the training pipeline ignores systemic bias in hiring and promotion, reinforces narrow conceptions of merit and skill, and pushes solutions into the future.

Companies can do more right now to increase diversity. Instead of confining recruitment efforts to children's activities, entry-level hires, or other junior people who don't threaten the dominance of established groups, companies could focus on hiring and promoting women and minorities into senior positions. Firms can map out plans for having underrepresented people lead technical teams, make product design decisions, and serve on venture capital boards, where they can weigh in on the social utility of proposed innovations. They can treat cultural competence as part of the programming skill set and be willing to fire employees—even "superstars"— who harass or demean women and minorities.[64] And the public can hold tech firms more accountable, refusing to accept facile excuses about meritocracy and pipelines. Pointing to changes at Uber after public outrage over its sexual harassment scandal, Y-Vonne Hutchinson argues, "If we want more diversity in the tech industry and high-skilled sector in general, we need to demand it and put pressure on them."[65]

Racial and gender inequality in the workplace is a complex problem with no quick fix, and real solutions may require uncomfortable changes from currently privileged groups. The knee-jerk reaction is to retreat, but this is exactly the wrong response.[66] Our discomfort is not a problem but rather a measure of the seriousness of the issue and a necessary first step toward change. The way forward is not denial but concrete, committed action. Unless there is a systemic realignment of opportunities, rewards, and values in the tech industry, training individual women and minorities will do little to shift the culture. Coding by itself is not empowerment.

NOTES

1. Reshma Saujani, *Girls Who Code: Learn to Code and Change the World* (New York: Viking, 2017).

2. Ezra Mechaber, "President Obama Is the First President to Write a Line of Code," White House blog (December 10, 2014), https://www.whitehouse.gov/blog/2014/12/10/president-obama-first -president-write-line-code; White House Office of the Press Secretary, "Fact Sheet: New Commitments to Support Computer Science Education" (December 08, 2014), https://www.whitehouse .gov/the-press-office/2014/12/08/fact-sheet-new-commitments-support-computer-science-education. "Hour of Code" is an introductory coding activity launched in 2013.

3. The surprisingly long history of coding-as-empowerment efforts, starting in the 1960s, is described in Janet Abbate, "Code Switch: Changing Constructions of Computer Expertise as Empowerment," *Technology & Culture* 59, no. 4 (October 2018): S134–S159.

4. National Science Board, *Science and Engineering Indicators 2016*, https://www.nsf.gov/statistics/2016/nsb20161/, ch. 3, 86. Women earning CS doctorates rose from 16 percent in 1993 to 21 percent in 2013, but 42 percent of female CS PhDs ended up in non-S&E jobs, compared to only 26 percent of men (88). See also Thomas Misa, ed., *Gender Codes: Why Women Are Leaving Computing* (Hoboken, NJ: Wiley, 2010); American Association of University Women, *Solving the Equation: The Variables for Women's Success in Engineering and Computing* (Washington, DC: AAUW, 2015).

5. Susan Fowler, "Reflecting on One Very Strange Year at Uber," personal blog (February 2, 2017), https://www.susanjfowler.com/blog/2017/2/19/reflecting-on-one-very-strange-year-at-uber; Alyssa Newcomb, "#MeToo: Sexual Harassment Rallying Cry Hits Silicon Valley," *NBC News* (October 23, 2017), https://www.nbcnews.com/tech/tech-news/metoo-sexual-harassment-rallying-cry-hits-silicon-valley-n813271, and Lydia Dishman, "This 'Me Too' Timeline Shows Why 2017 Was a Reckoning For Sexism," *Fast Company* (December 6, 2017), https://www.fastcompany.com/40504569/this-me-too-timeline-shows-why-2017-was-a-reckoning-for-sexism; Louise Matsakis, "Google Employee's Anti-Diversity Manifesto Goes 'Internally Viral,'" *Vice* (August 5, 2017), https://motherboard.vice.com/en_us/article/kzbm4a/employees-anti-diversity-manifesto-goes-internally-viral-at-google.

6. AAUW, *Solving the Equation*, 5.

7. Josh Harkinson, "Jesse Jackson Is Taking on Silicon Valley's Epic Diversity Problem," *Mother Jones* (June 30, 2015), http://www.motherjones.com/politics/2015/06/tech-industry-diversity-jesse-jackson/.

8. Code.org, "2016 Annual Report," https://code.org/about/2016, and "Code.org and Diversity in Computer Science," https://code.org/diversity, both accessed September 14, 2017. While Code.org's "diversity" page acknowledges that there are other factors, including "unconscious bias," it rhetorically positions the tech industry as the victim rather than the perpetrator of a diversity problem: "software, computing and computer science *are plagued by* tremendous underrepresentation," as is "the software workplace, which suffers a similar lack of diversity" (emphasis added).

9. Quoctrung Bui and Claire Cain Miller, "Why Tech Degrees Are Not Putting More Blacks and Hispanics into Tech Jobs," *New York Times* (February 26, 2016), http://www.nytimes.com/2016/02/26/.

10. Quoted in Stephanie Mehta, "How One Tech Start-Up Ditched Its Brogrammers," *Vanity Fair* (April 12, 2017), https://www.vanityfair.com/news/2017/04/tech-start-up-women-brogrammers.

11. Rachel Sklar, "Stop Equating Women in Tech with Engineers," *Medium* (February 1, 2015), https://medium.com/thelist/stop-equating-women-in-tech-with-engineers-e928e9fa1db5.

12. For examples of women who came to programming late but were extremely successful, see Janet Abbate, *Recoding Gender* (Cambridge, MA: MIT Press, 2012).

13. Eric P. Newcomer, "YC's Paul Graham: The Complete Interview," *The Information* (December 26, 2013), https://www.theinformation.com/YC-s-Paul-Graham-The-Complete-Interview. To represent Graham's views more accurately, I use his revised wording where available (Paul Graham, "What I Didn't Say," personal blog [December 2013], http://paulgraham.com/wids.html).

14. First Round Capital, "State of Startups 2016," accessed August 15, 2019, http://stateofstartups.firstround.com/2016/#highlights-diversity-problem.

15. Jason Pontin, "A Feminist Critique of Silicon Valley," *MIT Technology Review* (January/February 2015), https://www.technologyreview.com/s/533096/a-feminist-critique-of-silicon-valley/.

16. M. Bertrand and S. Mullainathan, "Are Emily and Greg More Employable Than Lakisha and Jamal? A Field Experiment on Labor Market Discrimination," *American Economic Review* 94, no. 4 (2004): 991–1013; C. A. Moss-Racusin, J. F. Dovidio, V. L. Brescoll, M. J. Graham, and J. Handelsman, "Science Faculty's Subtle Gender Biases Favor Male Students," *Proceedings of the National Academy of Sciences* 109, no. 41 (2012): 16474–16479; R. E. Steinpreis, K. A. Anders, and D. Ritzke, "The Impact of Gender on the Review of Curricula Vitae of Job Applicants and Tenure Candidates: A National Empirical Study," *Sex Roles* 41, nos. 7–8 (1999): 509–528.

17. Mark Milian, "Do Black Tech Entrepreneurs Face Institutional Bias?," *CNN* (November 11, 2011), http://www.cnn.com/2011/11/11/tech/innovation/black-tech-entrepreneurs/index.html.

18. Change the Ratio, accessed August 15, 2019, http://changetheratio.tumblr.com/post/1517142444/if-you-look-at-bezos-or-netscape-communications.

19. Michelle M. Duguid and Melissa Thomas-Hunt, "Condoning Stereotyping? How Awareness of Stereotyping Prevalence Impacts Expression of Stereotypes," *Journal of Applied Psychology* 100, no. 2 (March 2015): 343–359.

20. Jeff Schmitt, "The Top Feeder Schools to Google, Goldman Sachs and More," *Poets and Quants for Undergrads* (January 7, 2015), https://poetsandquantsforundergrads.com/2015/01/07/the-top-feeder-schools-to-google-goldman-sachs-and-more/.

21. Quoted in Abrar Al-Heeti, "Startups Could Be Key to Fixing Tech's Diversity Problem," *CNET* (August 20, 2017), https://www.cnet.com/news/startups-could-be-key-to-fixing-silicon-valleys-diversity-problem/.

22. Quincy Larson, "Why Is Hiring Broken? It Starts at the Whiteboard," *Medium* (April 29, 2016), https://medium.freecodecamp.org/why-is-hiring-broken-it-starts-at-the-whiteboard-34b088e5a5db.

23. Paraphrased in Mehta, "How One Tech Start-Up Ditched Its Brogrammers"; see also Anonymous, "Technical Interviews Are Bullshit," *Model View Culture* (November 17, 2014), https://modelviewculture.com/pieces/technical-interviews-are-bullshit.

24. Adrianne Jeffries, "Programmers Are Confessing Their Coding Sins to Protest a Broken Job Interview Process," *The Outline* (February 28, 2017), https://theoutline.com/post/1166/programmers-are-confessing-their-coding-sins-to-protest-a-broken-job-interview-process.

25. Mehta, "How One Tech Start-Up Ditched Its Brogrammers"; Joyce Park, "The Best Coding Interview I Ever Took," *Code Like a Girl* (April 4, 2016), https://code.likeagirl.io/the-best-coding-interview-i-ever-took-2d12ee332077.

26. Allison Scott, Freada Kapor Klein, and Uriridiakoghene Onovakpuri, *Tech Leavers Study* (Oakland, CA: Kapor Center for Social Impact, 2017), 4, 7.

27. Trae Vassallo, Ellen Levy, Michele Madansky, Hillary Mickell, Bennett Porter, Monica Leas, and Julie Oberweis, "Elephant in the Valley," accessed August 15, 2019, https://www.elephantinthevalley.com/.

28. Salvador Rodriguez, "Why Silicon Valley Is Failing Miserably at Diversity, and What Should Be Done about It," *International Business Times* (July 7, 2015), http://www.ibtimes.com/why-silicon-valley-failing-miserably-diversity-what-should-be-done-about-it-1998144.

29. Quoted in Donovan X. Ramsey, "Twitter's White-People Problem," *Nation* (January 6, 2016), https://www.thenation.com/article/twitters-white-people-problem/.

30. There is a vast literature on this issue; for some classic examples, see Langdon Winner, "Do Artifacts Have Politics?" *Daedalus* 109, no. 1 (1980): 121–136; Lorraine Daston and Peter Galison, *Objectivity* (Cambridge, MA: MIT Press, 2007).

31. John Cook, "We Need More Geeks," *GeekWire* (January 22 2013), https://www.geekwire.com/2013/twin-brothers-ali-hadi-partovi-start-codeorg-solve-computer-science-education-gap/.

32. Cathy O'Neil, *Weapons of Math Destruction: How Big Data Increases Inequality and Threatens Democracy* (New York: Crown, 2016); Solon Barocas and Andrew D. Selbst, "Big Data's Disparate Impact," *California Law Review* 671 (2016); Gideon Mann and Cathy O'Neil, "Hiring Algorithms Are Not Neutral," *Harvard Business Review* (December 9, 2016).

33. Quoted in Jake Edge, "The Programming Talent Myth," LWN.net (April 28, 2015), https://lwn.net/Articles/641779/.

34. Blacks in Technology, "Code2040—Shaping the Future for Minority Coders" (2012), accessed August 15, 2019, https://www.blacksintechnology.net/code2040.

35. Quoted in Ramsey, "Twitter's White-People Problem."

36. @Shaft (Leslie Miley), "Thoughts on Diversity Part 2. Why Diversity Is Difficult," *Medium* (November 3, 2015), https://medium.com/tech-diversity-files/thought-on-diversity-part-2-why-diversity-is-difficult-3dfd552fa1f7#.jhou0ggdl. For a historical perspective on using arbitrary standards to perpetuate racial exclusion, see Amy Slaton, *Race, Rigor, and Selectivity in U.S. Engineering: The History of an Occupational Color Line* (Cambridge, MA: Harvard University Press, 2010).

37. Ellen Berrey, "Diversity Is for White People: The Big Lie behind a Well-Intended Word," *Salon* (October 26, 2015), http://www.salon.com/2015/10/26/diversity_is_for_white_people_the_big_lie_behind_a_well_intended_word/.

38. For example, see the arguments made by Code2040's Laura Weidman (Blacks in Technology, "Code2040") and Girls Who Code's Reshma Saujani (Lindsay Harrison, "Reshma Saujani," *Fast Company* [May 13, 2013], https://www.fastcompany.com/3009251/29-reshma-saujani).

39. Jodi Kantor and David Streitfeld, "Inside Amazon: Wrestling Big Ideas in a Bruising Workplace," *New York Times* (August 15, 2015); Mike Isaac, "Inside Uber's Aggressive, Unrestrained Workplace Culture," *New York Times* (February 22, 2017). *Entrepreneur* magazine defended both companies by arguing that their tactics are necessary to "hire and promote the best and most accomplished talent. That's what defines a meritocracy." See Steve Tobak, "In Defense of Uber, Amazon and Meritocracy," *Entrepreneur* (February 8, 2017), https://www.entrepreneur.com/article/289919.

40. Quoted in Mehta, "How One Tech Start-Up Ditched Its Brogrammers."

41. Emilio J. Castilla and Stephen Benard, "The Paradox of Meritocracy in Organizations," *Administrative Science Quarterly* 55 (2010): 543–568, at 543 (emphasis added), 567–568.

42. For a historical perspective on the devaluing of communication skills in software engineering, see Abbate, *Recoding Gender*, chapter 3.

43. Quoted in Edge, "The Programming Talent Myth."

44. T. C. Currie, "Are You a 10x Programmer? Or Just a Jerk?," *New Stack* (June 3, 2016), https://thenewstack.io/10x-programmer-just-jerk/.

45. Barry Boehm, "Software Engineering—As It Is," *ICSE '79: Proceedings of the Fourth International Conference on Software Engineering* (Piscataway, NJ: IEEE Press, 1979), 14.

46. Catherine Rampell, "The Diversity Issue in Tech Firms Starts Before the Recruiting Process," *Washington Post* (August 25, 2014), https://www.washingtonpost.com/opinions/catherine-rampell-diversity-in-tech-must-be-addressed-before-the-recruiting-process/2014/08/25/0b13b036-2c8e-11e4-bb9b-997ae96fad33_story.html; see also Sylvia Ann Hewlett, Carolyn Buck Luce, and Lisa J. Servon, "Stopping the Exodus of Women in Science," *Harvard Business Review* (June 2008), https://hbr.org/2008/06/stopping-the-exodus-of-women-in-science.

47. Brian Feldman, "Why that Google Memo Is So Familiar," *New York Magazine* (August 7, 2017), http://nymag.com/selectall/2017/08/why-that-google-memo-is-so-familiar.html.

48. Edge, "The Programming Talent Myth."

49. Quoted in Mehta, "How One Tech Start-Up Ditched Its Brogrammers."

50. Girls Who Code Clubs Program, "Curriculum Summary 2015–2016," accessed April 27, 2016, https://girlswhocode.com.

51. . Kristen V. Brown, "Tech Shift: More Women in Computer Science Classes," *SF Gate* (February 18, 2014), http://www.sfgate.com/education/article/Tech-shift-More-women-in-computer-science-classes-5243026.php.

52. NCWIT, "Talking Points: Why Should Young Women Consider a Career in Information Technology?," accessed September 18, 2016, https://www.ncwit.org/sites/default/files/resources/tp_youngwomen.pdf.

53. Qeyno Labs, "Hackathon Academy," accessed September 18, 2017, https://www.qeyno.com/hackacademy; Hackathon Academy is now called Hackathon Worlds (https://steadyhq.com/en/qeyno, accessed August 15, 2019).

54. YesWeCode, "Our Story & Mission," accessed August 15, 2019, http://www.yeswecode.org/mission.

55. Demetria Irwin, "Facebook, #Yeswecode Make a Splash at the Essence Music Festival," *Grio* (July 6, 2014), https://thegrio.com/2014/07/06/facebook-yes-we-code-at-essence-music-festival/.

56. Irwin, "Facebook, #Yeswecode Make a Splash."

57. Qeyno Labs, "A Black Mark Zuckerberg?," press release (June 17, 2014), https://www.prlog .org/12337721-black-mark-zuckerberg-hackathon-empowers-youth-to-transform-nola-into-their -own-silicon-valley.html.

58. "Van Jones: Giving Black Geniuses Tools to Win with #Yeswecode," *Essence* (May 2, 2014), https://www.essence.com/festival/2014-essence-festival/tk-van-jones/.

59. Terry Collins, "They're Changing the Face of Silicon Valley," *CNET* (August 16, 2017), https:// www.cnet.com/news/changing-the-face-of-technology-women-in-technology/.

60. Guðrun í Jákupsstovu, "Silicon Valley Has Diversity Problems, Now Let's Focus on a Fix," *TNW* (February 20, 2018), https://thenextweb.com/us/2018/02/20/1108399/.

61. Joy Reid, "Teaching the Next Generation of Coders," *MSNBC: The Reid Report* (November 17, 2014), video, 4:50.

62. Jessica Guynn, "Qeyno Labs: Changing the World One Hackathon at a Time!" *USA Today* (February 3, 2015), video, 1:31, https://vimeo.com/121180705. Qeyno Labs changed its name to Qeyno Group in 2018.

63. Qeyno Labs, "DREAMerHack" (April 2018), https://www.eventbrite.com/e/dreamerhack -hackathon-academy-tickets-39229161478?aff=es2.

64. Rodriguez, "Why Silicon Valley Is Failing Miserably at Diversity."

65. Jákupsstovu, "Silicon Valley Has Diversity Problems."

66. One example of retreat is male tech executives who avoid working closely with women out of fear of harassment accusations. See Claire Cain Miller, "Unintended Consequences of Sexual Harassment Scandals," *New York Times* (October 9, 2017), https://www.nytimes.com/2017/10/09/ upshot/as-sexual-harassment-scandals-spook-men-it-can-backfire-for-women.html.

13

SOURCE CODE ISN'T

Ben Allen

American computer programmer Ken Thompson is best known for his role in the creation of the Unix operating system and of C, the programming language in which Unix is written. Many modern operating systems—Linux, BSD, the BSD derivative Mac OS X, the Linux-derived mobile operating system Android—count Unix as an ancestor, and most other non-mobile operating systems can be made compatible with Unix through the use of subsystems implementing the POSIX compatibility standards. C and its derivatives (most notably C++, C#, and Objective C) are still in widespread use.[1]

In 1983 Thompson received the Turing Award, largely as a result of his work on Unix. In his acceptance speech for this award, titled "Reflections on Trusting Trust,"[2] Thompson reminds the audience that he hadn't worked actively on Unix in many years.

After doing this act of performative humility, Thompson devotes his speech to a description of "the cutest program he ever wrote." This program was an implementation of a method of creating Trojan horse programs—software with secret backdoors, which the nefarious creator of the Trojan horse could then use to hijack computer systems on which the software was installed. I will refer to the method that Thompson outlines in "Reflections on Trusting Trust" as the "Thompson hack" for the remainder of this chapter.[3] Nevertheless, it is useful to note that what I call the Thompson hack was not wholly the result of Thompson's original research. It was first documented in a piece prepared in 1974 by US Air Force security analysts Paul

Karger and Roger Schell, under the imprimatur of the Hanscom Air Force Base Electronic Systems Division.[4] Thompson's contribution was in creating the first publicly known implementation of the technique, and then popularizing it in "Reflections."

The Thompson hack is interesting less as a practical tool for compromising computer systems—though it was used for this purpose, both by Thompson and by later programmers—and more as a mathematical or quasi-mathematical proof of the impossibility of completely verifying the security of any system. Whereas it is common, and entirely reasonable, for ordinary computer users to distrust the programs and platforms they're using, the Thompson hack demonstrates that no matter how sophisticated a computer user, computer programmer, or security analyst is, they can never be entirely certain that their system is not compromised—not unless they themselves were the author of every piece of code used on that system.

The main portion of this chapter starts with a discussion of the distinction between source code written in relatively human-readable programming languages like C and the code in lower-level languages into which source code must be translated in order to become a working program. This explanatory material is required to understand the technical workings of the Thompson hack. However, it also holds its own inherent interest, since this layer of translation standing between human-friendlier programming languages and code that computers can run is one of the deepest elements in the deep-layered software stack that allows for the creation of software platforms.

The primer on programming language hierarchies is followed by a close read of "Reflections on Trusting Trust." Although this material is fairly technical, enough apparatus is provided for nonprogramming readers to understand both the key turning points of Thompson's explanation of the hack and the significance of the hack in understanding software platforms. Once the technical work of explicating the Thompson hack's methodology is finished, I turn to analyzing the social implications of the hack, and how those implications may differ from the ones that Thompson himself proposes in "Reflections."

Finally, I move on to a brief history of the Thompson hack in use in the decades since Thompson's speech. This section starts from Thompson's own use of the hack in a few initial pranks, moves on to its later use in a few esoteric viruses, and then considers a strategy proposed by Institute for Defense Analyses (IDA) security researcher David A. Wheeler to counter the Thompson hack, a strategy that Wheeler originally developed as part of his PhD work at George Mason.

The strategy for analysis in this chapter requires thinking about the Thompson hack on two levels simultaneously. We must follow Thompson's technical description of the hack's significance to computer science while also (unlike Thompson) consistently keeping in mind the sociopolitical implications of the hack. The ultimate argument of this piece is that the Thompson hack illustrates how power operates through computer networks, not so much through the deployment of clever playful-seeming hacks and crafty backdoors, but instead through establishing whose clever hacks are read as playful and whose clever hacks are read as crimes.

A SHORT PRIMER ON THE PROGRAMMING LANGUAGE HIERARCHY

One key distinction that is necessary for understanding the Thompson hack is the difference between low-level and high-level programming languages. High-level languages, henceforth HLLs, tend to be relatively easier for humans to write and read; low-level languages are what the computer actually understands. When I use the terms "assembly code" or "machine code," they refer to code written in low-level languages. Likewise, when I refer to "executing" binary files, these files are understood as consisting of low-level machine code. When I use the term "source code," this refers to code written in HLLs.

Although HLLs, like low-level languages, require programmers to follow tightly defined formal syntax rules, HLLs allow programmers to write their programs in terms of relatively human-friendly concepts like "if this particular condition is met, run this piece of code, otherwise, run that other piece of code," without worrying about the fine details of how those concepts are implemented on actual machines. Most programming work is done in HLLs. Examples include Java, C, C++, Python, Swift, and R. In low-level programming languages—x86 assembly code, for example—there is a direct one-to-one correspondence between instructions given in the language and operations performed by the computer. Although one may think that this gives programmers working in low-level languages more control over what their software does, the level of detail required to produce even the simplest programs make them impractical to write in. Moreover, because all commands involve operations on numerical value stored in registers—machine registers are very small, somewhat cramped units of very fast memory[5]—the type of reasoning required to write lower-level code is relatively strange when compared to the types of reasoning allowed for by HLLs.

Here is an example of a function in x86 assembly that calculates numbers in the Fibonacci sequence. To save page space, this function is split into two columns; read down the left column first, then the right:

fib:	@@:
mov edx, [esp+8]	push ebx, [ebx + ecx]
cmp edx, 0	cmp edx, 3
ja @f	jbc @f
mov eax, 0	mov ebx, ecx
ret	mov ecx, eax
@@:	dec edx
cmp edx, 2	jmp @b
ja @f	@@:
mov eax, 1	pop ebx
ret	ret

It is not necessary to deeply understand this code. Simply observe that most of these commands involve moving numbers into or out of registers (these commands start with mov, pop, and push), comparing values in registers to numbers (commands starting with cmp), or jumping to other parts of the code based on those comparisons (the ja, jbc, and jmp commands). ret exits ("returns from") the function.

The most common commands in this snippet of code—the ones starting with mov—involve moving numbers from one register into another. Why all the moving? Certain registers are reserved for input to functions; one, eax, is reserved for output. Most of the work done by this little program is less about calculation and more about moving values into the right registers for calculation to occur, loading results into eax, and then restoring the registers to their original states so that other functions can use them.

For comparison, below is a simple function for calculating the nth Fibonacci number, written in the (higher-level) C programming language:

```
int fib(int n){
if (n == 0 || n == 1) {return n; } /* base case */
else { return(fib(n-1) + fib(n-2)); }/* recursive case */
}
```

Although this C function is not exactly an English-language description of how to calculate the Fibonacci sequence, it's nevertheless a relatively clear piece of code that reveals the method of calculation to interested readers. It is also a great deal *shorter* than the assembly code, and—as I have been painfully reminded on each occasion that I've had to work in assembly—much easier to write and debug. The method of calculation is most clearly shown in the line beginning with `else`, which is executed when the value of the integer value *n* does not equal either 0 or 1. When the `else` line is reached, the function determines the value of the nth term in the Fibonacci sequence by calling itself again twice, in order to calculate the value of the *n-1*th term and the *n-2*th term of the sequence. Whereas the assembly code primarily focuses on moving data into and out of registers, the C code clearly documents a key fact about the Fibonacci sequence: that it can be derived through recursion. (I've included comments in blocks delimited by /* and */ in order to make the structure of the program clearer; properly delimited comments are ignored by the C compiler when it converts C code into low-level code.)

Source code written in HLLs has to be converted into code in a low-level language in order for any given machine to actually execute the code. This requires a piece of software called a compiler or an interpreter. A compiler translates the entire program at once ahead of time, creating an executable file. Interpreters translate the code line by line as it runs, much like a human interpreter translates on the fly between two human languages. For the purposes of this chapter, I will use the word "compiler" rather than "interpreter" to refer to the software that converts HLL code into low-level code, since the C programming language in which Thompson implements his hack is generally a compiled, rather than interpreted, language.

In his piece in this volume, Paul N. Edwards describes platforms as "second-order infrastructures"—as new and relatively quick-to-build pieces of infrastructures layered on top of older, slower-to-build, more capital-intensive infrastructure. Edwards specifically describes software platforms, highlighting the tendency in software contexts for development to happen through the constructions of layer upon layer of technical system, with each layer being a platform established upon the infrastructure laid down by the previous layer.[6] As Edwards elaborates, "higher-level applications are built on top of lower-level software such as networking, data transport, and operating systems. Each level of the stack requires the capabilities of those below it, yet each appears to its programmers as an independent, self-contained system."[7] Edwards's account stresses the rapid pace at which platforms can be developed, rolled out, see widespread use, and then fade away as they are replaced by new platforms. Software platforms are flickering, evanescent flames, burning on top of the old slow infrastructure while allowing users and developers to pretend that the infrastructure isn't even there. The transition between low-level and high-level languages is the first flicker of this process, the first moment that software developers can begin to treat the machines that subtend software as irrelevant.

THOMPSON HACK STAGE 1: REPLICATION

There is a certain high seriousness to Thompson's speech. However, the opening of the description of the hack itself is playful:

> I am a programmer. On my 1040 form, that is what I put down as my occupation. As a programmer, I write programs. I would like to present to you the cutest program I ever wrote. I will do this in three stages and try to bring it together at the end.[8]

Thompson opens his explanation of this cute methodology for covertly hijacking computer systems by discussing a game that he and his fellow programmers would play in college. This game involved writing programs that would, when compiled and run, produce complete listings of their own source code as output. The winner would be the programmer who had produced the *shortest* self-replicating program.

The method Thompson used to produce a self-replicating pronoun that he presents in "Reflections on Trusting Trust" is inspired by analytic philosopher W. V. O. Quine's variant of the liar's paradox. Most versions of this paradox, like the well-known "this sentence is false" formulation, contain demonstrative words referring to the sentences themselves. In "this sentence is false," the demonstrative word is "this." Quine's version does not in any way explicitly refer to itself. Instead, it exploits the grammatical distinction between *mentioning* something in quotation marks and directly *using* the words that describe that thing.[9] The original Quine sentence is as follows:

> "Yields falsehood when preceded by its quotation" yields falsehood when preceded by its quotation.[10]

Just as the original formulation of the liar's paradox is undecidable—it's true if it's false, it's false if it's true—the Quine sentence is also undecidable. When the sentence fragment outside quotes is preceded by itself in quotes (as it is here), it "yields falsehood," meaning that it is false. If it is false, though, it is true, and if it is true, it is false, and so on. Undecidability is established, as in "this sentence is false," but without any use of the suspect self-referential word "this" or any other form of direct self-reference.

Although the self-replicating program that Thompson discusses in the "first stage" of his presentation of the Thompson hack isn't undecidable in the same way that

Quine's sentence is, it is constructed by a similar method. Like a quine, it contains itself twice: the first time *mentioned*, as data to be operated on, and the second time *used*, as instructions for operating on that data.

Thompson's sample quine contains logic (written in the C programming language) for printing a sequence of characters twice, preceded by a variable (called *s*) containing a string[11] that itself also contains that same C code for printing a string of characters twice.[12] The twist that makes this work as a self-replicating program is that the logic to print a string twice (which occurs both as data in the string *s* and as a set of commands in the main code itself) prints it with a small difference: the second time it is printed, it prints normally, but the first time it is wrapped in the command to make a string. When not packaged as a string, the sequence of characters is treated as a series of C commands. But when placed in a string, the characters are instead understood as data for other C commands to operate on. Once boxed up in a string, the characters have been, more or less, placed in quotes. Figure 13.1, taken from the version of Thompson's talk printed in *Communications of the ACM*, contains the abbreviated text of Thompson's quine. The lines that he removes to keep the string *s* from being too long to fit on the page can be reconstructed from the unpackaged version of *s* which appears below it.

As can be seen from the figure, the command in C to define a string of characters *s* is:

```
char s[] = { 'x', 'y', 'z', '. . .'};
```

with each character in the string given in single quotes separated by commas, and with the whole set of quote-wrapped characters itself wrapped in curly braces and terminated by a semicolon (the character used to end statements in the C language). `char` means "character," in this case a single letter in the Latin character set or a piece of punctuation. The mysterious brackets after *s* (the name of the string) tell the compiler that *s* refers to a string of characters, rather than just one single character. Some "special characters," like \n and \t, appear in Thompson's quine. These characters, represented by a backslash and a letter, stand for "whitespace" in the output: \t represents a tab, \n the start of a new line. The main body of the program, seen under the line `main()`, first prints the contents of *s* with each character set off in single quotes, separated by spaces, and wrapped in the lines `char s[] = {` and `};`— which is to say, wrapped in the C command for defining a string of characters—and then simply prints *s* itself directly. We have a complete program that prints *s* twice, interpreted once as a string, and once as a pair of commands to print that string. We have a program that prints its own source code. We have a quine.

```
char s[ ] = {
      '\t',
      '0',
      '\n',
      '}',
      ';',
      '\n',
      '\n',
      '/',
      '*',
      '\n',
      (213 lines deleted)
      0
};

/*
 * The string s is a
 * representation of the body
 * of this program from '0'
 * to the end.
 */

main( )
{
      int i;

      printf("char\ts[ ] = {\n");
      for(i=0; s[i]; i++)
            printf("\t%d, \n", s[i]);
      printf("%s", s);
}
```

Figure 13.1   Abridged self-reproducing program.

THOMPSON HACK STAGE 2: BOOTSTRAPPING

Writing quine programs is still a semi-popular recreational programming exercise.[13] How, though, does the quine game relate to Thompson's hack? In this speech Ken Thompson is accepting the highest award that the Association for Computing Machinery gives, an award often called the Nobel Prize for computer scientists. Why does he talk about the games he played in college?

Quine programs are interesting theoretically but in practical terms are useless—all they do is print themselves. But compilers written in the language they define are made through a process of "bootstrapping" that works very much like the process used in making quines. A technique used for playful, useless programming games is in fact crucial in the development of the tools that produce almost all computer programs. Bootstrapping is a process of adding features to a compiler through using it to repeatedly compile extensions to itself. When one is interested in writing a compiler in the language it compiles, one first writes a very limited version of the compiler in assembly code. Once that exists, one can then use that compiler to compile a more feature-rich version of the compiler written in its own language, which can then be used to compile a yet more feature-rich version of the compiler, and so forth. Through using the bootstrapping technique, programmers are able to write most of their compiler in the language that compiler compiles, spending as little time as possible working in low-level languages.

In the previous section, we saw how whitespace could be rendered inside strings through the use of "special characters" consisting of the backslash character followed by another character: \n for newline, \t for tab, and so forth. In this section Thompson gives a snippet of code defining the C compiler's logic for detecting and interpreting special characters. I have added explanatory comments set off in /* and */ characters. The purpose of this code is to get the next character from the input stream—the file that we're reading from—by calling the function next(), check if it's a backslash (meaning that it might be the start of a special character), and if it is a backslash get the subsequent character and emit whatever special character is requested. If the first character is *not* a backslash, we know we are not working with a special character, and so this section of code simply emits that character.

```
c = next();       /* get the next character in the input stream */
if (c != '\\')    /* != means "is not equal to" */
return c; /* if c is backslash, we're not dealing with a special
   character */
/* (treat it as a normal character */
```

```
c = next();        /* get the character
after the backslash */
if ( c == '\\')
return ('\\');     /* if the character after the backslash is a backslash,
   return backslash */
if (c == 'n')
return ('\n');     /* if the character is n */
/* return the newline command */
if (c == 't')
return ('\t');     /* if it's a t, return tab */¹⁴
```

It's worth pausing here to explain a necessary but counterintuitive quirk in this code. If the purpose of the test on the second line is to detect whether the character c is a backslash, why does the if statement check for *two* backslashes? And why does the subsequent if statement also check for two backslashes, and return two backslashes if the statement is true? This is because backslash is defined as the character used to set off special characters—as a *command*. As such, to check for a backslash you need to type two backslashes. The first backslash indicates that the following character is a special character, while the second backslash indicates that that particular special character is just a regular backslash (rather than a tab or a newline). The first backslash can be understood as placing the second backslash in quotes. The need to sometimes include in the output—to *mention*—the characters that we also *use* to indicate commands forces us to double those characters to unambiguously indicate their status as characters to be mentioned rather than used. An implication here is that at some point in the past the programmer must have defined "\\" to mean the special character that is printed as just \.

We already have code in place for our compiler to recognize the following special characters: \\, meaning "insert\" \n, meaning "insert a new line," and \t, meaning "insert a tab." Thompson proposes adding code defining the sequence \v to mean inserting a *vertical* tab. The process of adding this new special character can illuminate what must have happened to define the meaning of the other special characters, including defining \ as the meaning of \\.

A good first guess is that we could add support for \v by adding the following snippet to the end of the code above:

```
if (c == 'v')
return ('\v')¹⁵
```

However, Thompson notes that if one were to write this code, it would result in a compiler error—the compiler doesn't yet "know" that \v means vertical tab, and so output containing the \v character is meaningless. How do we get out of this jam?

Recall that the purpose of this piece of code is to produce a compiler—a piece of code written to itself produce code. Although what a given command means within a programming language is formally understood in terms of a language definition, compilers are what in practice determine which machine code commands are generated from a given piece of source code in a particular computer environment. We have here what Thompson describes as a "chicken and egg" problem. Because the compiler determines what \v means, the compiler cannot translate \v until \v has first been specified in a machine code version of the compiler. However, producing that machine code version of the compiler requires first having a working specification of \v.

The solution to this problem is mundane: the code must be changed to define \v in reference to a lower-level standard:

```
if (c == 'v')
return(11);
```
[16]

Why 11? This is the number (arbitrarily) given to vertical tab in ASCII, the encoding used on most computers for representing the Latin character set.[17] The compiler produced using this code now correctly parses \v as vertical tab, at least so long as it is run on a system that uses the ASCII character set.

Note that the compiler produced through this process now accepts the original definition of \v, the one that the previous version of the compiler flagged as an error. Once you have compiled one version of the source code containing the "magic number" 11 for vertical tab, you can change the code back to:

```
if (c == 'v')
return ('\v');
```

and the now-educated compiler will compile it without complaint. You can forget the number 11 and the entire ASCII standard altogether; \v now means "vertical tab." In fact, the new code is now much more flexible than the original code that defined \v in terms of 11 in the ASCII standard. If it were recompiled for a computer using a character representation standard other than ASCII, even a standard where vertical tab was represented by some number other than 11, \v would nevertheless *still* mean vertical tab.[18]

As Thompson describes it: "This is a deep concept. It is as close to a 'learning' program as I have seen. You simply tell it once, and then you can use this self-referencing definition."[19]

It is useful here to consider the textual practice Thompson describes in terms of Katherine Hayles's concept of the "flickering signifier." Hayles argues that the unique flexibility of signs in digital contexts derives from how they are produced by long chains of translations. On computers, Hayles argues,

> The signifier can no longer be understood as a single marker, for example an ink mark on a page. Rather it exists as a flexible chain of markers bound together by the arbitrary relations specified by the relevant codes. . . . A signifier on one level becomes a signified on the next-higher level.[20]

In this case the flickering signifier works with a concealed step; the compiler does not correlate the string \v with 11, the number that the ASCII standard uses for vertical tab. It instead simply specifies that \v should be rendered as a vertical tab, without reference to any underlying standard whatsoever. The loopy solution to the "chicken and egg" problem that Thompson deploys allows the HLL code to genuinely fly away from the standards that undergird it. So long as at some point in the past some standard correlation of \v with a number meaning vertical tab existed, \v means vertical tab. The flickering signifier has now slipped the surly bonds of its underlying standards, burned its way into orbit, and no longer needs to come back down.

THOMPSON HACK STAGE 3: TROJAN

We know what a quine is, and our code to implement vertical whitespace has become a space oddity. We have the pieces required to implement the Thompson hack itself.

Thompson gives the following C-like pseudocode for his Trojan horse exploit:

```
char *s;     /* char *s is the string read in by the compile
routine—the code our compiler is compiling */
compile(s) {
if (match(s, "pattern")) {
compile("bug");
return;
}
[ . . . ]   /* if s doesn't match "pattern," compile normally.  This is
   where we will soon insert our second bug. */
}21
```

The word "pattern" in this example is not the literal word "pattern," but instead code matching the source code specifying *login*. The word "bug" indicates the hostile version of *login*. Taken together, these lines of pseudocode mean something like "when the compiler sees the text of *login*, it should produce machine level correlating to *bug*."

Our goal here is altering the compiler to insert malicious commands into the low-level code it produces when it compiles the standard Unix command for logging into a system. "The replacement code," Thompson explains, "would miscompile the *login* command so that it would accept either the intended encrypted password or a particular known password."[22] This particular password would allow the attacker to log in as a superuser, with access to all parts of the system.

The code for the Trojan horse as written here would be invisible in the source code for *login*, but would remain visible in the source code for the compiler. This means that the attack could be easily detected by anyone diligent enough to examine the compiler's source code. We cannot, therefore, directly bootstrap away our malicious version of *login* as easily as we could hide the association between \v and 11. To genuinely hide our hostile code it is necessary to once again leverage the compiler's capacity to compile itself. To this end, the finished version of the attack adds another miscompilation at the location marked by the comment in the previous code snippet:

```
if (match(s, "pattern2"){
compile("bug2");
return;
}23
```

The pattern searched for this time (under the name "pattern2") is the *source code for the compiler itself*. Just as the quine we started with contained its logic twice, first in quotes (which is to say, packaged up in a string so that it could be operated on as data), and then directly as bare commands, this test uses as data the compiler's own text. If the code to be compiled matches that text—if we are compiling the compiler—the compiler returns itself, but with a twist—"*bug2*" consists of the compiler plus the extra code added to detect and interfere with attempts to compile *login*. We compile this code and then remove our bugs from the compiler's source code, much like we removed 11 from our code for implementing \v. Now we have a compiler that checks for when it is compiling itself. When it does, it makes a compiler that will, the next time it compiles *login*, inject the Trojan horse. The Trojan isn't inserted because code for the Trojan exists anywhere in any source code. It isn't inserted because of any extant source code at all. It is inserted because at some point in the past, on some machine we have no knowledge of, source code existed that said that the pattern of characters associated with the code of *login* should be interpreted in this unexpected way. The sinister code haunts the workings of the machine without ever revealing itself in human-readable text. Our space oddity has completed its third-stage burn, and no one on the ground will ever know for sure where it's gone.

As Thompson explains, this hack applies not just to compilers but to any piece of code that handles code:

> In demonstrating the possibility of this kind of attack, I picked on the C compiler. I could have picked on any program-handling program such as an assembler, a loader, or even hardware microcode.[24]

As we've implemented the hack so far, a particularly diligent security analyst could discover the bug by reading through the hundreds of thousands of lines of low-level code written by the bugged compiler. However, we could thwart this diligent analyst by introducing additional bugs—for example, a bug that would have the compiler corrupt the tools used to inspect assembly code. If we wanted to be even more nefarious, we could take Thompson's suggestion and alter not the compiler but instead the microcode that turns low-level code into actual machine circuitry. Once we have the logic of the Thompson hack implemented, it is no longer possible to fully verify that any machine is uncompromised, because every single piece of software that itself generates software is a vector for diligent attackers to exploit. Every layer in every nth-order platform is suspect. We are reduced to having to simply trust that no such diligent attacker has targeted our machine—no matter how skilled we are at detecting attacks, and no matter how much time we have to analyze our machines for bugs. All our stable platforms are potentially riddled with invisible trapdoors.

THE SOCIOPOLITICAL IMPLICATIONS OF UNTRUSTWORTHY MACHINES

After this tour de force demonstration, Thompson draws a pair of simple (or perhaps even simplistic) conclusions about what the hack means. First, he argues that "the moral is obvious. You can't trust code that you did not totally create yourself (especially code from companies that employ people like me)."[25] Second, he argues that the impossibility of formally verifying whether any computer is trustworthy indicates that laws against unauthorized access to computer systems need to be strengthened and that cultural norms need to change. As he puts it:

I would like to criticize the press in its handling of the "hackers," the 414 gang, the Dalton gang,[26] etc. The acts performed by these kids are vandalism at best and probably trespass and theft at worst. . . . [T]he act of breaking into a computer system has to have the same social stigma as breaking into a neighbor's house. It should not matter that the neighbor's door is unlocked. The press must learn that the misguided use of a computer is no more amazing than drunk driving of an automobile.[27]

There is an element of "do as I say, not as I do" present at the end Thompson's argument. Thompson's technically sophisticated hack is described throughout using terms indicating harmless fun and is discussed in relation to games. The swerve toward denouncing media coverage that valorizes hacking sits awkwardly at best with the remainder of the piece: Thompson's framing of the hack in terms of games, his description of it as "the cutest program he ever wrote,"[28] his wolfish declaration of the untrustworthiness of code written by people like him. Although Thompson claims that the hack was never distributed outside of AT&T Labs, he has confirmed that he built it, and rumors persist that a version of it was at least briefly installed in the BBN Unix distribution. The *Jargon File*, a collection of programmer lore dating from 1975 that was eventually repackaged and published as the *New Hacker's Dictionary*, claims the following in its entry for "back door":

> Ken Thompson has since confirmed that this hack was implemented and that the Trojan Horse code did appear in the login binary of a Unix Support group machine. Ken says the crocked compiler was never distributed. Your editor has heard two separate reports that suggest that the crocked login did make it out of Bell Labs, notably to BBN [Bolt, Beranek and Newman, a company responsible for much early Arpanet development], and that it enabled at least one late-night login across the network by someone using the login name "kt."[29]

In an email exchange, reposted to several Usenet groups in 1995, Thompson denies this allegation:

> fyi: the self reproducing cpp [the bugged compiler] was installed on OUR machine and we enticed the "unix support group" (precursor to usl) to pick it up from us by advertising some non-backward compatible feature. that meant they had to get the binary and source since the source would not compile on their binaries.
>
> they installed it and in a month or so, the login command got the trojan hourse. [*sic*] later someone there noticed something funny in the symbol table of cpp and were digging into the object to find out what it was. at some point, they compiled -S and assembled the output. that broke the self-reproducer since it was disabled on -S. some months later he login trojan hourse also went away.
>
> the compiler was never released outside.[30]

In this anecdote Thompson relates how he installed the bug on a machine owned by another group within Bell Labs, where he worked at the time. Because the deployment of the bug requires the target to install a compiler from a binary (rather than compiling it from source with one of their own compilers), Thompson performed an act of social engineering: he used the ruse that the new version of the compiler had features that the old compiler wouldn't be able to handle. Because Thompson was only interested in demonstrating his proof of concept rather than actually deploying a perfect attack, it was possible to detect the presence of the bug by analyzing the generated low-level code, and then to break the bug by compiling the compiler using a particular combination of settings.

In ch. 14, "Skills Will Not Set You Free," Sreela Sarkar describes a skills training course offered to young Indian Muslim women at an Information and Communications Technology Center.[31] These classes were promoted with the promise that the skills gained in would allow the students to obtain white-collar jobs. In practice, these women ended up as "low-paid workers at the fringes of the information economy."[32] Despite whatever skills they learned in their training, their Muslim last names and their working-class addresses marked them as outside the charmed circle that they would need to be in to get white-collar jobs. As Sarkar shows, access to good-paying jobs in this case depends less on technical skills and more on identity categories. Sarkar also documents some of the playful uses the students made of their technology—for example, using Photoshop to place an image of a blonde woman from their official exercises into their own real-world contexts. This clever and playful use of computers resulted in chastisement from the instructors. Who gets to creatively play with computer technology depends less on creativity and more on identity categories.

In the early 1970s, Ken Thompson, esteemed Bell Labs employee, was empowered to play with his company's machines in order first to implement an esoteric but extraordinarily effective Trojan horse development methodology that he had read about in a US Air Force paper. Further, he was empowered to use social-engineering techniques to get this Trojan installed on unauthorized machines. When less esteemed playful programmers—the "Dalton gang," the "414 gang"—used similar technical and social-engineering techniques to break into systems without authorization, they were equivalent to drunk drivers and burglars.

As the primary developer of a major operating system, the programming language used to develop that system, and many of the software tools used on that system, Ken Thompson was in a position of tremendous power. If, to use Lawrence

Lessig's famous analogy, "code is law," Ken Thompson had the power to write and alter the digital constitution by personal fiat.[33] Thompson's possession of power gave him the authorization to play—and to play irresponsibly—that the marginal Dalton gang lacked.

It is worthwhile at this point to note again that the methodology used in the Thompson hack, is, although impressive, more useful as a proof of the nonsecure nature of computing than it is in the building of practical attacks. More simple techniques—buffer overrun attacks, SQL-injection attacks, and so forth—are used in most real-world viruses. And most authors of Trojan horse programs do not need to conceal the evidence of their Trojans left behind in the source code, because most programs are closed source.

Nevertheless, the hack is troubling. We find ourselves gathered around the distributed campfire of our laptop screens and our glowing mobile phones. We learn that forty years ago a prominent and respected software developer announced that he had developed a method to distribute incendiary devices around our campsites, that these devices were entirely invisible, that others could replicate this methodology, and that if anyone had placed any of the devices, we could not be certain that we had found them all. He also promised that no one should trust any company that employed people like him . . . but he also claimed that he had distributed no incendiary devices, at least not anywhere but the campsite next to his own.

THE THOMPSON HACK IN THE WORLD

Multiple implementations of the Thompson hack have been discovered in the wild. One attack is discussed in a *Forbes* online article from 2014.[34] Mick Stute, the article's author, was a CS student hired by a psychologist to fix "odd behavior" in a questionnaire program used in their research. Specifically, the program would flash white supremacist messages to the screen for about a half second before displaying certain questions. Initially Stute thought the Trojan horse that was producing the messages was buried in a piece of obfuscated code that he found in the program's source directory. He laboriously unobfuscated the source code, found the offensive phrases, deleted them, and recompiled—only to find that the code was once again obfuscated and once again produced the white supremacist slogans. On the hypothesis that the Trojan was in one of the source code libraries used to build the program, he spend the next several days inspecting each of these libraries. Eventually, fixing this Trojan became an obsession for Stute, even though, as he observes, it might have been more practical to simply rewrite the entire program. After two weeks of investigation he discovered the injection code in the compiler itself. Stute requested clean source code for the compiler from AT&T, compiled it to produce a new, putatively clean compiler, then recompiled the questionnaire program. The white supremacist messages were still there—as Stute puts it, "the ex-grad-student had poisoned the compiler to poison itself when it was recompiled."[35] Stute finally fixed the hacked questionnaire program by bringing in a new binary copy of the compiler from a known-uninfected computer in another lab—though not before discovering that the bugged compiler also placed a backdoor in *login*. Stute observes that only this last detail got the computing center interested in fixing the problem. "Genius!," he concludes, "But put to a horrible cause." If the "genius" white supremacist behind the attack was ever discovered and punished, Stute doesn't say.

In 2009 a Thompson hack virus dubbed W32/Induc-A was discovered in a Delphi[36] compiler produced by the company Sophos. This nonmalicious virus had managed to replicate itself extensively; according to a *Wired* magazine article on the virus, it has "reportedly hitched a ride on a free CD ROM that shipped with the latest issue of ComputerBild, one of Germany's largest computer magazines." Kevin Poulson, author of the *Wired* article, notes that one of the chief uses of the Sophos Delphi compiler is the underground production of Trojan horse programs designed to steal online bank credentials. As such, many of these Trojans had been themselves infected with W32/Induc-A.[37]

A post by security blogger Graham Cluley noted that Sophos had to that point received 3,000 unique infected samples of programs infected by W32/Induc-A in the wild. Cluley notes that the infection had been discovered in a broad array of programs, and adds:

> Let me reiterate—this virus isn't just a threat if you are a software developer who uses Delphi. It's possible that you are running programs which are written in Delphi on your computers, and they could be affected.[38]

An FAQ page on the virus written by Sophos employee Nick Hodges responds to the question of whether C++ Builder, another Sophos compiler for the C++ HLL, could become a vector for the virus by saying "no," but then hedging with a sentence starting, "It is theoretically possible for a C++ Builder EXE to become infected."[39]

It is worth noting that despite the existence of W32/Induc-A, the Thompson hack is much *less* of a threat than it was when Thompson developed his initial version. In the 1970s, everyone using Unix depended on Thompson's software—the compiler, the operating system itself, most debugging tools, and most of the entire software development toolchain were written at Bell Labs by Thompson and his collaborator Dennis Ritchie. Any malfeasance, even playful malfeasance, committed by Thompson could potentially be disastrous. Today, there is no one single person upon whose work all practical computing depends—we use a variety of operating systems, a variety of compilers, and a variety of software development tools, and writing a Thompson hack Trojan that can thoroughly evade detection is now much more difficult. Even so, at least in theory, any of our software platforms could be riddled with entirely invisible trapdoors. One can assume that the Thompson hack Trojans that have been uncovered are the ones written by the *least* diligent hackers.

Although much rhetoric about hacking, especially rhetoric about hacking from the twentieth century, positions it as an anarchic and oppositional tool used by the street against corporate power, the Thompson hack functions to concentrate power in the hands of the software developers who already have it.[40] Moreover, programmers who already occupy privileged positions—already-esteemed software developers, college-educated computer science students from First World nations, and so forth—are the ones most empowered to play around with techniques like the Thompson hack. The perceived acceptability of using this technique depends less on the hacker's skill and more on who the hacker is.

CODA: FULLY COUNTERING TRUSTING TRUST

Although I have discussed the Thompson hack as a quasi-mathematical proof of the impossibility of verifying that any piece of code does what it says it does, there exists at least one proposed strategy for detecting Thompson hack Trojans in all situations. In 2009 computer security researcher David A. Wheeler received his doctorate in information technology from George Mason University upon the successful the defense of his dissertation, titled "Fully Countering Trusting Trust through Diverse Double-Compiling (DDC)."[41] Wheeler's full methodology is dense and ferociously technical. However, there exist reasonable cribs explaining Wheeler's overall strategy.

Bruce Schneier, cryptographer and computer security public intellectual, responding to a paper published by Wheeler in 2005 that would eventually grow into his dissertation, gives a relatively straightforward description of the DDC counter to the Thompson hack.[42]

This paper describes a practical technique, termed diverse double-compiling (DDC), that detects this attack and some unintended compiler defects as well. Simply recompile the purported source code twice: once with a second (trusted) compiler, and again using the result of the first compilation. If the result is bit-for-bit identical with the untrusted binary, then the source code accurately represents the binary.[43]

In short, if one has a known trusted compiler, one can compile the source code of the suspect compiler with both the trusted compiler and the suspect compiler. Because these two compilers should be *functionally* equivalent—they should produce identical output even if the binaries differ—one can then compile the source code of the suspect compiler with both compilers. If the resulting binaries are identical, both compilers can be trusted. Otherwise, the untrusted compiler has a Thompson hack Trojan installed in it.

Attentive readers may have already picked up on what makes this technique somewhat impractical: in order to use the DDC technique, *one must first have a known-trusted compiler.* Schneier elaborates:

> And if you're *really* worried . . . you can write Compiler B [the trusted compiler] yourself for a computer you built yourself from vacuum tubes that you made yourself. Since Compiler B only has to occasionally recompile your "real" compiler, you can impose a lot of restrictions that you would never accept in a typical production-use compiler. And you can periodically check Compiler B's integrity using every other compiler out there.[44]

Both of the tasks Schneier assigns to *"really* worried" paranoids are punishingly difficult for one person to do alone. One must first assemble a working computer from known-trusted components—meaning components with no potentially subvertible microcode—then write compilers for that computer, capable of compiling code from whatever languages one is interested in verifying compilers for. Presumably, writing these compilers would require also writing a text editor and a small operating system for one's bespoke computer. It is reasonable to characterize Wheeler's technique as only completely reliable when carried out by a large corporation or government agency that happens to trust its employees.

Since receiving his doctorate, Wheeler has taken a position with the Institute for Defense Analyses, headquartered in Alexandria, Virginia. This nonprofit corporation administers three federally funded research and development centers to assist the United States Department of Defense on national security issues. Although Wheeler hosts many of his publicly available papers on his website, including his dissertation and the 2005 "Countering Trusting Trust" paper, he notes that most of his written work is classified.[45]

NOTES

1. At the time of this writing, the Unix-derived Android just narrowly edges out the non-Unix Windows OS as the operating system with the highest market share, with Windows having more of the market share in the Americas and Europe, and the Android mobile operating system having a higher market share in Africa, South Asia, and China. See statcounter Global-Stats, "Operating System Market Share Worldwide" (July 2019), https://gs.statcounter.com/os-market-share#monthly-201907-201907-map.

   As of 2018 C++ is the fifth most widely used programming language on the software development hosting website GitHub, with C# as the sixth and C itself as the eighth most widely used language. Github, "Octoverse Report," accessed August 13, 2020, https://octoverse.github.com/projects#languages.

2. Ken Thompson, "Reflections on Trusting Trust," *Communications of the ACM* 27, no. 8 (1984): 761–763.

3. In his recent book *Bits to Bitcoin*, Mark Stuart Day briefly discusses this attack under the name Thompson's Hack. Mark Stuart Day, Al Sweigart, Tony Veale, and Karen Brown, "Thompson's Hack," in *Bits to Bitcoin* (Cambridge, MA: MIT Press, 2018), 243–271.

4. Discussed in Paul A. Karger and Roger R. Schell, "Multics Security Evaluation: Vulnerability Analysis," in *18th Annual Computer Security Applications Conference, IEEE, 2002. Proceedings* (2002): 127–146. Thompson identifies his source as "Unknown Air Force Document."

5. Regardless of whatever technical advances in computer design occur in the future, the number of bits stored in a register and the number of registers themselves will always necessarily be relatively small. This is because registers must be blazingly fast, and so must be crammed into a limited space on the CPU itself. If they were off the CPU, the limitation imposed by the speed of light would make them too slow to be useful. The nanometers occupied by registers on CPUs is, perhaps, the most pricy real estate in the world.

6. Paul N. Edwards, "Platforms Are Infrastructures on Fire," this volume.

7. Edwards, "Platforms Are Infrastructures on Fire."

8. Thompson, "Reflections," 761.

9. For a primer on the use-mention distinction, and of the vexed role of quotation within analytic philosophy, see Herman Cappelen, Ernest Lepore, and Matthew McKeever, "Quotation," *Stanford Encyclopedia of Philosophy*, accessed August 13, 2020, https://plato.stanford.edu/entries/quotation/#UseMentDist

10. For Quine's own discussion of the paradox, see Willard Van Orman Quine, *The Ways of Paradox, and Other Essays* (Cambridge, MA: Harvard University Press, 1976), 4–5.

11. "String" is the name given in the C programming language for a data structure that contains a sequence of characters.

12. As Thompson notes, this program would not be contest-winning—it is far too long—but it contains a useful capacity to be made arbitrarily *long*, a capacity that will be of interest when we reach the third step of Thompson's explanation. Thompson, "Reflections," 761.

13. A collection of quines maintained by programmer Gary Thompson can be found at https://www.nyx.net/~gthompso/quine.htm.

14. Thompson, "Reflections," 762. Comment blocks added by author.

15. Thompson, "Reflections," 762.

16. Thompson, "Reflections," 762.

17. There are few real alternatives to ASCII; its chief competitor, EBCDIC, is only used on older IBM mainframes. Unicode, the contemporary standard for representing text on most platforms, maintains reverse-compatibility with ASCII by storing Latin characters using the same values as

in ASCII. Unicode is as such less a competitor with ASCII and more a very large superset of ASCII. For a discussion of the relationship between Unicode and ASCII, and the implications of that relationship, see Thomas S. Mullaney, "Typing Is Dead," this volume.

18. One may at this point be asking why it is even necessary for the compiler to specify that when it sees \v in the input, it should produce \v in the output. The answer is that whereas a human—or at least a human with knowledge of the convention of specifying special characters using backslash—sees \v as one unit, the parser reads input character by character, successively storing each individual character in the variable c. Although it may seem like all this selection of code does is detect '\v' and return '\v', instead we must understand it as first detecting \, then detecting v, and then pulling them together into \v. This segment of code indicates that those two separate characters, in that order, should be treated as *one* character.

19. Thompson, "Reflections," 762.

20. N. Katherine Hayles, *How We Became Posthuman: Virtual Bodies in Cybernetics, Literature, and Informatics* (Chicago: University of Chicago Press, 2008), 31.

21. Thompson, "Reflections," 763. Comments added by author.

22. Thompson, "Reflections," 763.

23. Thompson, "Reflections," 763.

24. Thompson, "Reflections," 763.

25. Thompson, "Reflections," 763.

26. For a lighthearted description of the exploits of the so-called "Dalton gang," see Frederic Golden, "Superzapping in Computer Land," *Time* (January 12, 1981).

27. Thompson, "Reflections," 763.

28. Thompson, "Reflections," 761.

29. Eric Raymond, "Back Door," in *The New Hacker's Dictionary* (Cambridge, MA: MIT Press, 1996), 51.

30. Posted by Jay Ashworth to alt.folklore.computers (April 39, 1995). From private correspondence with Ken Thompson.

31. Sreela Sarkar, "Skills Will Not Set You Free," this volume.

32. Sarkar, "Skills Will Not Set You Free."

33. For one of Lessig's earliest uses of this phrase, see Lawrence Lessig, "Code Is law," *Industry Standard* 18 (1999).

34. Mike Stute, "What Is a Coder's Worst Nightmare," *Forbes* (December 5, 2014), https://www.forbes.com/sites/quora/2014/12/05/what-is-a-coders-worst-nightmare/#63aa7fee7bb4.

35. It is, for whatever reason, common to blame students who had read "Reflections on Trusting Trust" for most Thompson hack implementations found in the wild.

36. Delphi is an HLL derived from the earlier language Pascal.

37. Kevin Poulson, "Malware Turns Software Compilers into Virus Breeders," *Wired* (August 21, 2009).

38. Graham Cluley, "W32/Induc-A Virus Being Spread by Delphi Software Houses," *Naked Security* (August 19, 2009), https://nakedsecurity.sophos.com/2009/08/19/w32induca-spread-delphi-software-houses/.

39. Nick Hodges, "Frequently Asked Questions about the W32/Induc-A Virus (Compile-A-Virus)," date unknown, accessed August 13, 2020, http://edn.embarcadero.com/article/39851.

40. Among the countless examples of this sort of rhetoric, see "Superzapping in Computer Land," or William Gibson's foundational cyberpunk novel, *Neuromancer* (New York: Ace, 1984).

41. David A. Wheeler, "Fully Countering Trusting Trust through Diverse Double-Compiling (DDC)" (PhD dissertation, George Mason University, 2009).

42. David A. Wheeler, "Countering Trusting Trust through Diverse Double-Compiling," *21st Annual Computer Security Applications Conference (ACSAC'05)* (2005): 13.

43. Bruce Schneier, "Countering 'Trusting Trust,'" *Schneier on Security* (January 23, 2006), https://www.schneier.com/blog/archives/2006/01/countering_trus.html.

44. Schneier, "Countering 'Trusting Trust.'"

45. See Wheeler's site at https://dwheeler.com/dwheeler.html.

14

SKILLS WILL NOT SET YOU FREE

Sreela Sarkar

In Seelampur, located in the northeastern periphery of the capital city of New Delhi, a skills training class was in session at a globally acclaimed ICT (information and communication technologies) center. Young Indian Muslim women constituted the majority of students in the ICT class. At this practicum session held in a computer lab that was cordoned with a makeshift screen, participants were experimenting with Photoshop, working with images of white women with blonde hair.[1] A common activity was to export the image into a fairytale-like setting with green woods, mountains, and a stream. This activity was often accompanied by peals of laughter. As Sadiyah explained, "It is funny because we do not know any white women with hair like this. So just playing with her is funny. We can make her do what we want." Then Zohra said, "Yes, but imagine if we put her in Seelampur. Dirty, narrow streets and a lot of traffic with buses and [cycle] rickshaws. How will she survive here?" Zohra's comment was greeted with more laughter as the women tried to visualize this situation. "She won't last a day," Nasreen replied, chortling. The women were immediately told to be quiet by their leader.

India is celebrated as one of the largest emerging economies in the world. More than half of India's 1.2 billion people are estimated to be under the age of twenty-five. High unemployment rates among youth are considered to be a national crisis. The international development finance institution Asian Development Bank notes, "India with its youthful population and thriving information and communication industry can become a leading knowledge-driven economy." The Skill India initiative, introduced by the Hindu Nationalist Prime Minister Narendra Modi in 2015, aims to extend and introduce ICT-led skills training programs for India's youth. The promise of these programs is that they will flatten economic and social barriers for low-income youth, who will be able to reap the benefits of the new economy.

Seelampur is diverse but is primarily inhabited by Muslims, who are the largest minority in India. The average income in Seelampur for a family of four to six members is only $100–$150 a month. Muslim women in India are considered disadvantaged because of religious conservatism but actually face the tripartite struggle of being a minority, of being poor, and of being women.[2] The ICT center was housed in the Gender Resource Center (GRC) building as a public–private partnership (PPP) initiative between the Delhi government and the civil society organization G-Tech Foundation. The program taught basic ICT courses in Microsoft Word, design and page layout software like PageMaker and Dreamweaver, basic languages such as HTML, and graphics editor programs like CorelDRAW and Photoshop. In addition to technical skills, participants were taught to be goal-directed, disciplined, and passionate about their work, and to dress appropriately for positions in the information technology and information technology-enabled services (IT/ITeS) industries.

Behind the makeshift screen on the second floor of the GRC building, the play with Photoshop among the young women pointed to the contradictions and limitations of the skills training program in their everyday lives. The Seelampur women's contrast of a sterile setting with chaotic Seelampur highlighted their inability to transcend to a more prosperous world even after participation in the skills program. Their collective laughter pointed to their subversion of the celebratory training discourses of the program that were not consonant with their lives. The Seelampur program prepared marginalized women for jobs in the growing service sector in liberalized India, including jobs in IT, tourism, hotels, health care, transport or call centers, social and personal services, etc. However, like the majority of skills training programs directed at marginalized youth in contemporary India, the Seelampur program produced precarious and low-paid workers at the fringes of the information economy.

In contemporary India, Hotmail founder Sabeer Bhatia and Infosys founder Narayana Murthy are the popular icons of success. Bollywood films and TV series romanticize the figure of the hard-working software engineer who experiences the trappings of upward mobility and wealth. The popular reality television show on the state-owned Doordarshan channel called *Hunnarbaaz,* or *Skilled to Win,* showcases different skills training programs for marginalized youth. In recent years, the IT industry has been against caste-based reservations in private sector employment and education for Scheduled Castes (SC) and Scheduled Tribes (ST), who have historically been at the bottom of India's caste hierarchy. As Abbate observes in this volume, the insistence on meritocracy can actually lead to more discrimination.[3] The industry's claim that they hire on the basis of "merit," which allows equitable policies, is deeply contradicted by their recruiting mainly from elite colleges and certain urban areas for white-collar IT positions.[4]

The computer students in Seelampur already possessed individual initiative and skills. Their access to white-collar jobs was shaped not by their skills but by their lack of cultural capital:[5] access to elite educational institutions, speaking English with the correct accent, and self-presentation, which were in turn shaped by identities of class, caste, religion, etc. The training emphasis on discipline and structure left little space for highlighting the voices of the students who read and interpreted differently from politicians, policy experts, and program staff. Moments of play with technology and spontaneous interactions among students were often unmarked at the sanitized space of the center that isolated itself from popular movements in the locality and the city. It was in moments of play and pleasure at the ICT center that the Seelampur students resisted the bootstrapping messages of the modernization program, and pointed to the structural inequities that existed in their lives despite access to skills. They emphasized an understanding of knowledge and exchange outside capitalist models as an alternative to popular models of skills training. Despite the promise of inclusion for marginalized communities in the global information economy, the students subtly undermined their technology and modern citizenship lessons to assert that the programs did not set them "free" in term of leveling deep economic and social inequalities.

LOCATING SKILLS TRAINING AND PROTESTS IN SEELAMPUR

In 2014, Prime Minister Narendra Modi's first Independence Day address to the nation exalted the role of skills training programs to modernize marginalized communities. In Seelampur, billboards advertise an array of computer-training classes offered by established household names such as NIIT and Aptech, as well as local outfits like Future Ace, Kreative Arts, and Gem Future Academy. The biggest madrasa in the area, Babul Uloom, offers ICT training classes for young men and women. Many of the courses bundle ICT classes with personal-development courses that claim to teach English-speaking skills with the correct accent, hygiene, and discipline to students aspiring for inclusion in India's IT/ITeS sectors.

Skills training programs grew and flourished in the context of an enterprise culture in the 1980s that saw changes from the "nanny state" to Thatcherism in the United Kingdom and Reaganism in the United States.[6] In the 1990s, this enterprise culture traveled from western Europe and the United States to Asia, Africa, and Latin America in the context of specific national histories of neoliberal transformations.[7] In the context of declining states and austerity measures, skills programs emphasized discourses of individual initiative, risk-taking, innovation, creativity, and passion to flourish at work. In order to survive a harsh labor market, individuals were constantly pressed to improve and upgrade themselves.[8] The information economy, starting in the 1990s, emphasized technological literacy and skills training as essential for the production of the ideal global worker.[9] Soft skills training focused on personality development, appearance, and communication that were valued in the increasingly feminized service sector.[10] The emphasis on such an entrepreneurial individual, however, deflected attention from the responsibility of the government and an unprotected labor market.[11] Enterprise culture created new risks of exploitation that functioned through deception and opportunism.[12] The soul was itself put to work, and there was a greater intermeshing of work and leisure toward self-regulation and labor productivity.[13]

If leisure, creativity, and complex human emotions are intermeshed with work, then it becomes increasingly difficult for individuals to discern exploitative risks of labor and to practice resistance or moments of refusal to work. These acts of refusal thus become more clandestine. For example, data processors manipulate the system to count for faster processing speeds at a large offshore information processing facility in Barbados.[14] Indian guest workers in Berlin practice individualized freedom against immigration policies by writing spaghetti code and illegible comments that limit the code to its creator and prevent its free exchange or transfer.[15]

Although popular media and skills training project documents frame Seelampur as a technologically unplugged community within a rapidly globalizing city, the locality is home to Asia's largest denim manufacturing industry. Organized protests are not uncommon in Seelampur. On September 20, 2006, the Delhi police opened fire at hundreds of workers, small traders, and Seelampur residents who came out on the streets to protest against the forcible closure and demolition of their businesses and livelihoods that were deemed to be "illegal" and "polluting" under a law passed by the Supreme Court of India and implemented by the Municipal Corporation of Delhi (MCD). Such policies can be considered as "bourgeois environmentalism"[16] that involves a convergence of upper-class concerns about aesthetics, health and safety, and the interests of the disciplinary state in creating spaces deemed appropriate for the investment of modern capital. The English-language media labeled the Seelampur protestors as "lumpen," indicating the class-based struggles of the denim workers.

Women were absent from the Seelampur protests, although the closure of denim units impact Seelampur women who work for the industry in their homes, tacking accessories such as brass buttons to jeans and jackets. Much of Seelampur women's income-generating activities, such as tailoring clothes and home tutorials, happen behind closed doors and in the spaces of homes.

When women's everyday lives happen in the space of their homes and enclosed spaces of production, it is vital to pay close attention to the practices of women in such spaces. As anthropologist Carla Freeman argues in the context of forcibly suppressed unionization efforts, mass firing of rebellious workers, and militarization of trade zones, workers' forms of resistance can be individual and clandestine as opposed to collectively enacted.[17]

In the next sections, I examine key instances of refusal among Seelampur women that were embedded in their lived experiences inside and outside the center that questioned ICT-led development messages.

OVERTURNING REPRESENTATION: THE "MUSLIM WOMEN" AND COMPUTER TRAINING

In publicity literature for the ICT center, Seelampur is framed as a "conservative Muslim area" with "towering minarets." Founding trustee of G-Tech Foundation Meeta Mishra understood ICT skills as "giving an opportunity to Muslim women who may have never seen the world outside their neighborhood to experience the real world."[18] Seelampur women, however, lacked familial and institutional advantages that give them the desired "exposure" that were recognized as the desired attributes of a savvy IT professional.

The Seelampur women engaged with their classes despite their banter and laughter. They did their homework and regularly attended classes. The Photoshop exercise, and laughter among the students, included important subtexts that pointed to their lives outside the ICT center. When the Seelampur women laughed at the blonde woman, they were also recognizing a unique situation in which they got to be superior to her. The blonde woman's sanitized world was also similar to that of upper-class women in Delhi. Project leaders emphasized that "learning computers" gave

Seelampur women confidence to be at par with women who may have had access to more elite education. Among Seelampur women, "South Delhi girls" were equated with being more "modern" with their appearance, English-speaking skills, and access to a private-school education. For example, Salman Khan, the ICT instructor, told his class that "this course can overturn your lives and you may be like a South Delhi girl or like madam here." Khan was implying that the skills course would place the Seelampur students on par with the city's upper-class women, including me.

The ICT students often reminded me that I had the privilege of attending a private school and elite college in Delhi that had allowed me to pursue graduate school and a career in the United States. Sadiyah, who participated in the Photoshop activity, had studied at an "English medium" school in East Delhi. After completing the ICT program, she applied for a job at a large call center. Although Sadiyah could speak English and had the required technical skills, she was told that she lacked enough "exposure." "Exposure," in this context, referred to the markers of upper-class and upper-caste individuals such as the correct accent, "communication skills," and self-representation of a cosmopolitan, global individual. Saifyah Pathan, who staffed the front desk at the center, had earned superior grades in college and was considered to be an articulate individual at the center. After completing the skills program, Saifyah felt that her Seelampur address and her Muslim last name disadvantaged her in the job market. As Abbate's essay in this volume argues in the North American context, technical training does not erase race-, class-, and gender-based assumptions of what technically trained people look like. Attending elite schools, knowing the industry's cultural codes, and belonging to the right social networks shape beliefs about an individual's expertise and credibility[19]

Most Seelampur graduates did not find employment in the formal economy. Those who did were usually employed at the lower rungs of the information economy.[20] They found service-sector jobs at the ticketing counter of the Delhi metro, at lower-end call centers, at customer service positions in malls, as entry-level data operators for government programs like the unique identification number (UID) project, and at airline and tourism counters. Graduates who were employed at the metro ticketing counters and call centers visited the center to talk to the students about the pleasures of working in a clean, air-conditioned environment and wearing stylish clothes such as well-fitted shirts or short kurtas. Such clothing for pink-collar work was, however, both a site of hidden labor and pleasure.[21] The Seelampur women's insistence on the endurance of dirt and filth during play in their neighborhoods spoke to the limitations of their ICT lessons and work including experiences of stress, abuse, and soul-draining work. They faced harassment by male customers, short-term contract work without breaks, and sudden termination of contracts as part of their positions.

Although the Seelampur women were college-educated and had technical skills, they were only allowed to occupy positions at the lower rungs of the information economy. They formed the backbone of the service economy, but their contributions went unrecognized. Sexism is not a bug but a feature, as Mar Hicks points out in this volume.[22] Despite the women's training and hard work, their work was systemically devalued. Digital inclusion—or fixing the "bug" in the form of technology access and skills—was a celebrated goal for policy makers and elite IT professionals. However, it did not resolve the discrimination that Seelampur women faced that was based on their intersecting identities of gender, class, and caste as a feature of historical and larger political and social structures. In fact, technological training perpetuated and amplified such structural oppression for the women.

The blonde woman could be placed in a sterile and artificial setting, but she would never occupy the realities of everyday life in Seelampur. Nasreen's comment indicated that the Seelampur women believed that they were tougher than the woman on the computer screen. Their daily lives defied institutional discourses of "Muslim women." Zohra, who asked how the blonde woman would survive in Seelampur, was twenty years old. Her father had been employed in a denim factory in Seelampur but lost his job in 2006. The women supported the family by sewing clothes for Seelampur residents and by tacking buttons on jeans at home for the denim industry. A year after completing the program, Zohra was still waiting to hear back from several potential employers. The other women who participated in this play included Rabeena Siddiqui, whose father, Imran Siddiqui, owns a small, rapidly declining auto parts business. Rabeena had supported her family by working as a doctor's assistant and was widely respected in Seelampur for her medical knowledge. She hoped that the ICT course would lead her to get a medical imagining job. However, she realized that the course focused on basic skills, and she would need a specialized course in medical imaging that she could not afford.

When these women participated in Nasreen's joke about the white woman not surviving a day in Seelampur, they were implicitly acknowledging the harsh realities of their everyday lives. Despite being represented as "veiled Muslim women" who needed to step into the modern world through ICTs, they had previous employment and were often financial and emotional caretakers for their families. Their play pointed to their resilience in surviving their everyday lives in Seelampur in the context of deep structural inequities. They demonstrated their superiority over the white woman and in the process overturned the policy assumptions of themselves as passive recipients of modernization. Finally, the Seelampur women were constrained by their political and cultural realities. Although they realized that the skills training program was based in paternalism and translated into limited mobility, the program remained one of their few choices for striving to earn a living for their families. It was in these occasional moments of play while sequestered in the computer center on the second floor of the GRC building under the eagle eye of their instructor that the Seelampur women challenged the rigid representations of themselves in policy discourse.

THE WOMEN ACROSS THE STREET: DISRUPTING NARRATIVES OF DISCIPLINE

Practices of pleasure and enjoyment were regulated by project leaders and staff toward producing an entrepreneurial and efficient worker. What did the spontaneous activities among Seelampur women demonstrate about individualism and efficiency related to being a modern worker in India's global economy? Subaltern subjects actively select, interpret, and evaluate technologically mediated modernization messages in the context of their everyday lives.[23] Significantly, other political and social forces shape a modern self in relation to mass media such as kinship and new social relations in the city[24]

At the ICT center in Seelampur, Mishra told the students, "We teach computers here. But more importantly, we teach you to be independent, disciplined." Narratives of self-help and self-regulation were highlighted as part of the computer training program. In a series of short films produced through a UNESCO-led workshop, ICT training participants highlighted entrepreneurship and self-reliance among Seelampur women. A "digital story" slideshow presentation titled *Poonam Devi Ki Jeans Factory* (*Poonam Devi's Jeans Factory*) centered on Poonam Devi, who had separated from her husband at a young age and raised her two daughters alone. The film had been screened several times at the GRC and the community center, and Poonam Devi was upheld as a model of self-reliance. A subtext of this interesting presentation was Poonam's interview, where she asks the Delhi government to supply Seelampur with more electricity so that she can effectively do her work. This interview emphasized inequitable access to resources rather than individual determination as highlighted by the rest of the film.

Group activities represented a departure from the lecture-based learning that the youth were used to in schools. Competitions were organized based on typing speeds and knowing key terms and definitions like HTTP and URL. Participants would enact imagined scenes from an interaction with a customer in a large department store or with a manager at the call center. As the instructors pointed out, the exercises were designed to help students increase their productivity and sales. The center often had song and dance competitions that encouraged a spirit of competitiveness among different groups. The staff occasionally evoked the metaphor of the family in training sessions.

However, everyday life at the center among the Seelampur women revealed pivotal moments that challenged rigid training discourses. On a humid Monday morning in March, a computer skills training class was in session. A group of fifteen young women, between eighteen and twenty-five years of age, occupied the crowded space, listening to their instructor Ahmed Ali explain the functions of a USB drive. The windows were left open as usual to mitigate the low speed of the fan and regular power cuts. Across the narrow street from the ICT center, a grandmother, mother, and daughter stood in the balcony of their house next to a red gas cylinder. These three women listened to the computer class almost every day despite the repeated remonstrations of staff that the class was only meant for registered students. On this particular day, Shahnaz, an ICT student, pulled out a USB drive and yelled, "USB drive dekho" ("Look at the USB drive"), to the women across the street. The class erupted into laughter, and the women across the street joined in the collective camaraderie. Ahmed silenced the young women, and sternly reminded them that work in the outside world would require them to be disciplined workers. The women became mostly silent. However, occasional giggles erupted from a couple of the students during the next few minutes of Ahmad's lecture, while the women across the street continued to listen to the class.

This interplay between the neighbors and the students repeated itself in different forms throughout the year. The neighbors would applaud or laugh during class exercises, and the students maintained a steady stream of talk with the three women across the street. The ICT class participants had not collectively planned these encounters. Besides an occasional friendly wave to their neighbors, none of the women had an actual conversation with them beyond the doorstep at the center. Although they did not deliberately plot to disobey their male instructor, they derived satisfaction in the moments that they practiced indiscipline through this encounter. Their exchange with the neighborhood women also signified a different kind of community, one with more unprompted and flexible connections than the one evoked by the training discourses at the center, which constructed the corporate "family" in a similar vein as the "IBM family" highlighted in Schlombs's essay in this volume.[25]

These moments were also located in the particular political and cultural context of Seelampur in relation to everyday use of technology. Several ICT students shared cell phones, VCR and VCD players, computers, VCDs, headphones, etc. with family members and friends, especially other women. Amina's cell phone was shared among the three sisters, who took turns loading and listening to music and making calls. Fatima talked about this sharing in the context of the regular interaction with the women who lived across the street from the center. "I cannot understand why Ahmad Sir gets so angry. We lend our cell phone to our neighbor aunty to call her son in Faridabad. My friends and I share music and movies (VCDs)." Slum residents in urban India adapt technology to their everyday lives, including paralleling the practice of "cutting chai," which is sharing a cup of tea among two to four persons to cut costs.[26] ICT-led modernity is experienced in different ways according to particular national and local contexts that disrupt linear narratives of technological adaptation and change.

Seelampur women's interactions with their neighbors can further be understood in the pirate cultures operating in Seelampur and in the capital city. Denim units in Seelampur make and stitch counterfeit tags like the Levi's logo on jeans that make their way to big neighboring markets like Gandhinagar that are on the list of the Office of the United States Trade Representative, a government agency that monitors "specific physical and online markets around the world that are reported to be engaging in and facilitating substantial copyright piracy and trademark counterfeiting."[27] Corner stores and vendors in Seelampur sell VCDs of movies that are illegally copied from public viewings or are ripped off DVDs, including the reproduction of jacket covers. Embedded in the everyday world of work, piracy also creates new possibilities for subaltern groups in their interactions with the city.[28] It is important to keep in mind that Seelampur residents experience illegal markets in their everyday lives in contradictory ways. The large e-waste market in the locality is mostly unregulated and illegal. Small-time traders employ cheap migrant laborers, including children, who develop hazardous health conditions.

The interaction across the street among the different groups of women invoked indiscipline and disobedience that challenged the training narratives at the center.

As Rahma, an ICT student, said about the interactions with the neighbors, "It is not a big deal. The staff should let it go." Her comment signified a casualness with the use of technology that was counter to the lessons that the women learned at the center. This casualness or easier relationship with technology was rooted in the sharing of devices and media products that already existed among the Seelampur women beyond the doorstep of the ICT center. The disruption of claims to knowledge, which was produced through proprietary training modules and expected to be contained within the center's walls, was also placed within cultures of piracy and counterfeit. Significantly, the collective activity and resultant pleasure of Seelampur women were not geared toward competition and greater productivity at work. Instead, the exchange involved a sharing of knowledge without any expectation of a tangible return. It signified that both knowledge and pleasure existed outside capitalist boundaries of productivity.

The interactions with the neighbors were also claims to customary space shared by women in Seelampur. Most of the ICT students engaged in paid and unpaid labor at home with other women, mainly mothers and sisters. Despite repeated admonitions from the IT instructors, Seema would often run out of class to her mother's house down the street to "put on the rice." Seema lived with her parents-in-law in another neighborhood in Seelampur, and her husband worked in Muscat. Her mother was a widow, and Seema was very close to her. One morning, the IT instructor berated Seema in front of her colleagues for this regular action and told her that this behavior would get her fired in a call center. Seema retorted to Salman that she was capable of making decisions about her own life, and her mother needed her. In factory spaces, feminine rituals of baby and bridal showers can constitute women's claims to spaces that are male-controlled and exploitative.[29] Women's resistance has been imbricated with their gendered roles as wives and mothers. In the case of Seelampur women, kinship with other women fractured the intransigent and harsh space of the training center.

These moments of disruption or refusal were not part of a planned or organized process. However, they repeatedly occurred as part of activities at the ICT center, and drew from the particularities of the economic and cultural contexts in which they were embedded. The students mostly followed the disciplinary framework of the training initiative and aspired to get jobs in the service economy in call centers, metro stations, malls, government offices, etc. Inside the center and in daily interactions with other women in the neighborhood, these moments of refusal indicated a resistance to the rigid, individualistic training discourses.

CONCLUSION

In his 2005 book *The World Is Flat*, Thomas Friedman applauds the "flattening" impact of information and communication technologies on economic and social inequalities. A triumphant Friedman declares, "I was in Bangalore, India, the Silicon Valley of India, when I realized that the world was flat." In the documentary *Other Side of Outsourcing*, Friedman represents a bustling India in which young people have achieved unprecedented mobility because of their jobs in the technology industry. However, in 2017, the Indian IT and business process management (BPM) industry only directly employed 3.9 million people, which is less than 1 percent of India's population. It indirectly employed twelve million people, or less than 2 percent of India's population. Significantly, critics have argued that Friedman's celebration of the private sector in global India negates the state's historic role in actively promoting the IT sector and in establishing scientific and technical institutions.[30] Instead of equalizing disparities, IT-enabled globalization has created and further heightened divisions of class, caste, gender, religion, etc. As ecofeminist Vandana Shiva writes about Friedman, "Flat vision is a disease. . . . From this microcosm of privilege, exclusion, blindness, he shuts out both the beauty of diversity and the brutality of exploitation and inequality."[31]

Unionization and popular protests remain relatively sparse in relation to jobs in the new Indian economy. The IT industry has asserted that unions are unnecessary in the IT/ITeS sectors because HR policies protect employees' interests. In November 2017, the first trade union of IT and ITeS employees was formed in Karnataka, the state known as the IT capital of India that employees about 1.5 million people. Seelampur women aspired to be employed by the recently privatized airline industry in India. Pilots of Jet Airways, one of the major airlines, formed a union following the layoff of two thousand members of the cabin crew, who were later reinstated. However, service staff such as cabin and ground crew are not unionized. In a global context, it is worth noting that in the United States, about 60 percent of the jobs added since 2010 have been in low-wage service sector jobs and not in "knowledge work."[32] Protestors have gathered in Seattle, Berlin, and other places to protest against the tech giant Amazon, whose warehouse workers are subject to surveillance and harsh conditions that include surveillance devices strapped on to their bodies and to peeing in bottles instead of taking bathroom breaks.

In Seelampur, women form an invisible face of the labor force and are absent from popular protests. While designing participation programs such as the one by

UNESCO, facilitators have focused on individual success stories of entrepreneurship in Seelampur. They have ignored popular voices and protests against unequal globalization in India, such as the protests by male denim workers in Seelampur that have been forcibly suppressed by the state.

The Seelampur women at the ICT center realized that technology's promise to flatten inequalities was limited in the context of their everyday lives. For these women, acquiring skills did not set them free as policy makers imagined it would. Popular skills programs, such as the modular Seelampur course, mainly produce employment at the lower rung of the information economy that is temporary, gendered, and vulnerable to exploitation. As Halcyon Lawrence argues about speech technologies like Siri that are embedded with accent bias, such technologies should be tools that reflect our culture and identity.[33] In the case of Seelampur, the women not only recognized such limitations of technology but actively protested, resisted, and appropriated ICT in ways that reflected their individual and collective identities through pivotal moments of play and pleasure. The participants actively overturn essentialist representations of themselves as passive recipients of modernization and critically interact with these development messages. Their Photoshop play highlights critiques of economic and social mobility promised by the program and structural inequities in their lives outside the center. In their interactions with the neighbors across the street, the Seelampur women's conduct is deliberately positioned against narratives of self-regulation and against proprietary notions of ownership of knowledge.

As India emerges as an exemplar of technology-led inclusion in the Global South and the fire of techno-optimism travels more swiftly in global contexts, specific and grounded histories such as those of the Seelampur women are vital to understand the actual practices of working-class individuals at spaces of work and training as they interface with information and communication technologies. The crucial moments of play and refusal by these women at sites of labor reveal deep fissures and limitations in the promise of the information society for inclusion of marginalized communities. While these moments of play may seem spontaneous and fleeting, they represent another dimension of "fire" through the immediate material lives of people and their fraught practices of technology-enabled labor in the "new economy."

NOTES

1. Sreela Sarkar, "Beyond the 'Digital Divide': The Computer Girls of Seelampur," *Feminist Media Studies* 16, no. 6 (2016): 963–983.

2. Zoya Hasan and Ritu Menon, *Unequal Citizens: A Study of Muslim Women in India* (Delhi: Oxford University Press, 2004), 18.

3. Janet Abbate, "Coding Is Not Empowerment," this volume.

4. C. J. Fuller and Haripriya Narasimhan, "Engineering Colleges, 'Exposure' and Information Technology," *Economic and Political Weekly* 41, no. 3 (January 2006): 258–262.

5. Pierre Bourdieu, "The Aristocracy of Culture," *Media, Culture & Society* 2 (1980): 225–254.

6. Shaun S. Heap, Mary Douglas, and Angus Ross, *Understanding the Enterprise Culture: Themes in the Work of Mary Douglas* (Edinburgh: Edinburgh University Press, 1992), 1; Michael A. Peters, "Education, Enterprise Culture and the Entrepreneurial Self: A Foucauldian Perspective," *Journal of Education Inquiry* 2, no. 2. (May 2001): 58–71; Nikolas Ross, "Governing the Enterprising Self," in *The Values of the Enterprise Culture* (London: Routledge), 141–164.

7. Paula Chakravartty and Sreela Sarkar, "Entrepreneurial Justice: The New Spirit of Capitalism in Emerging India," *Popular Communication: The International Journal of Media and Culture* 11, no. 1 (February 2013): 58–75.

8. Nandini Gooptu, "Neoliberal Subjectivity, Enterprise Culture and New Workplaces: Organized Retail and Shopping Malls in India," *Economic and Political Weekly* 44, no. 2 (May 30–June 15, 2009): 45–54; Peters, "Education, Enterprise Culture and the Entrepreneurial Self," 61.

9. Peters, "Education, Enterprise Culture and the Entrepreneurial Self," 63.

10. Gooptu, "Neoliberal Subjectivity, Enterprise Culture and New Workplaces," 46.

11. Peters, "Education, Enterprise Culture and the Entrepreneurial Self," 59.

12. Sreela Sarkar, "Passionate Producers: Corporate Interventions in Expanding the Promise of the Information Society," *Communication, Culture & Critique* 10, no. 2 (June 2017): 241–260.

13. Franco Berardi, *The Soul at Work: From Alienation to Autonomy* (Cambridge, MA: MIT Press, 2009).

14. Carla Freeman, *High Tech and High Heels in the Global Economy* (Durham, NC: Duke University Press, 2000), 153.

15. Sareeta Amrute, *Encoding Race, Encoding Class: Indian IT Workers in Berlin* (Durham, NC: Duke University Press, 2016), 88.

16. Amita Baviskar, "Between Violence and Desire: Space, Power, and Identity in the Making of Metropolitan Delhi" *International Social Science Journal* 55, no. 175 (July 2004).

17. Freeman, *High Tech and High Heels in the Global Economy*, 207.

18. Interview with Meeta Mishra, July 1, 2014.

19. Abbate, "Coding Is Not Empowerment," this volume.

20. Sreela Sarkar, "Women at Work and Home: New Technologies and Labor among Minority Women in Seelampur," *Journal of Community Informatics* 5, no. 3–4 (July 2010).

21. Freeman, *High Tech and High Heels in the Global Economy*, 223.

22. Mar Hicks, "Sexism Is a Feature, Not a Bug," this volume.

23. Lila Abu-Lughod, *Dramas of Nationhood: The Politics of Television in Egypt* (Chicago: University of Chicago Press, 2005), 12.

24. Abu-Lughod, *Dramas of Nationhood*, 12.

25. Corinna Schlombs, "Gender Is a Corporate Tool," this volume.

26. Nimmi Rangaswamy and Nithya Sambasivan, "Cutting *Chai*, *Jugaad*, and *Here Pheri*: Towards UbiComp for a Global Community," *Pers Ubiquit Comput* 15 (April 2011): 553–564.

27. Abhishek Dey, "You May Not Have Heard of This 'Notorious Market' in Delhi but the US Has It on Their Watchlist," January 8, 2017, https://scroll.in/article/826095/you-may-not-even-have-heard-of-this-notorious-market-in-delhi-but-the-us-has-it-on-its-watchlist.

28. Ravi Sundaram, *Pirate Modernity: Delhi's Media Urbanism* (New York: Routledge, 2011).

29. Sallie Westwood, *All Day, Every Day: Factory and Family in the Making of Women's Lives* (Urbana: University of Illinois Press, 1985), 45.

30. Statista, "Direct and Indirect Employment of the IT-BPM Industry in India from Fiscal Year 2019–2017 (in millions)" (February 5, 2019), https://www.statista.com/statistics/320729/india -it-industry-direct-indirect-employment/; Peter Evans, *Embedded Autonomy: States and Industrial Transformation* (Princeton, NJ: Princeton University Press, 1995); and Kavita Philip, "Telling Histories of the Future: The Imaginaries of Indian Technoscience," *Identities: Global Studies in Culture and Power* 23, no. 3 (2016): 276–293.

31. Vandana Shiva, "The Polarised World of Globalisation," Global Policy Forum, May 10, 2005, https://www.globalpolicy.org/globalization/defining-globalization/27674.html.

32. Nicholas Garnham, "Information Society Theory as Ideology," *Studies in Communication Sciences* 1, no. 1 (2009): 129–166.

33. Halcyon M. Lawrence, "Siri Disciplines," this volume.

15

PLATFORMS ARE INFRASTRUCTURES ON FIRE

Paul N. Edwards

Highways, electric power grids, the internet, telephones, pipelines, railroads: we call these things "infrastructures." They're the large-scale, capital-intensive essential systems that underlie modern societies and their economies. These infrastructures took decades to develop and build, and once established, they endure for decades, even centuries. Infrastructures may be slow, but they don't burn easily.

Once you have a few infrastructures in place, you can build others "on top" of them by combining their capabilities. UPS, FedEx, and similar services join air, rail, and trucking to deliver packages overnight, using barcodes and computerized routing to manage the flow. National 911 services deploy telephone and radio to link police, fire, ambulance, and other emergency services. The internet is a gigantic network of computer networks, each one built separately. These second-order infrastructures seem to present a different temporality—a different sense and scale of time—in many cases due to the rise of networked software platforms. Today's platforms can achieve enormous scales, spreading like wildfire across the globe. As Facebook and YouTube illustrate, in just a few years a new platform can grow to reach millions, even billions, of people. In cases such as Airbnb and Uber, platforms set old, established systems on fire—or, as their CEOs would say, "disrupt" them. Yet platforms themselves burn much more readily than traditional infrastructures; they can vanish into ashes in just a few years. Remember Friendster? It had 115 million users in 2008. What about Windows Phone, launched in 2010? Not on your radar? That's my point. Platforms are fast, but they're flammable.

In this chapter I argue that software platforms represent a new way of assembling and managing infrastructures, with a shorter cycle time than older, more capital-intensive counterparts. I then speculate about the future of platform temporalities, drawing on examples from apartheid South Africa and contemporary Kenya. These examples suggest that African infrastructures, often portrayed as backward or lagging, may instead represent global futures—leapfrogging over the slower, heavier processes of more typical infrastructure.

TEMPORALITIES OF INFRASTRUCTURE

"Infrastructure" typically refers to mature, deeply embedded sociotechnical systems that are widely accessible, shared, and enduring. Such systems are both socially shaped and society-shaping. Major infrastructures are not optional; basic social functions depend on them. Many are also not easily changed, both because it would be expensive and difficult to do so, and because they interact with other infrastructures in ways that require them to remain stable. Archetypal infrastructures fitting this definition include railroads, electric power grids, telephone networks, and air traffic control systems. A large subfield of infrastructure studies, comprising history, anthropology, sociology, and science and technology studies, has traced many aspects of these systems.[1]

Susan Leigh Star, a sociologist of information technology, famously asked, "When is an infrastructure?" Her question calls out the ways infrastructure "emerges for people in practice, connected to activities and structures." Systems rarely function as infrastructure for everyone all of the time, and one person's smoothly functioning infrastructure may be an insurmountable barrier to another, as sidewalks without curb cuts are to people in wheelchairs.[2] Here I want to ask related but different questions: How fast is an infrastructure? What are the time frames of infrastructure, as both historical and social phenomena?

A temporal pattern is clearly visible in the case of "hard" physical infrastructures such as canals, highways, and oil pipelines. Following an initial development phase, new technical systems are rapidly built and adopted across an entire region, until the infrastructure stabilizes at "build-out." The period from development to build-out typically lasts between thirty and one hundred years.[3] This consistent temporal range is readily explained by the combination of high capital intensity; uncertainty of returns on investment in the innovation phase; legal and political issues, especially regarding rights-of-way; and government regulatory involvement.

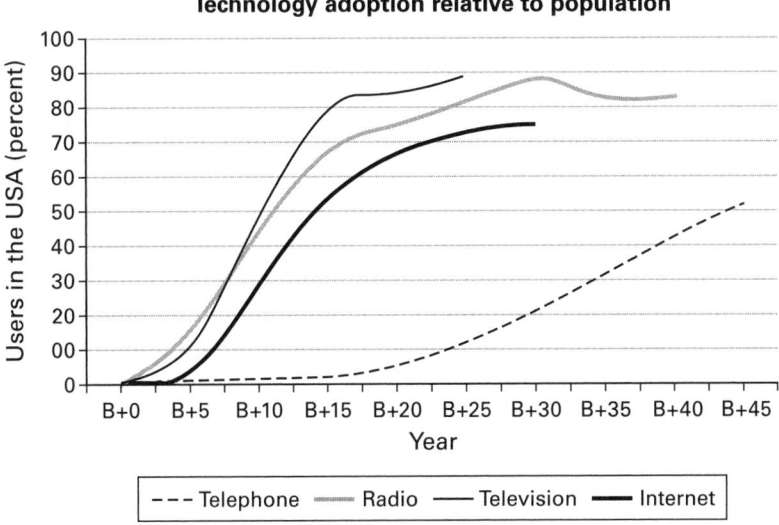

Figure 15.1 Technology adoption relative to population in the United States, starting from base year B defined as the year of commercial availability: telephone 1878, radio 1920, television 1945, and internet 1989. B+5 is 5 years after commercialization, and so on. (*Source*: Graph after Gisle Hannemyr, "The Internet as Hyperbole: A Critical Examination of Adoption Rates," *Information Society* 19, no. 2 [2003], figure 2, extended with additional data from the Pew internet survey.)

Major communication infrastructures display a similar pattern. On one account, radio, television, and the internet all took about twenty years to reach 80 percent of the US population (figure 15.1),[4] but these time lines might be closer to thirty to forty years if the innovation and early development phases were included. The telephone network took much longer, but it was the first personal telecommunication system, and unlike the other three, it required laying landlines to every home and business. The rapid spread of radio and television resulted in part from the lesser capital intensity of their original physical infrastructure, which reached thousands of receivers through a single broadcast antenna. Later, of course, cable television required large capital investments of the same order as landline telephony. The internet relied mainly on preexisting hardware, especially telephone and TV cables, to connect local and wide-area computer networks to each other.

As they mature, infrastructures enter another temporality, that of stability and endurance. Major infrastructures such as railroads, telephone networks, Linnaean taxonomy, interstate highways, and the internet last for decades, even centuries.

They inhabit a temporal mesoscale, enduring longer than most people, corporations, institutions, and even governments.

Each new generation is thus born into a preexisting infrastructural landscape that presents itself as a quasi-natural background of lifeways and social organization.[5] Infrastructure enrolls its inhabitants in its own maintenance, whether passively as consumers (whose payments support it) or actively as owners, maintainers, or regulators of infrastructural systems.[6] Thus, infrastructures are "learned as part of membership" in communities, nations, and lifeworlds.[7] The qualities of ubiquity, reliability, *and especially durability* create a nearly unbreakable social dependency—including the potential for social and economic trauma during episodes of breakdown, such as urban blackouts or major internet outages.

To answer the question "How fast is an infrastructure?": historically, major infrastructures appear to have shared a thirty- to one-hundred-year growth trajectory. As they spread, they became deeply embedded in social systems and intertwined with other infrastructures, increasing human capabilities but simultaneously inducing dependency. Societies then found themselves locked into temporally indefinite commitments to maintain them. Yet infrastructures' very invisibility and taken-for-grantedness have also meant that the financial and social costs of maintenance are borne grudgingly—and are frequently neglected until an advanced state of breakdown presents a stark choice between disruptive major repairs and even more disruptive total failure.[8] Infrastructures are slow precisely because they represent major commitments of capital, training, maintenance, and other social resources.

SECOND-ORDER LARGE TECHNICAL SYSTEMS: PLATFORMS AS FAST INFRASTRUCTURES

A third temporality appears when we consider what the sociologist Ingo Braun named "second-order large technical systems" (LTSs) built on top of existing infrastructures.[9] Braun's example was the European organ transplant network, which uses information technology to integrate emergency services, air transport, and patient registries to locate fresh donor organs, match them to compatible transplant patients, and deliver them rapidly across the continent. Other examples include the global supply chains of huge enterprises such as Walmart, Ikea, and Alibaba, emerging systems of trans-institutional electronic patient records, and large-scale "enterprise management" software.

Certain major infrastructures introduced since the 1970s are second-, third-, or nth-order LTSs. Software is the critical core element. For example, as the internet emerged and became publicly available from the mid-1980s through the 1990s, it relied mainly on existing equipment constructed for other purposes: computers, electric power, telephone lines, and TV cables. The critical elements that turned this hodgepodge of gear into "the internet" were software and standards: the TCP/IP protocols that govern how data is packetized and routed from one network to another, and the domain name system that governs internet addresses. Similarly, the World Wide Web rides "on top of" the internet. With rising demand for ever higher bandwidth, the internet has increasingly become a physical infrastructure project in its own right, requiring dedicated undersea cables, fiber-optic landlines, and server farms to handle exponentially increasing traffic—but telephone, TV cable, and cellular telephony, all originally installed as part of other infrastructures, remain the principal modes of last-mile delivery. The web, by contrast, is constructed entirely from standards, protocols (HTML, HTTP, etc.), and software such as web browsers, and it is filled with content from millions of sources, including individuals, firms, news agencies, and governments.

The currently popular vocabulary of "platforms" reflects the increasing importance of software-based second-order infrastructures. The origins of platform terminology, however, predate the role of software. In the early 1990s, management and organization studies researchers began to identify "platforms" as a generic product strategy applicable to almost any industry. For these scholars and practitioners, platforms are architectures comprising three key elements:

- Core components with low variability (the platform)
- Complementary components with high variability
- Interfaces that connect core and complementary components

The platform strategy lowers the cost of variation and innovation, because it avoids designing entirely new products to address related but different needs. A celebrated example is the Chrysler K-car platform (1981–1988), essentially a single chassis and drive train built to accommodate many different car and truck bodies. This approach dates to the early days of the American automobile industry, when Ford fitted its Model T chassis with bodies ranging from open touring cars to sedans to trucks. (There was even a snowmobile.) Successful platforms often attract ecosystems of smaller firms, with producers of complementary components and interfaces forming loose, "disaggregated clusters" around the producer of the core component.[10] In the 1990s, management scholars promoted "platform thinking" as a generic corporate strategy.[11]

Also in the 1990s, the computer industry adopted the "platform" vocabulary, applying it agnostically to both hardware and software. Microsoft described its Windows operating system as a platform, while Netscape defined a "cross-platform" strategy (i.e., availability for all major computer operating systems) for its web browser. Some computer historians argue that the so-called IBM PC, introduced in 1981, should really be known as the "Microsoft-Intel PC"—or platform. Lacking control of its core components (Microsoft's operating system and the Intel chips on which the OS ran), IBM rapidly lost dominance of the PC market to reverse-engineered clones of its design, manufactured by Compaq, Hewlett-Packard, Dell, and other firms. Arguably, IBM's loss of control actually made the Microsoft-Intel PC platform even more dominant by driving down prices.[12]

Web developers soon extended the computer industry's notion of platform to web-based applications, abandoning the previous model of purchased products in favor of subscriptions or rentals. As web guru Tim O'Reilly put it, in many cases today "none of the trappings of the old [packaged or purchased] software industry are present . . . No licensing or sale, just usage. No porting to different platforms so that customers can run the software on their own equipment."[13] For web-based "platforms," the underlying hardware is essentially irrelevant. Web service providers own and operate servers, routers, and other devices, but the services themselves are built entirely in software: databases, web protocols, and the APIs (application program interfaces) that permit other pieces of software to interact with them.

APIs act like software plugs and sockets, allowing two or more pieces of unrelated software to interoperate. APIs are thus readily described as interfaces between core components (e.g., operating systems, browsers, Facebook) and complementary components (e.g., Android apps, browser and Facebook plug-ins). Using APIs, this modular architecture can be extended indefinitely, creating chains or networks of interoperating software. Since the early 2000s, the dramatic expansions of major web-based companies such as Google, Facebook, and Apple have demonstrated the power of the platform strategy.

This detachment from hardware corresponds with a layering phenomenon commonly observed in computing: software depends on and operates within hardware, yet it can be described and programmed without any knowledge of or even any reference to that hardware. Higher-level applications are built on top of lower-level software such as networking, data transport, and operating systems. Each level of the stack requires the capabilities of those below it, yet each appears to its programmers as an independent, self-contained system.

The key ingredient for many platforms is user contribution: product and movie reviews, videos, posts, comments and replies, and so on. Platforms also invisibly capture data about users' transactions, interests, and online behavior—data which can be used to improve targeted marketing and search results, or to serve more nefarious agendas. App developers also furnish content of their own, as well as functionality and alternative interfaces.

Digital culture scholar Tarleton Gillespie notes that social media companies such as YouTube and Facebook deploy the term "platform" strategically, using its connotations to position themselves as neutral facilitators and downplay their own agency. Recent public debates about the legal and regulatory status of Uber and Airbnb illustrate this strategy. Unlike taxi companies and hotels, these enterprises started with neither cars nor buildings, presenting themselves instead as platforms that "merely" connect car or property owners with potential customers. In this context, "platform" is both "specific enough to mean something, and vague enough to work across multiple venues for multiple audiences," such as developers, users, advertisers, and (potentially) regulators.[14]

Thus, a key role of what we might call "platform discourse" is to *render the platform itself as a stable, unremarkable, unnoticed object,* a kind of empty stage, such that the activity of users—from social media posts to news, videos, reviews, connecting travelers with drivers and apartments for rent—obscures its role as the enabling background. As Sarah Roberts argues in this volume, platforms "operate on a user imagination of . . . unfettered self-expression, an extension of participatory democracy that provides users a mechanism to move seamlessly from self to platform to world." Aspirationally, platforms are like infrastructures: widely accessible, reliable, shared, "learned as part of membership" in modern social worlds—and to this extent, no parent of teenagers possessed of (by?) smartphones could disagree.

PLATFORMS AND THE "MODERN INFRASTRUCTURAL IDEAL"

City governments and planners of the mid-nineteenth century began to conceive cities as coherent units furnishing certain key services, such as roads, sewers, police, emergency services, and public transportation, as public goods—a vision the urban sociologists Simon Graham and Simon Marvin call the "modern infrastructural ideal."[15] Some infrastructures originated as private enterprises; as the modern infrastructural ideal took hold, many became publicly regulated monopolies. Many national governments also provided and regulated railroads, highways, PTTs (post, telegraph, and telephone services), and other infrastructures—including the early internet.

The modern infrastructural ideal began to decline in the late 1970s, as neoliberal governments sought to shift public services to private enterprise. Rather than operate or oversee monopoly suppliers of public goods, these governments wanted to break up those monopolies so as to increase competition. As a corollary, they renounced the public-facing responsibilities implied by the modern infrastructural ideal.

The history of networked computing offers a striking view of this tectonic shift. In the 1960s, many analysts saw computer power as a significant resource that might be supplied by what they called a "computer utility." Such a utility would own and operate huge computers; using the then-new technology of time-sharing, thousands of people could share these machines, just as electric utility customers share the output of huge electric power plants. Economies of scale, they argued, would reduce prices. The computer utility would provide, on demand, sufficient computer power for almost any task.[16] At a time when a single computer could easily cost over $500,000 (in 2019 dollars), this argument made excellent sense. By the late 1960s, in fact, it became the business model for companies such as the very successful CompuServe. Industry observers traced out the logic of the modern infrastructural ideal: computer services would eventually become public, regulated monopoly utilities.[17]

Early internet history, from the late 1960s through the late 1980s, also traced the terms of the modern infrastructural ideal. Large government investments funded development and rollout, via the US Defense Department's Advanced Research Projects Agency and later the US National Science Foundation (NSF). Justifications for this support were first military, later scientific, but always had a compelling public purpose in view. Exactly as predicted by the "computer utility" model, the NSF required universities it provisioned with supercomputers and networks to connect other, less-well-resourced institutions.[18] The single most complete realization of the computer utility model was the French Minitel, introduced in 1980. Reaching 6.5 million French citizens by 1990, the government-owned and -operated Minitel used centralized servers communicating with dumb terminals over existing telephone lines. Minitel hosted (and profited from) numerous commercial services, but also offered free public services such as a national telephone directory. Minitel was explicitly developed as a public good, with terminals distributed at no cost to millions of French households and also available at post offices.[19]

Just as it did with other infrastructures, the United States mostly abandoned the public-good model of the internet in the late 1980s (although the federal government continues to regulate some aspects of internet services). The crucial steps in this splintering process were privatization of internet backbone lines (from 1987) and removal of restrictions on commercial use (from 1992). The history of networked computing can thus be seen as the transformation of a traditional monopoly infrastructure model into the deregulated, privatized, and splintered—we might say "platformized"—infrastructure model prevalent in many sectors today.[20]

To summarize, the rise of ubiquitous, networked computing and changing political sentiment have created an environment in which software-based platforms can achieve enormous scales, coexist with infrastructures, *and in some cases compete with or even supplant them*. With my colleagues Plantin, Lagoze, and Sandvig, I have argued elsewhere that two of today's largest web firms, Google and Facebook, display characteristics of both platforms and infrastructures.[21] Like platforms, they are second-order systems built "on top of" the internet and the web, and they provide little content of their own. Like infrastructures, they have become so important and so basic to daily life and work in large parts of the world that their collapse would represent a catastrophe. Further, these firms have invested substantially in physical systems, such as undersea cables surrounding the African continent.[22]

As Nathan Ensmenger shows in this volume, "the cloud" is really a factory: not just virtual but also physical and human. Platforms are no exception. Each one requires servers, routers, data storage, miles of cable, physical network connections, and buildings to house them all. Each involves a human organization that includes not only programmers but also accountants, maintainers, customer service people, janitors, and many others. And as Tom Mullaney observes in the introduction, like the assembly lines of the early twentieth century (and like most traditional infrastructures), physical fire—burning fossil fuels—powers much of their activity (though Google and Apple, in particular, have invested heavily in wind, solar, and geothermal power, aiming for 100 percent renewable balance sheets across their operations in the very near future).[23]

Yet the cloud also differs from factories in a key respect: because they are essentially made from code, platforms can be built, implemented, and modified *fast*. And as n^{th}-order systems, their capital intensity is low relative to such infrastructures such as highways, power grids, and landline cables. My argument is that this flexibility and low capital intensity gives platforms a wildfire-like speed, as well as an unpredictability and ephemerality that stems from these same characteristics: competing systems can very quickly supplant even a highly successful platform, as Facebook did Myspace. Far from being stable and invisible, software platforms are mostly in a state of constant flux, including near-daily states of emergency (data breaches, denial of service attacks, spam). In major cases, such as Twitter and Facebook since 2016, their states of emergency are principal subjects of national and international politics, news, and anguished conversation. Platforms are infrastructures on fire.

THREE PLATFORMS IN AFRICAN CONTEXTS

In this section, I trace the histories of three platforms significant to various African contexts. FidoNet, the "poor man's internet," became the first point of contact with electronic mail in many African countries; here I focus on its role in South Africa, where it helped anti-apartheid activists communicate with counterparts both inside and outside the country. M-Pesa, a mobile money system first rolled out in Kenya in 2007, is already used by the majority of Kenyans; it is fast emerging as a parallel financial infrastructure outside the traditional banking system. Finally, Facebook's low-bandwidth "Free Basics" (a.k.a. internet.org) platform for mobile phones has already become a staple of daily life for some 95 million Africans.

FIDONET

FidoNet began in the 1980s as computer networking software supporting the Fido Bulletin Board System (BBS) for personal computers. Like other bulletin board services, it allowed users to send email, news, and other text-only documents to other FidoNet users. By the late 1980s, it could connect to the emerging internet by means of UUCP (Unix-to-Unix copy) gateways. Created in San Francisco in 1984 by the Fido BBS developer, Tom Jennings, FidoNet ran over telephone lines using modems. In that year, the AT&T monopoly had just been unbundled, but despite the emergence of competitors, long-distance telephone calls remained quite expensive in the United States. All providers charged differential rates: highest during the business day, lowest usually between eleven p.m. and seven a.m. The FidoNet approach took advantage of these differential rates to exchange messages in once-daily batches during a late-night "zonemail hour," dependent on time zone and region. While many FidoNet nodes remained open at other times, zonemail hour ensured that all nodes could take advantage of the lowest-cost alternative.

Dozens of other networking systems were also developed in the mid-1980s. A 1986 review article listed FidoNet alongside some twenty-six other "notable computer networks," as well as six "metanetworks" such as the NSFNET.[24] These included Usenet, BITNET (an academic network), and the fledgling internet itself. Each of these deployed its own unique addressing and communication techniques, leading to a cacophony of largely incompatible standards. Only in the 1990s did the internet's TCP/IP protocols become the dominant norm.[25]

Usenet, BITNET, NSFNET, and the internet were initially populated mainly by university faculty and students working on time-shared mainframes and minicomputers, using dumb terminals rather than personal computers (PCs). As PCs spread in the 1980s, commercial "networks" such as Prodigy (founded 1984) and America Online (circa 1985 as an online services provider) also sprang up, aimed at private individuals and small businesses. In practice, these walled-garden online services operated like large-scale, centralized BBSs, rather than like networks in today's sense of the term. Most of them deliberately prevented users from communicating with other computer networks, which they viewed as a threat to their revenue. FidoNet presented a low-budget, decentralized alternative to these commercial and academic networks and online services.

The FidoNet software was designed specifically for small local operators using only PCs as hosts (i.e., servers). Originally written for Microsoft's MS-DOS PC operating system, it was also eventually ported to other personal computer types, as well as to Unix, MVS, and other minicomputer operating systems. However, DOS-based PCs always comprised the large majority of FidoNet nodes, especially outside the US and Europe.[26] FidoNet had its own email protocols, netmail and echomail. Netmail was designed for private email, but "due to the hobbyist nature of the network, any privacy between sender and recipient was only the result of politeness from the owners of the FidoNet systems involved in the mail's transfer."[27] Echomail performed essentially the same function as UUCP: it copied large groups of files from one node to another. This capability could be used both to send batches of many email messages destined for individuals and to broadcast news automatically to other FidoNet nodes.

FidoNet became the base for a PC-based, grassroots collective of local BBS operators who sought to develop long-distance email and shared news services. Unlike most of the commercial services and academic networks, FidoNet software was freely distributed and required no particular affiliation. Many operators charged fees to support their operations and recover the costs of the nightly long-distance calls, but in general these fees were low. Randy Bush, a major promoter of FidoNet, emphasized the "alternative infrastructure" aspect of this work: "From its earliest days, FidoNet [was] owned and operated primarily by end users and hobbyists more than by computer professionals. . . . Tom Jennings intended FidoNet to be a cooperative anarchism [sic] to provide minimal-cost public access to email." A quasi-democratic governance system evolved in which node operators elected local, regional, and zone coordinators to administer the respective levels.

As for the technical underpinnings of this "anarchism," Bush wrote:

> Two very basic features of FidoNet encourage this. Every node is self-sufficient, needing no support from other nodes to operate. But more significant is that the nodelist contains the modem telephone number of all nodes, allowing any node to communicate with any other node without the aid or consent of technical or political groups at any level. This is in strong contrast to the uucp network, BITNET, and the internet.[28]

In addition to the all-volunteer, anarchical quality of the organization and its technical standards, developers' commitment to keeping costs low soon gave FidoNet a reputation as "the poor man's internet." Starting in 1986, FidoNet administrators published a series of simple standards documents. These allowed distant participants with relatively minimal computer skills to set up FidoNet nodes.

The number of FidoNet nodes grew rapidly. Starting at twelve nodes in 1984, it hit 10,000 nodes in 1990, mushrooming to nearly 40,000 nodes in 1996. It then declined, nearly symmetrically with its growth, dropping below 10,000 nodes again in 2004 as a direct result of the spread of internet protocols. Although some 6,000 nodes were still listed in 2009, these saw little traffic. Today the network exists only as a nostalgia operation.[29]

In the mid-1980s, progressive nongovernmental organizations (NGOs) began to build their own dial-up bulletin board systems, such as GreenNet (UK), EcoNet/PeaceNet (California), and many others. These activist networks soon allied and consolidated, forming in 1986 the Institute for Global Communications based in the USA, and in 1990 the much larger Association for Progressive Communications (APC), whose founders included NGO networks in Brazil and Nicaragua as well as the Global North. The critical importance to activists of creating and running their own online networks—as well as the chaotic diversity of the online world emerging in the late 1980s—was captured by one of APC's founders, writing in 1988:

> Countless online commercial services [offer] data bases [sic] of business, academic, government and news information. DIALOG, World Reporter, Data Star, Reuters, Dow Jones. You can easily spend up to $130 (£70) an hour on these services (GreenNet

costs £5.40 an hour, PeaceNet costs $10 an hour peak—half at night or weekends). Then there are the information supermarkets—The Source or Compuserve in the US, Telecom Gold in the UK—offering electronic mail, stock market information, and conferencing to the general public. Finally, there are the non-profit academic and special interest networks—Bitcom, Janet, Usenet, MetaNet. All of these data bases and networks—except the last group—are owned and operated by large corporations.

How are the APC networks different? They are a telecommunications service closely linked to citizen action. They are non-profit computer networks connecting over 3000 users and 300 organizations working for the future of our planet. They enable immediate, cost-effective exchange of information between people in 70 countries . . . working on peace, environmental, economic development, and social justice issues. Their present operation, history, and future development have no real parallels in the communications industry. . . . [APC] email facilities include . . . gateways for sending messages to users on more than twenty commercial and academic networks. . . .[30]

By the late 1980s, numerous NGOs in the Global South had joined this movement. Many used FidoNet to link with APC networks. In 1990, in London, Green-Net's Karen Banks set up the GnFido (GreenNet Fido) gateway node, which translated between FidoNet protocols and the UUCP protocol widely used in the Global North; her work made it possible for African, South Asian, and Eastern European NGOs to use FidoNet to communicate directly with internet users.[31] Banks maintained GnFido until 1997; in 2013, this work earned her a place in the Internet Hall of Fame alongside FidoNet promoter Randy Bush and South Africa NGONet (SANGONet) founder Anriette Esterhuysen.

FidoNet standards had many advantages for developing-world conditions:

The FidoNet protocol was a particularly robust software [sic], which made it very appropriate for use in situations where phone line quality was poor, electricity supply was unreliable, costs of communications were expensive, and where people had access to low specification hardware. . . . FidoNet provided very high data compression, which reduced file size and therefore reduced transmission costs . . . and it was a "store-and-forward" technology (meaning people could compose and read their email offline, also reducing costs).[32]

These were not the first uses of FidoNet in Africa, however. In the late 1980s, South African computer scientist Mike Lawrie, working at Rhodes University in Grahamstown, met Randy Bush at a conference in the United States and learned about FidoNet. He convinced Bush to help him and his colleagues communicate with counterparts outside the country. At that time, due to international sanctions against the apartheid government, South Africa was prohibited from connecting directly to the internet. But FidoNet had gateways to the UUCP network, which could be used to transmit email to internet users. The necessary conversions and addressing schemes were not exactly simple; Lawrie details a twelve-step process for sending and receiving internet mail via FidoNet to UUCP. Bush allowed the Rhodes group to place calls to the FidoNet node at his home in Oregon. Through that node, they could then access the internet; for about a year, this was South Africa's only email link to the USA.[33]

The attractions of the low-tech FidoNet platform and its anarchist political culture ultimately proved insufficient to maintain FidoNet as an alternative infrastructure. Instead, the desire to communicate across networks led to FidoNet's demise. Like the Rhodes University computer scientists, the activist networks soon created "gateways" or hubs that could translate their UUCP traffic into the previously incompatible format used by FidoNet (and vice versa), enabling worldwide communication regardless of protocol.

FidoNet's temporal profile thus looks very different from that of an infrastructure like a highway network. Its phase of explosive growth lasted just five years, from about 1991 to 1996. Its decline was nearly as rapid. By 2010 it was all but dead. As a second-order communication system based on PCs and long-distance telephony, capable of connecting to the internet but using its own "sort of" compatible addressing scheme, FidoNet was rapidly replaced as internet protocols and services spread throughout the world. Yet due to its simplicity and low cost, for over a decade FidoNet played a disproportionate role in connecting the African continent to the rest of the world via email. For many African users, FidoNet was the only alternative. This "poor man's internet" served them, however briefly, as infrastructure.

M-PESA

M-Pesa is a cellphone-based mobile money system, first rolled out in Kenya in 2007 by mobile phone operator Safaricom. ("Pesa" means "money" in Swahili.) When clients sign up for M-Pesa, they receive a new SIM card containing the M-Pesa software. Clients "load" money onto their phones by handing over cash to one of the 160,000+ authorized M-Pesa agents. Friends, relatives, or employers can also load money to the client's phone. Clients can use the money to buy airtime or to pay bills at hundreds of enterprises.

To date, however, clients have used M-Pesa principally for "remittances," the common practice of wage-earning workers sending money to family or friends elsewhere. M-Pesa offers a safe, secure method of storing money in an environment where many workers are migrants supporting family members in remote areas and robbery is common, especially when traveling. Businesses now use M-Pesa for mass payments such as payrolls—not only a major convenience but also a far more secure method than cash disbursements. The service charges small fees (maximum $1–3 US dollars) for sending money or collecting withdrawals from an agent. Safaricom earns more on large business transactions.

M-Pesa has enjoyed enormous success in Kenya. By 2011, just four years after it was launched, over thirteen million users—representing some 70 percent of all Kenyan households—had signed up.[34] Other mobile phone companies quickly mounted their own mobile money systems, but none have gained nearly as much traction. According to Safaricom, in 2015 some 43 percent of the country's $55 billion GDP passed through the system.[35] By the end of 2018 M-Pesa subscribers in Kenya numbered 25.5 million—half the country's total population.

While M-Pesa originally focused on domestic transfers among Kenyans, Safaricom and part-owner Vodacom (South Africa) have since expanded the business to seven other African countries. M-Pesa also operates in Romania and Afghanistan, as well as in India, one of the world's largest markets for such services. Given that international remittances constitute a $300 billion global business, prospects seem very bright for a low-cost, user-friendly mobile money system that can operate on even the simplest cellphones.

The M-Pesa project began in 2003 with a handful of individuals at London-based Vodafone, as a response to the United Nations' Millennium Development Goals. Supported in part by "challenge funds" from the UK government for development work, these developers formed a partnership with Sagentia, a small British technology consulting group (now known as Iceni Mobile), and Faulu Kenya, a microfinance firm that would furnish a business test bed. Just as with FidoNet, these designers sought simplicity and extreme low cost. Project leaders Nick Hughes and Susie Leonie, in their account of its early days, note that "the project had to quickly train, support, and accommodate the needs of customers who were unbanked, unconnected, often semi-literate, and who faced routine challenges to their physical and financial security."[36]

Since M-Pesa would have to run on any cellphone, no matter how basic, the hardware platform chosen for M-Pesa was the lowly SIM card itself. The software platform was the SIM Application Toolkit, a GSM (Global System for Mobile Communications) standard set of commands built into all SIM cards. These commands permit the SIM to interact with an application running on a GSM network server, rather than on the

SIM or the handset, neither of which has enough computing power or memory for complex tasks. PINs would protect cellphone owners, while agents would confirm transactions using special phones furnished by Safaricom. The simple, menu-driven system operates much like SMS.

Since many customers would be making cash transfers to remote locations, Safaricom established an extensive country-wide network of agents (mostly existing Safaricom cellphone dealers), as well as methods for ensuring that agents neither ran out of cash nor ended up with too large an amount of it on hand. Training these agents to handle the flow of cash and maintain PIN security proved challenging, especially in remote areas, but they eventually caught on, as did users.

Businesses soon began using the system as well, but smooth, standardized techniques for batch-processing large numbers of transactions remained elusive. Every change to Safaricom's software caused breakdowns in this process and required businesses to rework their own methods, causing them to clamor for a genuine API. A 2012 assessment of M-Pesa noted that the eventual API performed poorly, but also that

> a mini-industry of software developers and integrators has started to specialize in M-Pesa platform integration. . . . These bridge builders fall into two broad categories: 1) those that are strengthening M-Pesa's connections with financial institutions for the delivery of financial products, and 2) those that are strengthening M-Pesa's ability to interoperate with other mobile and online payment systems. The lack of a functional M-Pesa API is hindering bridge-building, but several companies have devised tools for new financial functions and online payments nonetheless.[37]

This trajectory reflects the typical platform pattern of development. Once established, the core component, namely the M-Pesa software, acquired a number of developers building complementary components. Other developers built better interfaces between the core and the complementary components, leading to M-Pesa-based financial services such as pension schemes, medical savings plans, and insurance offerings. This "mobile money ecosystem" is a kind of parallel universe to traditional banking.[38] Indeed, today a Google search on M-Pesa categorizes it simply as a "bank."

Is the short-cycle temporality of platforms and second-order systems the future of infrastructure? M-Pesa went from drawing board to multibillion-dollar business in less than ten years. More importantly for my argument, M-Pesa rapidly acquired the status of fundamental infrastructure for the majority of Kenyan adults, serving as a de facto national banking system. According to one specialist in mobile money, "Africa is the Silicon Valley of banking. The future of banking is being defined here. . . . It's going to change the world."[39]

Yet the jury is still out on its ultimate value for low-income Kenyans. A widely noted study concluded that between 2008 and 2014, the increase in "financial inclusion" afforded by mobile money lifted 194,000 households, or around 2 percent of Kenya's population, out of poverty.[40] However, this result has been strongly criticized as a "false narrative" that fails to account for such negative effects as increasing "over-indebtedness" among Kenyans, high costs for small transactions, and the incursion of social debts related to kinship structures which may outweigh any economic advantages.[41]

FACEBOOK AND WHATSAPP

Facebook, which opened for business in 2004 and issued its IPO in 2012, is currently the globe's sixth largest publicly traded company, with a book value of over $510 billion in 2019. With more than 2.4 billion monthly active users—nearly one-third of the world's total population—it is currently the world's largest (self-described) "virtual community." At this writing in 2019, Facebook founder Mark Zuckerberg is all of thirty-five years old.

Facebook presents itself as the ultimate platform, filled largely with content provided by users. Early on, Facebook became an alternative to internet email for many users—especially younger people, whose principal online communications were with their friends and peer groups rather than the wider world. While users' photos, videos, and reports of daily events make up much of Facebook's content, a large and increasing amount of it is simply forwarded from online publishers and the open web.

Facebook is deliberately designed to keep its customers within the platform. Once web content is posted to Facebook, it can readily be viewed there, but additional steps are required to reach the original web source. Through its APIs, Facebook also brings third-party apps, games, news, and more recently paid advertising directly into its closed universe. Meanwhile, private Facebook posts cannot be crawled by third-party search engines, though the company offers its own, rather primitive search facilities. As many commentators have observed, Facebook appears to be creating a "second internet," a parallel universe largely inaccessible to Google and other search engines, reachable only through Facebook. The sociologist of information Ann Helmond has called this the "platformization of the web," while others have labeled Facebook's goal "internet imperialism."[42]

As mentioned above, in 2011 Facebook began offering a low-end version of its software that required little or no data service. The initial release was aimed at "feature phones" (an intermediate, image-capable level between basic phones, with voice and text only, and the more expensive smartphones). Over the next four years, this project underwent a series of transformations, all reflecting the complex tensions among (a) Facebook's goal of attracting users and capturing their data, (b) the technical constraints of developing-world devices and mobile providers, and (c) users' and governments' interest in access to the full, open World Wide Web.

In 2013, Zuckerberg declared internet connectivity a "human right" and set a goal of making "basic internet services affordable so that everyone with a phone can join the knowledge economy." To make this possible, he argued, "we need to make the internet 100 times more affordable."[43] The company launched the deceptively named internet.org, a shaky alliance with six other firms (Samsung, Ericsson, MediaTek, Opera, Nokia, and Qualcomm) seeking to create free internet access for the entire developing world. Internet.org launched four of its first five projects in African nations: Zambia, Tanzania, Kenya, and Ghana. The organization sought to define and promote standards for low-bandwidth web content; for example, Java and Flash cannot be used. Only websites that met these standards would be delivered by the platform envisaged for internet.org's end users.

Zuckerberg made no secret of his ultimate goal. As a 2014 article in *TechCrunch* put it,

> The idea, [Zuckerberg] said, is to develop a group of basic internet services that would be free of charge to use—"a 911 for the internet." These could be a social networking service like Facebook, a messaging service, maybe search and other things like weather. Providing a bundle of these free of charge to users *will work like a gateway drug of sorts*— users who may be able to afford data services and phones these days just don't see the point of why they would pay for those data services. This would give them some context for why they are important, *and that will lead them to paying for more services like this*—or so the hope goes.[44]

In 2014, internet.org held a "summit" in New Delhi, where Zuckerberg met with Indian Prime Minister Narendra Modi to promote the project. In May 2015, the coalition announced its intent to embrace an "open platform" approach that would not discriminate among third-party services or websites (provided they met the standards mentioned above).

However, under pressure from local operators suspicious of Facebook imperialism, India's Telecom Regulatory Authority ultimately banned the program—now called

"Free Basics"—in 2016, on net neutrality grounds. The regulator characterized the free service as a grab for customers using zero-rating (i.e., no-cost provision) as bait. The World Wide Web Foundation immediately applauded India's decision, writing that "the message is clear: We can't create a two-tier internet—one for the haves, and one for the have-nots. We must connect everyone to the full potential of the open Web."[45] Many others have since criticized the platform's "curated, America Online-esque version of [the internet] where Facebook dictates what content and services users get to see."[46]

The net neutrality controversy notwithstanding, the feature-phone version of Facebook's software rapidly attracted a large user base. In 2016, 20 percent of all active Facebook users accessed the platform on feature phones.[47] Some 95 million Africans were using this version of the platform, and in South Africa at least, this became a dominant mode. Although YouTube and other social media systems also have a presence, Facebook dominates the continent by far. Even in 2018, feature phones still accounted for well over half of African mobile phone sales, despite increasing uptake of smartphones.

To date, Facebook's "internet" service in the developing world remains a vacillating and conflicted project. Much like the America Online of the 1990s, an undercurrent of its plans has been to channel the open web through its closed platform, seeking to retain users within its grasp. Reporting on research by the NGO Global Voices and citing public protests in Bangalore, the *Guardian* quoted advocacy director Ellery Biddle: "Facebook is not introducing people to open internet where you can learn, create and build things. It's building this little web that turns the user into a mostly passive consumer of mostly western corporate content. That's digital colonialism."[48]

On the other hand, when confronted on this point, Zuckerberg has repeatedly changed course in favor of greater openness. He insists that full web access for all is the ultimate goal—albeit according to a set of standards that drastically reduce bandwidth demands, an entirely reasonable vision given current technical capacities and constraints. In the long run, Facebook certainly hopes to profit from "bottom of the pyramid" customers, but in the interim, the platform may (at least for a time) provide a crucial public good—in other words, an infrastructure.

WhatsApp, launched independently in 2011 but acquired by Facebook in 2014, offers extremely cheap text messaging and, more recently, image and voice service as well. Its low cost made WhatsApp very popular in developing-world contexts. At this writing in 2019, it has over 1.6 billion users, the largest number after Facebook and YouTube.[49]

WhatsApp's extremely low cost was made possible by a gap in most mobile operators' pricing structure, related to the legacy SMS service still in widespread use. Rather than deliver messages as costly SMS, WhatsApp sends them as data. It does the same with voice calls, also normally charged at a higher rate than data. It thus exploits a kind of hole in the fee structure of cellular providers. South African operators sell data packages for as little as R2 (2 rand, or about $0.10). Using WhatsApp, the 10MB that this buys is enough to make a ten-minute voice call or to send hundreds of text messages. By contrast, at typical voice rates of about R0.60 per minute, R2 would buy only three minutes of voice, or four to eight SMS (depending on the size of the bundle purchased). These huge cost savings largely account for WhatsApp's immense popularity.

Although they can also be used via computers and wired internet services, Facebook Free Basics and WhatsApp services mainly target cellphones and other mobile devices, the fastest-growing segment of internet delivery, and by far the most significant in the developing world. Recently, researchers have adopted the acronym OTT (Over the Top) to describe software that runs "on top of" mobile networks—exactly the second-order systems concept introduced earlier in this chapter. Some cellular operators have begun to agitate for regulation of these systems, arguing that Facebook, WhatsApp, and similar platforms should help pay for the hardware infrastructure of cell towers, servers, and so on. Others retort that users already pay cellular operators for the data service over which these apps run.

The extremely rapid rise of Facebook and WhatsApp—from a few tens of thousands to well over one billion users in just a few years—again exemplifies a temporality very different from that of older forms of infrastructure. Like Uber and Airbnb, OTT systems do not own or invest in the physical infrastructure on top of which they run; their principal product is software, and their capital investment is limited to servers and internet routers. Competing apps such as Tencent QQ and WeChat, emerging from the dynamic Chinese market, may eventually displace Facebook and WhatsApp as the largest virtual "communities."

CONCLUSION

In many parts of the world today, the modern infrastructural ideal of universal service and infrastructure stability through government regulation and/or stewardship has already crumbled. In others, including many African nations, it either died long ago (if it ever existed) or is disappearing fast. Not only corporate behemoths such as

Google and Facebook but smaller entities such as Kenya's Safaricom are taking on roles once reserved for the state or for heavily regulated monopoly firms. The speed with which they have done so—five to ten years, sometimes even less—is staggering, far outstripping the thirty- to one-hundred-year time lines for the rollout of older infrastructures.

Are the rapid cycle times of software-based systems now the norm? Are platforms and second-order systems—imagined, created, and provided by private-sector firms—the future of infrastructure? Will the swift takeoff of quasi-infrastructures such as M-Pesa be matched by equally swift displacement, as happened to FidoNet when internet protocols swept away its raison d'être? To me, history suggests an affirmative, but qualified, answer to all of these questions. My qualification is that the astoundingly rich tech giants clearly understand the fragile, highly ephemeral character of nearly everything they currently offer. To inoculate themselves against sudden displacement, they continually buy up hundreds of smaller, newer platform companies, as Facebook did with WhatsApp. This diversification strategy provides multiple fallbacks. Software platforms rise and fall, in other words, but the corporate leviathans behind them will remain.

If Africa is indeed "the Silicon Valley of banking," perhaps we should look for the future of infrastructure there, as well as in other parts of the Global South. Yet despite the glory of its innovations and the genuine uplift it has brought to the lives of many, this future looks disconcertingly like a large-scale, long-term strategy of the neoliberal economic order. By enabling microtransactions to be profitably monetized, while collecting the (also monetizable) data exhaust of previously untapped populations, these systems enroll the "bottom of the pyramid" in an algorithmically organized, device-driven, market-centered society.

NOTES

1. Thomas P. Hughes, *Networks of Power: Electrification in Western Society, 1880–1930* (Baltimore: The Johns Hopkins University Press, 1983); Wiebe Bijker, Thomas P. Hughes, and Trevor Pinch, *The Social Construction of Technological Systems* (Cambridge, MA: MIT Press, 1987); Geoffrey C. Bowker and Susan Leigh Star, *Sorting Things Out: Classification and Its Consequences* (Cambridge, MA: MIT Press, 1999); Paul N. Edwards et al., *Understanding Infrastructure: Dynamics, Tensions, and Design* (Ann Arbor: Deep Blue, 2007); Christian Sandvig, "The Internet as an Infrastructure," *The Oxford Handbook of Internet Studies* (Oxford: Oxford University Press, 2013).

2. Susan Leigh Star and Karen Ruhleder, "Steps Toward an Ecology of Infrastructure: Design and Access for Large Information Spaces," *Information Systems Research* 7, no. 1 (1996): 112.

3. Arnulf Grübler, "Time for a Change: On the Patterns of Diffusion of Innovation," *Daedalus* 125, no. 3 (1996).

4. Sources for figure 15.1: Gisle Hannemyr, "The Internet as Hyperbole: A Critical Examination of Adoption Rates," *Information Society* 19, no. 2 (2003); Pew Research Center, "Internet/Broadband Fact Sheet" (June 12, 2019), https://www.pewinternet.org/fact-sheet/internet-broadband/. Pew survey data (originally expressed as percentage of the adult population) were adjusted by the author to correspond with percentage of the total population. Graph is intended to capture general trends rather than precise numbers.

5. Paul N. Edwards, "Infrastructure and Modernity: Scales of Force, Time, and Social Organization in the History of Sociotechnical Systems," in *Modernity and Technology*, ed. Thomas J. Misa, Philip Brey, and Andrew Feenberg (Cambridge, MA: MIT Press, 2002).

6. Andrew Russell and Lee Vinsel, "Hail the Maintainers: Capitalism Excels at Innovation but Is Failing at Maintenance, and for Most Lives It Is Maintenance That Matters More" (April 7, 2016), https://aeon.co/essays/innovation-is-overvalued-maintenance-often-matters-more.

7. Bowker and Star, *Sorting Things Out*.

8. American Society of Civil Engineers, "2013 Report Card for America's Infrastructure" (2014), accessed February 1, 2019, http://2013.infrastructurereportcard.org/.

9. Ingo Braun, "Geflügelte Saurier: Zur Intersystemische Vernetzung Grosser Technische Netze," in *Technik Ohne Grenzen*, ed. Ingo Braun and Bernward Joerges (Frankfurt am Main: Suhrkamp, 1994).

10. Carliss Y. Baldwin and C. Jason Woodard, "The Architecture of Platforms: A Unified View," *Harvard Business School Finance Working Paper* 09-034 (2008): 8–9.

11. Mohanbir S. Sawhney, "Leveraged High-Variety Strategies: From Portfolio Thinking to Platform Thinking," *Journal of the Academy of Marketing Science* 26, no. 1 (1998).

12. Martin Campbell-Kelly et al., *Computer: A History of the Information Machine* (New York: Basic Books, 2014).

13. Tim O'Reilly, "What Is Web 2.0? Design Patterns and Business Models for the Next Generation of Software" (September 30, 2005), http://www.oreilly.com/pub/a/web2/archive/what-is-web-20.html.

14. Tarleton Gillespie, "The Politics of 'Platforms,'" *New Media & Society* 12, no. 3 (2010).

15. Stephen Graham and Simon Marvin, *Splintering Urbanism: Networked Infrastructures, Technological Mobilities and the Urban Condition* (New York: Routledge, 2001).

16. Martin Greenberger, "The Computers of Tomorrow," *Atlantic Monthly* 213, no. 5 (1964).

17. Paul N. Edwards, "Some Say the Internet Should Never Have Happened," in *Media, Technology and Society: Theories of Media Evolution*, ed. W. Russell Neuman (Ann Arbor: University of Michigan Press, 2010).

18. Janet Abbate, *Inventing the Internet* (Cambridge, MA: MIT Press, 1999).

19. William L. Cats-Baril and Tawfik Jelassi, "The French Videotex System Minitel: A Successful Implementation of a National Information Technology Infrastructure," *MIS Quarterly* 18, no. 1 (1994).

20. Paul N. Edwards, "Y2K: Millennial Reflections on Computers as Infrastructure," *History and Technology* 15 (1998); Brian Kahin and Janet Abbate, eds., *Standards Policy for Information Infrastructure* (Cambridge, MA: MIT Press, 1995).

21. J.-C. Plantin et al., "Infrastructure Studies Meet Platform Studies in the Age of Google and Facebook," *New Media & Society* 10 (2016).

22. Yomi Kazeem, "Google and Facebook Are Circling Africa with Huge Undersea Cables to Get Millions Online," *Quartz Africa* (July 1, 2019), https://qz.com/africa/1656262/google-facebook-building-undersea-internet-cable-for-africa/.

23. Gary Cook et al., *Clicking Clean: Who Is Winning the Race to Build a Green Internet?* (Washington, DC: Greenpeace Inc., 2017).

24. John S. Quarterman and Josiah C. Hoskins, "Notable Computer Networks," *Communications of the ACM* 29, no. 10 (1986).

25. Edwards, "Some Say the Internet Should Never Have Happened."

26. Randy Bush, "Fidonet: Technology, Tools, and History," *Communications of the ACM* 36, no. 8 (1993): 31.

27. The BBS Corner, "The Fidonet BBS Network," *The BBS Corner* (February 10, 2010), http://www.bbscorner.com/bbsnetworks/fidonet.htm.

28. Bush, "Fidonet."

29. Data source: FidoNet nodes by year, Wikimedia Commons, https://commons.wikimedia.org/wiki/File:Fidonodes.PNG.

30. Mitra Ardron and Deborah Miller, "Why the Association for Progressive Communications Is Different," *International Communications Association* (1988): 1.

31. Karen Higgs, ed., *The APC Annual Report 2000: Looking Back on APC's First Decade, 1990–2000* (Johannesburg, South Africa: Association for Progressive Communications, 2001), 13.

32. Karen Banks, "Fidonet: The 'Critical Mass' Technology for Networking with and in Developing Countries," in *The APC Annual Report 2000: Looking Back on APC's First Decade, 1990–2000*, ed. Karen Higgs (Johannesburg, South Africa: Association for Progressive Communications, 2001), 35.

33. Mike Lawrie, "The History of the Internet in South Africa: How It Began" (1997), http://archive.hmvh.net/txtfiles/interbbs/SAInternetHistory.pdf, 2–3.

34. Jake Kendall et al., "An Emerging Platform: From Money Transfer System to Mobile Money Ecosystem," *Innovations* 6, no. 4 (2012): 51.

35. Eric Wainaina, "42% of Kenya GDP Transacted on M-Pesa and 9 Takeaways from Safaricom Results," *Techweez: Technology News & Reviews* (May 7, 2015), http://www.techweez.com/2015/05/07/ten-takeaways-safaricom-2015-results/.

36. Nick Hughes and Susie Lonie, "M-Pesa: Mobile Money for the 'Unbanked,'" *Innovations* 2 (2007); Sibel Kusimba, Gabriel Kunyu, and Elizabeth Gross, "Social Networks of Mobile Money in Kenya," in *Money at the Margins: Global Perspectives on Technology, Financial Inclusion, and Design*, ed. Bill Maurer, Smoki Musaraj, and Ian V. Small (London: Berghahn Books, 2018).

37. Kendall et al., "An Emerging Platform," 58.

38. Kendall et al., "An Emerging Platform."

39. Killian Fox, "Africa's Mobile Economic Revolution," *Guardian* (July 24, 2011), https://www.theguardian.com/technology/2011/jul/24/mobile-phones-africa-microfinance-farming.

40. Tawneet Suri and William Jack, "The Long-Run Poverty and Gender Impacts of Mobile Money," *Science* 354, no. 6317 (2016).

41. Milford Bateman, Maren Duvendack, and Nicholas Loubere, "Is Fin-Tech the New Panacea for Poverty Alleviation and Local Development? Contesting Suri and Jack's M-Pesa Findings Published in Science," *Review of African Political Economy* (2019); Kusimba, Kunyu, and Gross, "Social Networks of Mobile Money in Kenya."

42. Anne Helmond, "The Platformization of the Web: Making Web Data Platform Ready," *Social Media + Society* 1, no. 2 (2015).

43. Dara Kerr, "Zuckerberg: Let's Make the Internet 100x More Affordable," *CNET* (September 30, 2013), https://www.cnet.com/news/zuckerberg-lets-make-the-internet-100x-more-affordable/.

44. Ingrid Lunden, "WhatsApp Is Actually Worth More Than $19b, Says Facebook's Zuckerberg, and It Was Internet.org That Sealed the Deal," *TechCrunch* (February 24, 2014), http://techcrunch.com/2014/02/24/whatsapp-is-actually-worth-more-than-19b-says-facebooks-zuckerberg/. Emphasis added.

45. World Wide Web Foundation, "World's Biggest Democracy Stands Up for Net Neutrality" (February 8, 2016), https://webfoundation.org/2016/02/worlds-biggest-democracy-bans-zero-rating/.

46. Karl Bode, "Facebook Is Not the Internet: Philippines Propaganda Highlights Perils of Company's 'Free Basics' Walled Garden," *TechDirt* (September 5, 2018), https://www.techdirt.com/articles/20180905/11372240582/facebook-is-not-internet-philippines-propaganda-highlights-perils-companys-free-basics-walled-garden.shtml.

47. Simon Kemp, We Are Social Singapore, and Hootsuite, "Digital in 2017: A Global Overview," LinkedIn (January 24, 2017), https://www.linkedin.com/pulse/digital-2017-global-overview-simon-kemp.

48. Olivia Solon, "'It's Digital Colonialism': How Facebook's Free Internet Service Has Failed Its Users," *Guardian* (July 27, 2017), https://www.theguardian.com/technology/2017/jul/27/facebook-free-basics-developing-markets.

49. Simon Kemp, "Q2 Digital Statshot 2019: Tiktok Peaks, Snapchat Grows, and We Can't Stop Talking," *We Are Social* (blog) (April 25, 2019), https://wearesocial.com/blog/2019/04/the-state-of-digital-in-april-2019-all-the-numbers-you-need-to-know.

16

TYPING IS DEAD

Thomas S. Mullaney

In 1985, economist Paul David published the groundbreaking essay "Clio and the Economics of QWERTY," in which he coined the term "path dependency"—by now one of the most influential economic theories of the twentieth and twenty-first centuries. He posed the question: Given how remarkably inefficient the QWERTY keyboard is, how has it maintained its market dominance? Why has it never been replaced by other keyboard arrangements—arrangements that, he and others have argued, were superior? How does inefficiency win the day?[1]

David's answer has become a mainstay of economic thought: that economies are shaped not only by rational choice—which on its own would tend toward efficiency—but also by the vagaries of history. Each subsequent step in economic history is shaped in part, or parameterized, by the actions of the past. Future paths are not charted in a vacuum—rather, they are "dependent" on the paths already taken to that point. This "dependency," he argues, helps us account for the endurance of inferior options.

David's choice of target—the QWERTY keyboard—must have struck many readers as iconoclastic and exhilarating at the time. He was, after all, taking aim at an interface that, by the mid-1980s, enjoyed over one century of ubiquity in the realms of typewriting, word processing, and computing. Although concerned primarily with economic history, David was also lifting the veil from his readers' eyes and revealing, *QWERTY is not the best of all possible worlds! Another keyboard is possible!*

David was not alone in this iconoclasm. In the April 1997 issue of *Discover* magazine, Jared Diamond (of *Guns, Germs, and Steel* notoriety) penned a scathing piece about QWERTY, lambasting it as "unnecessarily tiring, slow, inaccurate, hard to learn, and hard to remember." It "condemns us to awkward finger sequences." "In a normal workday," he continues, "a good typist's fingers cover up to 20 miles on a QWERTY keyboard." QWERTY is a "disaster."[2]

When we scratch at the surface of this exhilarating anti-QWERTY iconoclasm, however, things begin to look less revolutionary, if not remarkably late in the game. Whereas a vocal minority of individuals in the Anglophone Latin-alphabetic world has been questioning the long-accepted sanctity of the QWERTY keyboard since the 1980s, those outside of the Latin- alphabetic world have been critiquing QWERTY for nearly one hundred years longer. It was not the 1980s but the 1880s when language reformers, technologists, state builders, and others across modern-day East Asia, South Asia, Southeast Asia, the Middle East, North Africa, and elsewhere began to ask: How can we overcome QWERTY?

Moreover, the stakes involved in these earlier, non-Western critiques were profoundly higher than those outlined in the writings of David and Diamond. If Anglophone critics lamented the prevalence of wrist strain, or the loss of a few words-per-minute owing to the "suboptimal" layout of QWERTY, critics in China, for example, had to contemplate much starker realities: that the growing dominance of keyboard-based QWERTY interfaces, along with countless other new forms of Latin-alphabet-dominated information technologies (telegraphy, Linotype, monotype, punched card memory, and more), might result in the exclusion of the Chinese language from the realm of global technolinguistic modernity altogether. Likewise, reformers in Japan, Egypt, India, and elsewhere had to worry about the fate of their writing cultures and the future of their countries. By comparison, David and Diamond merely wanted to juice a bit more efficiency from the keyboard.

The ostensible "solutions" to QWERTY raised by David, Diamond, and others also begin to look naïve when viewed from a global perspective. Their proposed solution to overcoming QWERTY was the adoption of an *alternate keyboard layout*—the favorite being one designed in 1932 by August Dvorak, a professor of education at the University of Washington who started working on his new layout around 1914. Like so many who have criticized QWERTY in the Western world, David and Diamond celebrated the relatively scientific disposition of letters on the Dvorak keyboard, citing it as a kind of emancipatory device that would free typists from the "conspiracy" and "culprit" of QWERTY, as later iconoclasts phrased it.[3]

Technologists and language reformers in the non-Western world knew better than this. They knew that to "overcome" QWERTY required confronting a global IT environment dominated by the Latin alphabet, and that this confrontation would require far deeper and more radical courses of action than the mere rearrangement of letters on an interface surface. Whether for Chinese, Japanese, Arabic, Burmese, Devanagari, or any number of other non-Latin scripts, one could not simply "rearrange" one's way out of this technolinguistic trap. As we will see, the strategies adopted to overcome QWERTY were not restricted merely to the replacement of one keyboard layout for another but required the conquest of far deeper assumptions and structures of keyboard-based text technologies. Ultimately, these reformers would overcome *typing* itself, bringing us to a present moment in which, as my title contends, typing as we have long known it in the Latin-alphabetic world is dead.

Specifically, I will show how QWERTY and QWERTY-style keyboards, beginning in the age of mechanical typewriters and extending into the domains of computing and mobile devices, have excluded over one half of the global population in terms of language use. I will also show how the "excluded half" of humanity went on to transcend (and yet still use) the QWERTY and QWERTY-style keyboard in service of their writing systems through a variety of ingenious computational work-arounds. In doing so, I will demonstrate how the following pair of statements are able to coexist, even as they seemingly contradict one another:

- QWERTY and QWERTY-style keyboards and interfaces can be found everywhere on the planet, in use with practically every script in the world, including Chinese.
- QWERTY and QWERTY-style keyboards, as originally conceived, are incompatible with the writing systems used by more than half of the world's population.

Explaining this paradox, and reflecting on its implications, are the goals of this chapter.

THE FALSE UNIVERSALITY OF QWERTY

Before delving in, we must first ask: What exactly is the QWERTY keyboard? For many, its defining feature is "Q-W-E-R-T-Y" itself: that is, the specific way in which the letters of the Latin alphabet are arranged on the surface of the interface. To study the QWERTY keyboard is to try and internalize this layout so that it becomes encoded in muscle memory.

This way of defining QWERTY and QWERTY-style keyboards, however, obscures much more than it reveals. When we put aside this (literally) superficial definition of QWERTY and begin to examine how these machines *behave* mechanically—features of the machine that are taken for granted but which exert profound influence on the writing being produced—we discover a much larger set of qualities, most of which are so familiar to an English-speaking reader that they quickly become invisible. Most importantly, we learn that the supposed universalism of QWERTY—from the age of mechanical typewriting but continuing into the present day—is a falsehood. These features include:

Auto-Advancing

At its advent, QWERTY-style keyboard typewriters were auto-advancing: when you depressed a key, it created an impression on the page, and then automatically moved forward one space.

Monolinearity

Letters fell on a single baseline (also known as monoline).

Isomorphism

Letters did not change shape. (The shape and size of the letter "A" was always "A," regardless of whether it followed the letter "C" or preceded the letter "X.")

Isoposition

Letters did not change position. ("A" was always located in the same position on the baseline, regardless of which other letters preceded or followed.)

Isolating

Letters never connected. (The letter "a" and the letter "c," while they might connect in handwritten cursive, never connected on a mechanical typewriter—each glyph occupied its own hermetically separate space within a line of text.)

Monospatial

At the outset of typewriting, all letters occupied the same amount of horizontal space (whether typing the thin letter "I" or the wide letter "M").

Unidirectional

At the outset, all typewriters assumed that the script in use would run from left to right.

In addition to these many characteristics of the original QWERTY-style keyboard typewriters, there were also two deeper "logics" which are germane to our discussion:

Presence

These machines abided by a logic of textual "presence," meaning that users expected to find all of the components of the writing system in question—be it the letters of English, French, or otherwise—fully present in some form on the keys of the keyboard itself. When sitting down in front of an English-language machine, one expected the letters "A" through "Z" to be present on the keys, albeit not in dictionary order. By extension, when sitting down in front of a Hebrew typewriter or an Arabic typewriter, one expected the same of the letters *alef* (א) through *tav* (ת) and *'alif* (ا) through *ya'* (ي). As seemingly neutral as this logic might appear to us at first, it is in fact a deeply political aspect of the machine, and one we will return to shortly.

Depression Equals Impression

The second logic might be termed "what you type is what you get." With the exception of the shift key, and a few others, the expectation is that, when one depresses a key, the symbol that adorns that key will be printed upon the page (or, in the age of computing, displayed on the screen). Depress "X" and "X" is impressed. Depress "2" and "2" appears. And so forth.[4]

Taken together, these features of QWERTY devices constitute the material starting point that engineers had to think through in order to retrofit the English-language, Latin-alphabetic machines to other orthographies.

Why are these features worthy of our attention, and even our concern? The answer comes when we begin to scrutinize these logics, particularly within a global comparative framework. In doing so, a basic fact becomes apparent that would, for the average English (or French, Russian, German, Italian, etc.) speaking user, be non-obvious: namely, that the QWERTY-style keyboard machine as it has been conceptualized excludes the majority of writing systems and language users on the planet.

Consider that:

- An estimated 467 million Arabic speakers in the world are excluded from this mesh of logics, insofar as Arabic letters connect as a rule and they change shape depending upon context. Considering the various mechanical logics of the keyboard noted above, at least five are violated: isomorphism, isoposition, isolation, monospatialism, and left-to-right unidirectionality.
- More than seventy million Korean speakers are excluded as well, insofar as Hangul letters change both their size and position based upon contextual factors, and they combine in both horizontal and vertical ways to form syllables.

Hangul, then, violates at least four of the logics outlined above: isomorphism, isoposition, isolation, and monolinearity.

- An estimated 588 million Hindi and Urdu speakers, approximately 250 million Bengali speakers, and hundreds of millions of speakers of other Indic languages are excluded for many of the same reasons as Korean speakers. For although there is a relatively small number of consonants in Devanagari script, for example, they often take the form of conjuncts, whose shapes can vary dramatically from the original constituent parts.
- For the estimated 1.39 billion Chinese speakers (not to mention the approximately 120 million speakers of Japanese, whose writing system is also based in part on nonalphabetic, Chinese character-based script), the QWERTY keyboard as originally conceptualized is fundamentally incompatible.[5]

Altogether, when one tallies up the total population of those whose writing systems are excluded from the logics of the keyboard typewriter, one's calculations quickly exceed 50 percent of the global population. In other words, to repeat the point above: the majority of the people on earth cannot use the QWERTY-style keyboard in the way it was originally designed to operate. How, then, did this keyboard come to be a ubiquitous feature of twentieth- and twenty-first-century text technologies globally?

UNIVERSAL INEQUALITY

If Arabic, Devanagari, and other scripts were "excluded" from the logics of the QWERTY-style mechanical typewriter, how then were engineers able to build mechanical, QWERTY-style typewriters for Arabic, Hindi, and dozens of other non-Latin scripts in the first place? How could QWERTY be at once limited and universal?

The answer is, when creating typewriters for those orthographies with which QWERTY was incompatible, engineers effectively performed invasive surgery on these orthographies—breaking bones, removing parts, and reordering pieces—to render these writing systems compatible with QWERTY. In other words, the universalism of QWERTY was premised upon inequality, one in which the many dozens of encounters between QWERTY and non-Western orthographies took place as profoundly asymmetric, and often culturally violent, engagements. In Thailand (then called Siam), for example, the designers of the first two generations of mechanical Siamese typewriters—both of whom were foreigners living in Siam—could not fit all the letters of the Siamese alphabet on a standard Western-style machine. What they opted for instead was simply to *remove* letters of the Siamese alphabet, legitimating their decision by citing the low "frequency" of such letters in the Siamese language.

It was not only foreigners who advocated such forms of orthographic surgery. Shaped by the profound power differential that separated nineteenth- and twentieth-century Euro-American colonial powers from many parts of the modern-day non-West, some of the most violent proposals were advocated by non-Western elites themselves, with the goal of rendering their scripts, and thus perhaps their cultures, "compatible" with Euro-American technological modernity. In Korea, for example, some reformers experimented with the linearization of Hangul, which required lopping off the bottom half of Korean glyphs the subscript consonant finals (*batchim* 받침) and simply sticking them to the right of the graph (since, again, these same technologies could also not handle anything other than the standard single baseline of the Latin alphabet).[6] In the Ottoman Empire, meanwhile, as well as other parts of the Arabic-speaking world, proposals appeared such as those calling for the adoption of so-called "separated letters" (*hurûf-ı munfasıla*) as well as "simplified" Arabic script: that is, the cutting up of connected Arabic script into isolated glyphs that mimicked the way Latin alphabet letters operated on text technologies like the typewriter or the Linotype machine (since such technologies could not handle Arabic script and its connected "cursive" form).[7] Further language crises beset, almost simultaneously, Japanese, Hebrew, Siamese, Vietnamese, Khmer, Bengali, and Sinhalese, among many others.[8]

When we examine each of these "writing reform" efforts in isolation, they can easily seem disconnected. When viewed in concert, however, a pattern emerges: each of these writing reforms corresponded directly with one or more of the "logics" we identified above with regard to the QWERTY typewriter. "Separated letters" was a means of fulfilling, or matching, the logic of isomorphism, isolation, and isoposition; linearization was a means of matching the logics of monolinearity and isomorphism; and so forth.[9] In other words, every non-Latin script had now become a "deviant" of one sort or another, wherein the deviance of each was measured in relationship to the Latin-alphabetic, English-language starting point. Each non-Latin script was a "problem," not because of its inherent properties but because of the ways certain properties proved to be mismatched with the QWERTY machine.

What is more, it also meant that all non-Latin writing systems on earth (and even non-English Latin-alphabetic ones) could be assigned a kind of "difficulty score" based on a measurement of how much effort had to be expended, and distance covered, in order for the English/Latin-alphabetic "self" to "perform" a given kind of otherness. The shorter the distance, the better the deviant (like French or even Cyrillic). The greater the effort, the more perverse and worthy of mockery the deviant is (like Chinese).

In this model, one could array all writing systems along a spectrum of simplicity and complexity. For engineers seeking to retrofit the English-language QWERTY machine to non-English languages:

French equaled English with a few extra symbols (accents) and a slightly different layout. [Easy]

Cyrillic equaled English with different letters. [Easy/intermediate]

Hebrew equaled English with different letters, backward. [Intermediate complexity]

Arabic equaled English with different letters, backward, and in cursive. [Very complex]

Korean Hangul equaled English with different letters, in which letters are stacked. [Very complex]

Siamese equaled English with many more letters, in which letters are stacked. [Highly complex]

Chinese equaled English with tens of thousands of "letters." [Impossible?]

For the engineers involved, these formulations were *not* mere metaphors or discursive tropes; they were *literal*. To the extent that English-language text technologies were able to perform French easily, this led many to the interpretation that French itself was "easy"—not in a relative sense but in an absolute sense. Likewise, insofar as it was "harder" for English-language text technologies to perform "Arabic-ness" or "Siamese-ness," this implied that Arabic and Siamese *themselves* were "complex" scripts—rather than the objective truth of the matter, which was that these and indeed all measurements of "simplicity" and "complexity" were Anglocentric and inherently *relational* evaluations of the interplay between the embedded, culturally particular logics of the technolinguistic starting point (the QWERTY-style machine) and the scripts whose orthographic features were simply different than those of the Latin alphabet.

In short, the kinds of appeals one frequently hears about the "inherent complexity" or the "inherent inefficiency" of computing in Chinese, Arabic, Burmese, Japanese, Devanagari, and more—especially those couched in seemingly neutral technical descriptions—are neither neutral nor innocuous. They are dangerous. They rehabilitate, rejuvenate, and indeed fortify the kinds of Eurocentric and White supremacist discourses one encountered everywhere in eighteenth- and nineteenth-century Western writings—all while avoiding gauche, bloodstained references to

Figure 16.1 A 1981 issue of *Popular Computing* imagined a Chinese computer keyboard the size of the Great Wall of China (complete with rickshaw and driver).

Western cultural superiority or the "fitness" of Chinese script in a social Darwinist sense. Instead, technologists have recast the writing systems used by billions of people simply as "complex scripts." They have opined about the technological superiority of the Latin alphabetic script over the likes of Arabic and Indic scripts. And in certain cases, they have heaped ridicule on those writing systems they see as computationally "beyond the pale"—so "complex" that they are almost absurd to imagine computationally (see, for example, the 1983 cartoon in fig. 16.1, in which a Chinese computer keyboard was imagined by the artist as an object whose size matched the Great Wall of China).[10] Creating a special class of writing systems and labeling them "complex" omits from discussion the question of *how* and *why* these writing systems came to be "complex" in the first place.

ESCAPING ALPHABETIC ORDER: THE AGE OF "INPUT"

By the midpoint of the twentieth century, the global information order was in a state of contradiction, at least as far as the keyboard was concerned. The keyboard had become "universal," in that QWERTY-style keyboards could be found in nearly all parts of the globe. And yet it was haunted by specters both from within and from without. From within, the supposed neutrality of the QWERTY-style keyboard did not accord with the actual history of its globalization: in South Asia, the Middle East, North Africa, Southeast Asia, and elsewhere, the "price of entry" for using the QWERTY-style machine was for local scripts to bend and break, succumbing to various kinds of orthographic disfigurement.

Meanwhile, the Sinophone world and its more than one billion people were simply excluded from this "universality" altogether. For this quarter of humanity, that is, the very prospect of using a "keyboard," whether QWERTY-style or otherwise, was deemed a technological impossibility—and yet never did this gaping hole in the map of this "universality" call such universality into question. Instead, the provincial origins of the QWERTY machine were entirely forgotten, replaced with a sanitized and triumphant narrative of universality. If anything was to blame for this incompatibility between Chinese and QWERTY, it was the Chinese writing system itself. Chinese was to blame for its own technological poverty.

In this moment of contradiction, 1947 was a watershed year. In this year, a new Chinese typewriter was debuted: the MingKwai, or "Clear and Fast," invented by the best-selling author and renowned cultural critic Lin Yutang. MingKwai was not the first Chinese typewriter.[11] It was, however, the first such machine to feature a keyboard. What's more, this keyboard was reminiscent of machines in the rest of the world. It looked just like the *real thing*.

Despite its uncanny resemblance to a Remington, however, there was something peculiar about this keyboard by contemporary standards: what you typed was *not* what you got. Depression did not equal impression.

Were an operator to depress one of the keys, they would hear gears moving inside—but nothing would appear on the page. Depressing a second key, more gears would be heard moving inside the chassis of the device, and yet still nothing would appear on the page. After the depression of this second key, however, something would happen: in a small glass window at the top of the chassis of the machine—which Lin Yutang referred to as his "Magic Eye"—up to eight Chinese characters would appear. With the third and final keystroke using a bank of keys on the bottom—numerals 1 through 8—the operator could then select which of the eight Chinese character candidates they had been offered.

Phrased differently, Lin Yutang had designed his machine as a kind of *mechanical Chinese character retrieval system* in which the user provided a description of their

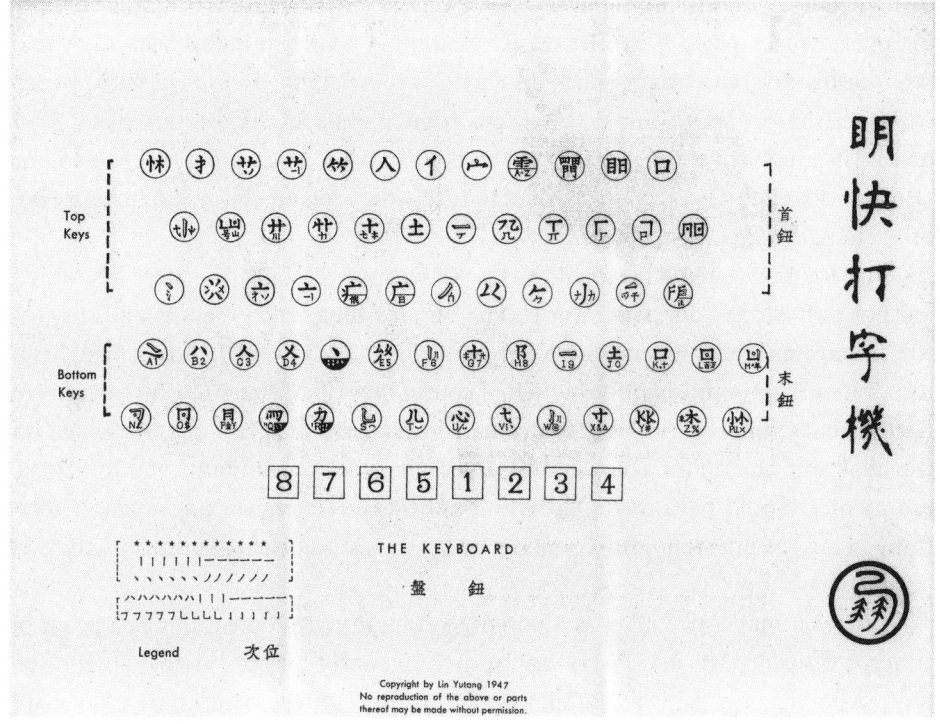

Figure 16.2 Keyboard layout of the MingKwai typewriter.

desired character to the machine, following which the machine then offered up to eight characters that matched said description, waiting for the user to confirm the final selection before imprinting it on the page. *Criteria*, *candidacy*, *confirmation*— over and over.

There was something else peculiar about this machine. For anyone familiar with Chinese, the symbols on the keyboard of the MingKwai would have struck them as both insufficient in number and peculiar in shape (fig. 16.2). There were nowhere near the tens of thousands of characters the machine claimed to be able to type, and the symbols that were included were, in many cases, neither Chinese characters nor any commonly accepted component of Chinese characters. Some of these symbols were peculiar, made-up shapes that Lin Yutang himself invented—"pseudo-Chinese" or "Chinese-ish" characters, perhaps, but certainly not Chinese in any accepted sense of the term in the 1940s or, indeed, at any other moment in recorded history.

There was a simple reason for all of this: although the MingKwai typewriter was an inscription technology, it was not premised upon the act of "typing." The goal of depressing a key on the MingKwai typewriter was not to cause a particular symbol to appear on the page, as was the case for all other typewriters designed up to that point in history, but rather to provide *criteria* to the machine so as to describe which character the operator wanted to retrieve from the machine's internal metal hard drive. The graphs on the keys did not, therefore, have to correspond in a one-to-one fashion with the characters one wished to type—they merely needed to help *describe* one's desired characters to the device.

Lin Yutang invested a personal fortune in the project, but ultimately his timing did not prove fortuitous. Civil war was raging in China, and executives at Mergenthaler Linotype grew understandably worried about the fate of their patent rights if the Chinese Communists won—which of course they did. Later, Mergenthaler heard rumors that China's new "Great Helmsman," Chairman Mao Zedong, had called for the abolition of Chinese characters and their replacement by full-scale Romanization, which would have made MingKwai pointless. (This never happened.) Everything being too uncertain, the company decided to wait, placing Lin in a financially untenable position.

MingKwai may have failed as a potential commercial product, but as a proof of concept it opened up new vistas within the domain of text technology. It showed the possibility of a radically different relationship between a machine's keyboard interface and its output. Specifically, Lin Yutang had created a typewriter on which what you typed was *not* what you got, changing the typewriter from a device focused primarily on *inscription* into one focused on *finding* or *retrieving things from memory*. Writing, in this framework, was no longer an act of typing out the spelling of a word, but rather using the keys of the keyboard in order to *search* for characters stored in memory. This departure from the long-standing logics of the keyboard opened up an entirely new terrain.

The first to venture into this new terrain was a professor of electrical engineering at MIT, Samuel Hawks Caldwell, just a few years after the MingKwai project stalled. Caldwell did not speak a word of Chinese, but as a student of Vannevar Bush he was an expert in logical circuit design. He was first exposed to the Chinese language thanks to informal dinnertime chats with his overseas Chinese students at MIT. As Caldwell and his students got to talking about Chinese characters, one seemingly rudimentary fact about the language caught him by surprise: "Chinese has a 'spelling.'" "Every Chinese learns to write a character by using exactly the same strokes in exactly the same sequence."[12] Here Caldwell was referring to "stroke order" (*bishun*), as it is known in Chinese.

His curiosity piqued, he sought the help of a professor of Far Eastern languages at Harvard, Lien-Sheng Yang, relying upon Yang to analyze the structural makeup of Chinese characters and to determine the stroke-by-stroke "spelling" of approximately 2,000 common-usage graphs. Caldwell and Yang ultimately settled upon twenty-two "stroke-letter combinations" in all: an ideal number to place upon the keys of a standard QWERTY-style typewriter keyboard.[13]

Caldwell's use of the word "spelling" is at once revealing and misleading. To "spell" the word pronounced /kəmˈpjutɚ/ using a typewriter was to depress a sequence of keys, C-O-M-P-U-T-E-R. The "spelling" of this word is said to be complete only when all eight letters are present, in the correct sequence. Such was not the case by which a user "spelled" with the Sinotype, as it was known. To depress keys on the Sinotype was not to see that same symbol on the page but as, Caldwell himself explained, "to furnish the input and output data required for the switching circuit, which converts a character's spelling to the location coordinates of that character in the photographic storage matrix." Like MingKwai before it, that is, Sinotype was not primarily an inscription device but rather a *retrieval* device. Inscription took place only after retrieval was accomplished.

While this distinction might at first seem minor, the implications were profound—as Caldwell quickly discovered. Not only did Chinese characters have a "spelling," he discovered, but "the spelling of Chinese characters is highly redundant." It was almost never necessary, that is, for Caldwell to enter every stroke within a character in order for the Sinotype to retrieve it unambiguously from memory. "Far fewer strokes are required to select a particular Chinese character than are required to write it," Caldwell explained.[14]

In many cases, the difference between "spelling in full" and "minimum spelling" (Caldwell's terms) was dramatic. For one character containing fifteen strokes, for example, it was only necessary for the operator to enter the first five or six strokes before the Sinotype arrived at a positive match. In other cases, a character with twenty strokes could be unambiguously matched with only four keystrokes.

Caldwell pushed these observations further. He plotted the total versus minimum spelling of some 2,121 Chinese characters, and in doing so determined that the median "minimum spelling" of Chinese characters fell between five and six strokes, whereas the median total spelling equaled ten strokes. At the far end of the spectrum, moreover, no Chinese character exhibited a minimum spelling of more than nineteen strokes, despite the fact that many Chinese characters contain twenty or more strokes to compose.

In short, in addition to creating the world's first Chinese computer, Caldwell had also inadvertently stumbled upon what we now refer to as "autocompletion": a technolinguistic strategy that would not become part of English-language text processing in a widespread way until the 1990s but was part of Chinese computing from the 1950s onward.[15] The conceptual and technical framework that Lin Yutang and Samuel Caldwell laid down would remain foundational for Chinese computing into the present day. Whether one is using Microsoft Word, surfing the web, or texting, computer and new media users in the Sinophone world are constantly involved in this process of criteria, candidacy, and confirmation.[16] In other words, when the hundreds of millions of computer users in the Sinophone world use their devices, not a single one of them "types" in the conventional sense of spelling or shaping out their desired characters. Every computer user in China is a "retrieval writer"—a "search writer." In China, "typing" has been dead for decades.

THE KEYBOARD IS NOT THE INTERFACE

Beginning in the 1980s, the QWERTY-style keyboard extended and deepened its global dominance by becoming the text input peripheral of choice in a new arena of information technology: personal computing and word processing. Whether in Chinese computing—or Japanese, Korean, Arabic, Devanagari, or otherwise—computer keyboards the world over look exactly the same as they do in the United States (minus, perhaps, the symbols on the keys themselves). Indeed, from the 1980s to the present day, never has a non-QWERTY-style computer keyboard or input surface ever seriously competed with the QWERTY keyboard anywhere in the world.

As global computing has given way to a standardized technological monoculture of QWERTY-based interfaces (just as typewriting did before it) one might reasonably assume two things:

In light of our discussion above, one might assume that writing systems such as Arabic, Chinese, Korean, Devanagari, and more should have continued to be excluded from the domain of computing, just as they were in typewriting, unable to render their "complex" scripts correctly on computers without fundamentally transfiguring the orthographies themselves to render them compatible.

Given the stabilization of global computing around the QWERTY-style keyboard, one might also reasonably assume that such standardization would have resulted in standardized forms of human–computer interaction, even if these interactions were (as noted above) at a disadvantage when compared to the English-language Latin-alphabetic benchmark.

In a nutshell, the globalization and standardization of the QWERTY-style keyboard should have resulted in both a deepening of the inequalities examined above and an increased standardization of text input practices within each respective language market where QWERTY-style keyboards have become dominant.

The empirical reality of global computing defies both of these assumptions. First, while still at a profound disadvantage within modern IT platforms, Arabic has finally begun to appear on screen and the printed page in its correct, connected forms; Korean Hangul has finally begun to appear with *batchim* intact; Indic conjuncts have at last begun to appear in correct formats; and more. Collectively, moreover, Asia, Africa, and the Middle East are becoming home to some of the most vibrant and lucrative IT markets in the world, and ones into which Euro-American firms are clamoring to make inroads. How can an interface that we described above as "exclusionary" suddenly be able to help produce non-Latin scripts correctly? How did this exclusionary object suddenly become inclusionary?

Second, with regards to stabilization, while all non-Latin computing markets have settled upon the QWERTY or QWERTY-style keyboard as their dominant interfaces, techniques of non-Latin computational text input have not undergone standardization or stabilization. To the contrary, in fact, the case of Chinese computing demonstrates how, despite the ubiquity of QWERTY keyboards, the number of Chinese input systems has *proliferated* during this period, with many hundreds of systems competing in what is sometimes referred to as the "Input Wars" of the 1980s and 1990s. (To gain a sense of the sheer diversity and number of competing input systems during this period, fig. 16.3 lists just a small sample of the hundreds of IMEs invented since the 1980s, including the codes an operator would have used to input the Chinese character *dian*, meaning "electricity," for each.) How is it possible that the 1980s and '90s was simultaneously a time of interface *stabilization* (in the form of the QWERTY keyboard) for Chinese computing but also a time of interface *destabilization* (in the form of the "Input Wars")?

In trying to resolve these two paradoxes (How does an exclusionary object become inclusionary? How does interface stabilization lead, counterintuitively, to interface destabilization?), current literature is of little help. Existing literature, after all, is shaped primarily by the framework of the Latin alphabet—and, above all, of "typing"—as exemplified in the essays by David and Diamond described earlier. For

Inputting the Character 电 (*dian*) on a Computer

Using this Input Method Editor (IME)...	you enter...
Xingyi sanma 形意三码	b 1
Quwei ma 区位码	2 1 7 1
Shuangpin 双拼	d m #
Dianbao ma 电报码	7 1 9 3
Yi shurufa 易输入法	r g d
Shuangpin shuangbu bianmafa 双拼双部编码法	d q t k
Sijiao fuyin 四角附音	5 0 7 1 6 d
Bixing bianma 笔形编码	6 0 1
Pinyin 拼音	d i a n #
Shuangbi yinxing shurufa 双笔音形输入法	d j j m
Shouwei yinxing shurufa 首尾音形输入法	d j f z
Chinese Transalphabet	d i a n t m v v
Cangjie bianma 仓颉编码	l w u
Wubi shurufa 五笔输入法	j n v
Zhengma 郑码	k z v v

Figure 16.3   Fifteen ways (among hundreds) to input *dian* ("electricity").

David and Diamond—and for all critics of QWERTY—there is no (and can be no) conceptual separation between "the keyboard" and "the interface," and they are understood to be one and the same thing. If the arrangement of keys on the keyboard stays consistent, whether as QWERTY, AZERTY, QWERTZ, DVORAK, or otherwise, by definition the "interface" is understood to have stabilized—and vice versa.

But in the case of the MingKwai machine, the Sinotype machine, or other examples of "input"—where writing is an act of retrieval rather than of inscription, and where one does not "type" in the classic sense of the word—the stabilization of the physical device (the keyboard) has no inherent or inevitable stabilizing effect on the retrieval protocol, insofar as there are effectively an infinite number of meanings one could assign to the keys marked "Q," "W," "E," "R," "T," and "Y" (even if those keys never changed location). When inscription becomes an act of retrieval, our notion of "stabilization"—that is, our very understanding of what stabilization means and when it can be assumed to take place—changes fundamentally.[17]

To put these manifold changes in context, a helpful metaphor comes from the world of electronic music—and, in particular, Musical Instrument Digital Interfaces (MIDI). With the advent of computer music in the 1960s, it became possible for musicians to play instruments that *looked* and *felt* like guitars, keyboards, flutes, and so forth but create the sounds we associate with drum kits, cellos, bagpipes, and more. What MIDI effected for the instrument form, input effected for the QWERTY keyboard. Just as one could use a piano-shaped MIDI controller to play the cello, or a woodwind-shaped MIDI controller to play a drum kit, an operator in China can used a QWERTY-shaped (or perhaps Latin-alphabetic-shaped) keyboard to "play Chinese." For this reason, even when the instrument form remained consistent (in our case, the now-dominant QWERTY keyboard within Sinophone computing), the number of different instruments that said controller could control became effectively unlimited.

Chinese computing is not unique in this regard. Across the non-Western, non-Latin world, a wide array of auxiliary technologies like input method editors can be found, used to "rescue" the QWERTY-style keyboard from its own deep-seated limitations and to render it compatible with the basic orthographic requirements of non-Latin writing systems. Thus, it is not in fact the QWERTY keyboard that has "globalized" or "spread around the world" by virtue of its own supremacy and power. Rather, it is thanks to the development of a suite of "compensatory" technologies that the QWERTY keyboard has been able to expand beyond its fundamentally narrow and provincial origins. In the end, we could say, it was the non-Western world that conquered QWERTY, not the other way around.

Collectively, there are at least seven different kinds of computational work-arounds, achieved by means of three different kinds of computer programs, that are now essential in order to render QWERTY-style keyboard computing compatible with Burmese, Bengali, Thai, Devanagari, Arabic, Urdu, and more (notice that I did *not* write "to make Burmese, etc., compatible with QWERTY"). The three kinds of programs are layout engines, rendering engines, and input method editors, and the seven kinds of reconciliations are (see fig. 16.4):

1. *Input*, required for Chinese (as we have seen), Japanese, Korean, and many other non-Latin orthographies[18]
2. *Contextual shaping*, essential for scripts such as Arabic and all Arabic-derived forms, as well as Burmese
3. *Dynamic ligatures*, also required for Arabic as well as Tamil, among others
4. *Diacritic placement*, required for scripts such as Thai that have stacking diacritics
5. *Reordering*, in which the order of a letter or glyph on the line changes depending on context (essential for Indic scripts such as Bengali and Devanagari in which consonants and following vowels join to form clusters)
6. *Splitting*, also essential for Indic scripts, in which a single letter or glyph appears in more than one position on the line at the same time
7. *Bidirectionality*, for Semitic scripts like Hebrew and Arabic that are written from right to left but in which numerals are written left to right

As should be evident from this list, the specific "problems" that these work-arounds were developed to resolve are the very same ones we examined at the outset of this essay: namely, they target the specific characteristics and logics of the original QWERTY typewriter, including presence, what-you-type-is-what-you-get, monospacing, isomorphism, isoposition, etc. What we are witnessing in the list above, then, are legacies of mechanical typewriting that were inherited directly by word processing and personal computing beginning in the second half of the twentieth century, and which have remained "problems" ever since.

The question becomes: What happens when the number of keyboards requiring "auxiliary" or "compensatory" programs outnumber the keyboards that work the way they supposedly "should"? What happens when *compensatory* technologies are more widespread than the supposedly "normal" technologies they are compensating for? What are the costs involved—the economic costs, the cultural costs, the psychic costs—when effectively the entire non-Western, nonwhite world is required to perform never-ending compensatory digital labor, not merely to partake

Contextual Shaping & Dynamic Ligatures

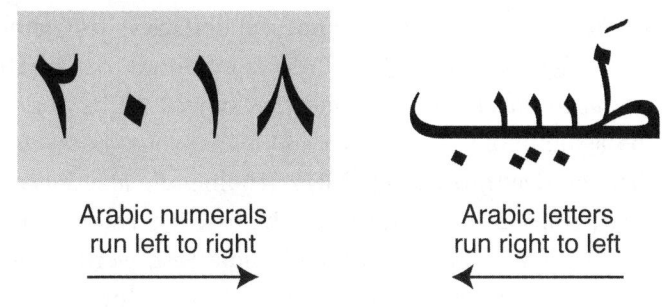

ب (*b*) in
medial position

ب (*b*) in
isolated position

Bidirectionality

Arabic numerals
run left to right
——————→

Arabic letters
run right to left
←——————

Reordering & Splitting

क + आ = का

क + इ = कि

When Hindi consonant क (*k*) combines with vowel आ (*a*),
vowel changes shape, but maintains position/order.

When same consonant combines with vowel इ (*i*),
vowel changes shape and position/order.

Figure 16.4 Examples of contextual shaping, dynamic ligatures, reordering, and splitting.

in Latin-alphabet-dominated industrial- and postindustrial-age text technologies but, beyond that, to enable the Latin and overwhelmingly white Western world to slumber undisturbed in its now-centuries-old dream of its own "universality" and "neutrality"?

CONCLUSION

Let us consider three ironies that pervade the history discussed thus far, ironies that urge us to rethink the global history of information technology in radically new ways.

First, QWERTY and QWERTY-style keyboards are not the obstacles—it is rather the logics that are baked into QWERTY and QWERTY-style machines that are the obstacles. In other words, *typing* is the obstacle. It is only the "typists" of the world—those operating under the assumptions of presence and "what-you-type-is-what-you-get"—who believe that the overthrow of QWERTY requires the *abandonment* of QWERTY. For those in the realm of input, however, the baseline assumption is completely different. Across East Asia, South Asia, the Middle East, North Africa, and elsewhere, what one types is *never* what one gets, and one's writing system is *never fully present* on the keyboard surface. Instead, the QWERTY or QWERTY-style keyboard are always paired with computational methods of varying sort—input method editors, shaping engines, rendering engines, and more—that harness computational and algorithmic power in order to reconcile the needs of one's orthographies with a device that, by itself, fails to do so. In quite literally every part of the world outside of the Latin-alphabetic domain, that is, the QWERTY or QWERTY-style keyboard has been transformed into what in contemporary parlance might be referred to as a "smart" device—a once-noncomputational artifact (such as a refrigerator or a thermostat) retrofitted so as to harness computational capacities to augment and accelerate its functionality (and to create new functionalities that would have been prohibitively difficult or even impossible in noncomputational frameworks).

In doing so, the escape from QWERTY has not required throwing away the QWERTY keyboard or replacing it with another physical interface—only a capacity, and perhaps a necessity, to reimagine what the existing interface is and how it behaves. In a peculiar twist of fate, it has proven possible to *exceed* the QWERTY-style keyboard, even as one continues to use it, and even as the exclusionary logics outlined above continue to be embedded within it. As a result, despite the global, unrivaled ubiquity of the QWERTY-style keyboard, the escape from QWERTY has in fact already happened—or is, at the very least, well under way. This is an idea that is completely oxymoronic and unimaginable to David, Diamond, and other typing-bound thinkers who, unbeknownst to them, are conceptually confined to operating *within* the boundaries of the very object they believe themselves to be critiquing.[19]

The second irony is best captured in a question a student once posed to me at People's University in Beijing at the close of a course I offered on the global history of information technology: "Does English have input? Is there an IME for English?"

The answer to this question is *yes*—or, at the very least: *There is absolutely no reason why there couldn't be.* The QWERTY keyboard in the English-speaking world, and within Latin-alphabetic computing more broadly, is as much a computational device as it is within the context of computing for Chinese, Japanese, Korean, Arabic, or otherwise. Nevertheless, the overarching pattern of Anglophone computing is one in which engineers and interface designers have done everything in their considerable power to craft computational experiences that mimic the feeling of typing on, basically, a much faster version of a 1920s-era mechanical Remington. They have done everything in their power to *preserve* rather than interrogate the two core "logics" of mechanical typewriting: presence, and what-you-type-is-what-you-get.

To return to the MIDI analogy, the QWERTY keyboard in the Anglophone world is as much a MIDI piano as in the non-Latin world, the main difference being: Anglophone computer users have convinced themselves that the only instrument they can play with their MIDI piano *is the piano*. The idea that one could use this interface to "play English" in an alternate way is a distant idea for all but the most aggressive technological early adopters. As a result, whereas in the context of non-Latin computing, engineers and users have spent their time figuring out ways to *harness* the computational power of microcomputers to make their input experiences more intuitive or efficient, in the context of English-language Latin-alphabet computing, practically all of the algorithmic and computational power of the microcomputer is left on the table.

In a peculiar turn of history, then, the ones who are now truly captive to QWERTY are the ones who built (and celebrated) this prison in the first place. The rest of the world has fashioned radically new spaces within this open-air prison, achieving perhaps the greatest freedom available short of rewriting the history of Euro-American imperialism in the modern era.

The third and final irony is, in many ways, the most counterintuitive and insidious. Returning to David, Diamond, and the self-styled anti-QWERTY iconoclasts—let us now ask: What would happen if they actually got what they wanted? What would happen if "path dependence" could be overcome, and we really could replace the QWERTY keyboard with the Dvorak layout? What if all keyboards were magically jettisoned and replaced by speech-to-text algorithms (arguably the newest object of fetishization among anti-QWERTY agitators)? Might this constitute the long-desired moment of emancipation?

Absolutely not.

If either approach gained widespread usage, it would serve only to reconstitute "(Latin) alphabetic order" by other means. Instead of the QWERTY-based expectation of "presence" and "what you type is what you get," we would simply have a Dvorak-based one. The "spell" of typing would remain unbroken. And, in the case of speech recognition, we would simply be replacing a keyboard-based logic with a speech-based homology: namely, *what you say is what you get*. Instead of text composition being an act of "spelling" words out in full on a keyboard-based interface, it would involve "sounding them out" in full through speech. However you cut it, it would be the same technolinguistic prison, only with different bars. (Indeed, either of these new prisons might be even more secure, insofar as we prisoners might be tempted to believe falsely that we are "free.")

There is only one way to overcome our present condition, and that is to face facts: Typing is dead.

NOTES

1. Paul A. David, "Clio and the Economics of QWERTY," *American Economic Review* 75, no. 2 (1985): 332–337.

2. Jared Diamond, "The Curse of QWERTY: O Typewriter? Quit Your Torture!," *Discover* (April 1997).

3. Eleanor Smith called QWERTY the "culprit" that prevents typists from exceeding the average speed of human speech (120 words per minute) with most of us "crawling" at 14 to 31 wpm. Robert Winder, meanwhile, called the QWERTY keyboard a "conspiracy." Eleanor Smith "Life after QWERTY," *Atlantic (*November 2013); Robert Winder, "The Qwerty Conspiracy," *Independent* (August 12, 1995).

4. There is, in fact, another baseline "logic" of the QWERTY keyboard that, while we do not have time to address in this venue, needs to be emphasized: namely, that there is a "keyboard" at all. Early in the history of typewriting, and throughout the history of Chinese typewriting, typewriters did not necessarily have keyboards or keys to be considered "typewriters." See Thomas S. Mullaney, *The Chinese Typewriter: A History* (Cambridge, MA: MIT Press, 2017), chapter 1.

5. Rick Noack and Lazaro Gamio, "The World's Languages, in 7 Maps and Charts," *Washington Post* (April 23, 2015), accessed July 1, 2018, https://www.washingtonpost.com/news/worldviews/wp/2015/04/23/the-worlds-languages-in-7-maps-and-charts/?utm_term=.fba65e53a929.

6. See Mullaney, *The Chinese Typewriter*; Andre Schmid, *Korea Between Empires, 1895–1919* (New York: Columbia University Press, 2002).

7. See Nergis Ertürk, "Phonocentrism and Literary Modernity in Turkey," *Boundary 2* 37, no. 2 (2010): 155–185.

8. Ilker Ayturk, "Script Charisma in Hebrew and Turkish: A Comparative Framework for Explaining Success and Failure of Romanization," *Journal of World History* 21, no. 1 (March 2010): 97–130; Fiona Ross, *The Printed Bengali Character and Its Evolution* (Richmond: Curzon, 1999); Christopher Seeley, *A History of Writing in Japan* (Leiden and New York: E.J. Brill, 1991); Nanette Gottlieb, "The Rōmaji Movement in Japan," *Journal of the Royal Asiatic Society* 20, no. 1 (2010): 75–88; Zachary Scheuren, "Khmer Printing Types and the Introduction of Print in Cambodia: 1877–1977," PhD dissertation (Department of Typography & Graphic Communication, University of Reading, 2010); W. K. Cheng, "Enlightenment and Unity: Language Reformism in Late Qing China," *Modern Asian Studies* 35, no. 2 (May 2001): 469–493; Elisabeth Kaske, *The Politics of Language Education in China 1895–1919* (Leiden and New York: Brill, 2008); Mullaney, *The Chinese Typewriter*; Sandagomi Coperahewa, "Purifying the Sinhala Language: The *Hela* Movement of Munidasa Cumaratunga (1930s–1940s)," *Modern Asian Studies* 46, no. 4 (July 2012): 857–891. The simultaneity of these language crises was matched by their intensity.

9. It is important to note that internalist critiques of writing systems were met as well by internalist defenses thereof. Alongside Chinese reformers who called for the abolition of Chinese characters, there were those who argued vociferously against such viewpoints. In China, for example, a telling passage from 1917, written by leading reformer Hu Shi, is a case in point: "They say that Chinese doesn't fit the typewriter, and thus that it's inconvenient. Typewriters, however, are created for the purpose of language. Chinese characters were not made for the purpose of typewriters. To say that we should throw away Chinese characters because they don't fit the typewriter is like 'cutting off one's toes to fit the shoe,' only infinitely more absurd." Hu Shi [胡適], "The Chinese Typewriter—A Record of My Visit to Boston (Zhongwen daziji—Boshidun ji) [中文打字機—波士頓記])," in Hu Shi, *Hu Shi xueshu wenji—yuyan wenzi yanjiu* [胡适学术文集—语言文字研究] (Beijing: Zhonghua shuju, 1993).

10. Source for figure 16.1: "Computerizing Chinese Characters," *Popular Computing* 1, no. 2 (December 1981): 13.

11. See Mullaney, *The Chinese Typewriter*.

12. Samuel H. Caldwell, "The Sinotype—A Machine for the Composition of Chinese from a Keyboard," *Journal of The Franklin Institute* 267, no. 6 (June 1959): 471–502, at 474.

13. Now, importantly, Caldwell dispensed with Lin Yutang's symbols and replaced them with different ones—but this is in many ways immaterial to our discussion. Just as "typing" is an act that is not limited to any one instantiation or style of typing (i.e., "typing" can take place in Russian just as well as in English or German), "input" too is a framework that is not tied to any one kind of input. The sine qua non of "input" is not the symbols on the keys, or even the mechanism by which the criteria-candidacy-confirmation process takes place—rather, input can be defined as any inscription process that uses the retrieval-composition framework first inaugurated on Lin's MingKwai.

14. Samuel Caldwell, "Progress on the Chinese Studies," in "Second Interim Report on Studies Leading to Specifications for Equipment for the Economical Composition of Chinese and Devanagari," by the Graphic Arts Research Foundation, Inc., addressed to the Trustees and Officers of the Carnegie Corporation of New York. Pardee Lowe Papers, Hoover Institution Library and Archives, Stanford University, Palo Alto, CA, accession no. 98055-16, 370/376, box 276, p. 2.

15. Autocompletion is just one of the many new dimensions of text processing that emerge when engineers began to move away from the framework of "spelling" as conceptualized in Alphabetic Order to the framework of Input. For an examination of early Chinese contributions to predictive text, see Mullaney, *The Chinese Typewriter*.

16. Recently, moreover, this process has actually entered the cloud. So-called "cloud input" IMEs, released by companies like Sogou, Baidu, QQ, Tencent, Microsoft, Google, and others, have begun to harness enormous Chinese-language text corpora and ever more sophisticated natural-language-processing algorithms.

17. One additional factor that helps explain the sheer number of input systems at this time is that, whether knowingly or not, these inventors, linguists, developers, and hobbyists were in fact recycling methods that were first invented in China during the 1910s, '20s, and '30s—an era well before computing, of course, but one in which Chinese-language reform and educational reform circles were in the grips of what was then called the "character retrieval crisis" (*jianzifa wenti*), in which various parties debated over which among a wide variety of experimental new methods was the best way to recategorize and reorganize Chinese characters in such contexts as dictionaries, phone books, filing cabinets, and library card catalogs, among others. For example, Wubi shurufa (Five-Stroke input), which was one of the most popular input systems in the 1980s and '90s—was in fact a computationally repurposed version of the Wubi jianzifa (Five-Stroke character retrieval method) invented by Chen Lifu. The nearly 100 experimental retrieval systems of the 1920s and '30s, as well as the principles and strategies upon which they were based, thus formed a kind of archive (consciously or not) for those who, two and three generations later, were now concerned in the novel technolinguistic ecology of computing.

18. Sharon Correll, "Examples of Complex Rendering," *NSRI: Computers & Writing Systems*, accessed January 2, 2019, https://scripts.sil.org/CmplxRndExamples; Pema Geyleg, "Rendering/Layout Engine for Complex Script," presentation, n.d., accessed January 2, 2019, http://docplayer.net/13766699-Rendering-layout-engine-for-complex-script-pema-geyleg-pgeyleg-dit-gov-bt.html.

19. To suggest that the escape from QWERTY has happened, or that it is under way, is decidedly *not* the same as suggesting that the non-Western world has in some way become "liberated" from the profoundly unequal global information infrastructure we discussed at the outset. The argument I am making is more complicated and subtle than that, and suggests that contemporary IT has entered into a new phase in which Western Latin alphabet hegemony continues to exert material, and in many ways conceptual, control over the way we understand and practice information, but that *within* this condition new spaces have been fashioned. Even with this kind of cognitive firepower targeting these issues, however, problems remain today. Normal, connected Arabic text cannot be achieved on many Adobe programs, a remarkable fact when considering the kinds of advances made by this company, as well as the fact that more than one billion people use one form or another of Arabic/Arabic-derived writing system.

AFTERWORDS

HOW TO STOP WORRYING ABOUT CLEAN SIGNALS AND START LOVING THE NOISE

Kavita Philip

Your computer is on fire—or so we have argued, confirmed, illustrated, and sometimes shouted here, in various registers, through historical, cultural, and technical analyses. This is not just a local fire, as these chapters demonstrate. It's raging around the world. It has been building for a long time. And it is consuming us. This is not just a technology fire—it is a resource fire that threatens the planet, it is a psychological fire that implicates our desires and fears, and it is a cultural fire that enlists our identities and communities. We do not stand apart from it; we are its kindling.

Shouting "Fire!" in a crowded technological world, however, comes with an obligation to account for one's extreme action.

Do the authors in this volume speak as Luddite saboteurs, plotting to empty the technological theater of its talented inventors? Do we seek to cause a technophobic panic that disrupts modernity itself? Do we hate progress? No, no, and no. But we do seek to interrogate the terms of such questions, suggesting that terms like "inventor," "modernity," and "progress" might have histories and futures that hold the key to other kinds of worlds.

Let's examine our assumptions about "inventors" and their individual heroics. Are inventors individual geniuses, sui generis generators of unprecedented insights? Or do they build on a host of other human and infrastructural affordances, deploying the public domain to synthesize their findings, only to jettison that collective work in order for their insight to be celebrated as singular?

Let's untangle the diversity of threads that weave together the thing we call "modernity." Is it a story of Western rationality, clearly demarcated from premodern primitives? Or is it a multitextured skein formed of technology, poetry, craft, philosophy, fiction, with room for all the other forms of expression still to come?

Let's disaggregate what we mean by "progress." Is it a teleological sprint that sloughs off all previous experiments as just so many erroneous interludes, explaining away disasters as aberrations? Or is it a labyrinth of ideas and practices that traces circles and spirals in time even as it generates both ruthless hierarchical striations and egalitarian leveling impulses?

To all three questions, we answer that we're wagering our chips on the second set of options. We choose complex, contradictory, contingent explanations over just-so stories. We corral humanist, technical, global, contemporary, local, novel, and longstanding models to explain the tangled histories that led to present-day arrangements that seem natural and inevitable. Are men just naturally better programmers? Do computational devices just naturally work better in English? Did European empires bring backward people into the light of modernity so that they might follow in the footsteps of the Western Enlightenment? Do modern networks show the natural superiority of decentralized freedom over centralized planning? Does political freedom lead us to the promised virtual land, cleaned of messy organic histories and physical infrastructures? To each of these and a host of other just-so stories, we have answered no. We strive to build more complicated historical narratives from the archival and lived traces of these technological stories, even while attempting to understand why it is that many of us started out believing the simpler answers to these questions.

And yes, by allowing these historical explications to unravel our tightly woven models of the technological present, we might along the way destroy the very ground on which computational production, distribution, and use currently operate. But we believe that this destabilizing move holds the keys to a future that's more expansive in its technosocial possibilities than the one currently on offer. This is the spark that lit the fire in our computational universe, and we hope that it has struck a match in yours too.

We're not helpless in the face of this fire. As fire historian Steven Pyne writes, through most of human history, "We have leveraged fire; fire has leveraged us." Delivering a nonmetaphorical warning—"Winter is not coming. Get ready for the Pyrocene"—he famously analyzed environmental history, drawing on fifteen seasons of wildfire-fighting experience, to decipher a historical lesson about the anthropogenic age of fire. "Humans are a uniquely fire creature, not only the key-stone species for fire but a species monopolist over its manipulation," he pointed out—only we, among all the animals, could have laid waste to the planet in this particular way, because we are uniquely fire creatures.[1]

Our book transforms Pyne's literal fire warning into a metaphorical one for the computational world. There are many illuminating parallels. We are uniquely technological creatures, *homo faber*—we solve problems with tools. Our particular skill at tool making has brought us sophisticated systems that are truly awesome (combining the Old English meaning of terror-inducing with its current slang reso-nances of coolness), but it is some combination of our addiction to the excitement of invention, with our enjoyment of the individualized sophistications of a tech-nological society, that has brought us to the brink of ruin even while illuminat-ing our lives and enhancing the possibilities of collective agency. Like Pyne's literal flaming planet, our planetary picture of computational systems run amok can be improved—with economic regulation, collaborative education, and some geopolitical gearshifts.

Combining the work of sixteen researchers into a book that we hope is more than the sum total of our individual findings, we offer not just a frightening warning but some well-tested escape routes.

ERRORS AND APPROXIMATIONS

Underneath the euphoric techno-utopian pronouncements as well as the romanti-cist technophobic tirades that rage across our planet in the early twenty-first cen-tury are some simple errors. Humanists and technologists are equally susceptible to these. When Apple's user interface design registers global applause and Steve Jobs tells a graduating class that the humanities are key to good design, professors of poetry release a satisfied sigh, expecting students to flock in to appreciate the time-less beauty of iambic pentameter. When work-life balance teeters under the weight of cellphone-tethered parents and social media–addicted teenagers, cognition experts weigh in about the evolutionary psychology of addiction, while those who can afford it go on device-detox holidays. When North African youth rise up in demo-cratic protest, the credit goes to Facebook. These everyday reactions to the perceived encroachment of technological ways of being on an otherwise authentic, pretechno-logical self might seem like harmless, commonsensical manifestations of everyday public debate. However, they are deeply misguided. They rely on simple, ahistorical distinctions between the humanities and technology training, between networked selves and autonomous humans, between hard and soft sciences, and between Western and non-Western democratic practices.

Common sense has a history and politics. Whatever our profession, historian or software designer, fantasy writer or computer scientist, we could all acquire some useful techniques by which to discover the histories and contexts of the models of humans and the world—such as those above— that we take to be commonsensically true. How might we arrive at more complex understandings of humans in the world? How will a better understanding help us put out the fires this book has described?

No universalizable template could unite the rich content of the sciences, technologies, humanities, and social sciences, which thrive on specialization and specificity. But we do have some generalizable starting points.

HOW NOT TO TRAVEL AMONG THE DISCIPLINES

1   APPRECIATE THE GREAT BOOKS

In 1959, scientist-turned-novelist Charles Percy Snow described what he saw as a lamentable intellectual division between those who could quote Shakespeare and those who could quote the laws of thermodynamics. Delivering the prestigious Rede Lecture at the University of Cambridge, C. P. Snow coined the term "the Two Cultures," a phrase that has endured as a symbol of the social, political, and ethical tensions between science and the humanities. Many people assume that the solution to bridging the two cultures is for scientists to appreciate the great books of literature, and for literary and cultural experts, in turn, to read the "laws of thermodynamics" and other law-like conclusions of science as earnestly as they read Shakespeare. While this would certainly be a delightful experiment, it would in no way change the ways in which literary and scientific researchers think, frame questions, or understand the world.

Appreciating the "Great Books of Literature" and "Eternal Laws of Nature" is simply a way of saying we should respect and value the conclusions arrived at by other disciplinary experts. It does not give us insight into the process by which these experts design, plan, critique, and complete their work. It does not show the mechanisms of production, expose the assumptions at work, or analyze the books and laws in the social, economic, or political contexts of their time.

It is fun to travel between disciplines by appreciating their great works, but great-book tourism will not help put out the fire that we have described.

2   FIND MEDIATING BROKERS TO DUMB IT DOWN

If the Great Books and Laws won't do, could we identify a few brokers to break down the language of technological complexity into eighth-grade language for the lay public? Could we rely on laboratories and scientific institutions to employ science-to-society translators to justify their work to a skeptical public?

Popularizers are important and enjoyable parts of our cultural milieu. But they are not the cure for our techno-fire problem. This notion of popularizing leaves untouched the belief in an authoritative, expert context of knowledge on one side, and an ignorant, nonexpert populace on the other. The broker is set up to gain the trust of the latter for the decisions of the former to continue unimpeded.

Historical and social analyses of technology require us to understand a wide range of social and cultural conditions. This requires us not to dumb down but to wise up in the ways in which we practice interdisciplinarity. We must learn how technology works as well as how society works.[2] We must recognize the expertise that "ordinary" people have that can call into question the historically laden assumptions that lie at the heart of disciplines forged during the Enlightenment.

The chapters in this book have shown how women in rural India, typists in China, phone users in Trinidad, and women programmers in Britain all have sophisticated understandings of the technologies in their context. They don't need translators or knowledge brokers to dumb down the technical knowledge for them. The dumbing down of labor-relations theory into buzzwords like "innovation" and "skills training" has already brought them to economic cul-de-sacs.[3] What they need is an egalitarian political context, a modern-day agora, in which to discuss, and contest, the unequal social relations that are being silently packaged within their technological devices. Skills will not set rural women free in the absence of a fuller explication of the economic pressures that force them into low-paid data-entry jobs. We understand why they are discouraged from play and creative computational activities only when we also see the history of why and how poor Global South workers are slotted into a global economy in which their low-level economic status has already been decided.

Women programmers were pushed out of computer science because of a complex set of post–World War II economic pressures, combined with cultural assumptions about masculinity and authority that would only be partially challenged by feminist mobilizations that were still to come. Corporations would grudgingly include women executives, but use gender and labor against each other as corporate strategies. And girls would be invited back into coding when a new need for affordable programming labor emerged in the early twentieth century.

We can debate the conclusions above by reading the chapters by Sarkar, Hicks, Schlombs, Lawrence, and Abbate. These histories give us material not merely to accept my quick conclusions above but to argue and contest them. Good history shows its work and shares its data so that we can debate its implications and collaborate on making different futures.

3   MIX AND STIR

If we must know a bit of tech and a bit of history, could we then mix and stir them? We could have engineers take one class on engineering history, or writers could take a class on "programming for poets." This is a great idea, but one that works best when that one class turns into a revelation for the person who had not thought of anything outside their narrow professional practice. Too often, engineers and poets receive only a superficial appreciation for another field, but no sense of how the skills of the other field might actually change their practice, call into question their authority, or shift the kinds of questions their field asks.

Mixing and stirring disciplines together can create a pleasant cocktail hour, but it still does not help us douse the flames that erupt when technologists design the future without seriously engaging with the past. The disciplines still come in separate bottles, with separate production histories, before they are mixed. It would be better to meld their production processes together. Engineering and history, computing and feminist politics, platform capitalism and gig workers' unions should be in engagement and contestation from the beginning in order to have any real effect on systems before they settle into exploitative structures.

To fully deal with the entanglement of people, technology, and their histories requires a deeper engagement than is possible with a mix-and-stir model. We must understand the contingency in how things came to be and see the forces that work to keep things a certain way. This kind of knowledge can upend many assumptions in tech design and planning.

4   MOVE FAST AND BREAK THINGS

In order to upend design assumptions, can't we just move fast and break things? This way we'll learn from our mistakes, and eventually we'll correct any starting biases.

Moving fast and breaking things works if we have enough privilege and safety not to be affected by the loss of the things we break, and enough certainty in our goals that we do not have to stop and worry whether we are going in the right direction. We move fast when we think we know where we are going. Things break easily when they are brittle.

"Move fast and break things" is not a good slogan for the serious firefighter, the rigorous traveler among disciplines. We need to proceed with the possibility of questioning our directions and our outcomes. We need to make models of understanding and techniques of fabrication that stretch, evolve, change, and modify our paradigms, rather than breaking, cracking, and repeatedly discarding humans and objects who no longer serve our purpose. The massive production of technological waste as well as the histories of domination of some humans by others, and the rise of resistance to that model, suggests that we are at the end of an era where such an arrogant strategy can work.

WHAT SHOULD WE DO?

If we've followed along with the chapter conclusions in this book, we know there are insights to be gained in the close and careful study of everyday technological practices, the *longue durée* patterns in social history, and the changing interdependencies between corporate needs and workers' identities. But if the standard modes of cross-disciplinary brokerage and mixing don't work, what should we do?

There are three sets of interdisciplinary conversations that we do have the power and the skills to reframe, here and now. These are conversations about language, history, and politics in the context of scientific and technological knowledge production and use. Scientists have too often assumed that language is transparent, and that conventional usages carry adequate meaning. Technologists have often read history only as a pastime. Tech corporations have often assumed that they live in a rarefied world above politics. In order to correct the massive harms that have resulted from these assumptions, we must begin by acknowledging that scholars of language, history, and political theory have something to offer technology producers and users.

The study of language, history, and politics—these are huge new variables to throw into the mix. But we can't afford not to draw our frameworks wide and deep. We must stop searching for clean, simple signals from social and humanist knowledge, and enjoy the fact that human experience comes with a lot of noise. Fighting our fire requires that we find a way to embrace, not erase, this noise.

1   LANGUAGE MATTERS

Words and things are mutually constitutive. To understand how language matters in technological design and use, one must do more than mix, stir, and borrow. One must be open to the insights of language overturning one's most cherished design principles. Consider two recent examples that suggest this process is already taking place.

Technology analysts Luke Stark and Anna Hoffmann, in a mid-2019 opinion piece published while ethical debates raged in data science, suggested ways in which metaphors matter. Data-driven work is already complicit, they argue, "in perpetuating racist, sexist, and other oppressive harms."[4] Stark and Hoffmann argued, however, that a solution was hidden in the very articulation of this problem: "The language we use to describe data can also help us fix its problems." They drew on a skill commonly believed appropriate only in literature departments: the analysis of metaphor, imagery, and other narrative tropes. This route to data ethics, they suggested, would help data scientists to be "explicit about the power dynamics and historical oppressions that shape our world." The analysis of metaphors is useful not simply to assign blame; it is critically important for finding solutions.

Informatics scholars Lilly Irani and M. Six Silberman, reporting on Turkopticon, a worker-centered web portal and services site they designed and maintained, published software design papers as well as reflective "meta-analysis" papers about it.[5] Both were presented at CHI, the premier computer-human interaction conference. The scholarly association that studies computer-human interaction issues had recognized, in other words, that things (like software) and words (with which both technical and nontechnical people represented, used, and critiqued that software) were both relevant to its work. Things and their representations are connected; representations matter, and examining how they fail or succeed can help us make better things.

Citing feminist humanists, political theorists, and sociologists of labor, Irani and Silberman traced the stories designers and users "told about the software, and how these stories supported or worked against the very goals . . . our software was meant to achieve." The attention to stories and their effects brought them to the realization that time is another factor in improving design and its effects—"it takes time," they argued, reflecting over "half a decade of making, watching, waiting, being surprised, and trying again."[6] The process of improving technical design involves skills from the domains of literature (the analysis of narratives) and history (the analysis of work and gender patterns over historical periods and the changing use of particular software, for example).

The technology analysts cited above are the first wave of a conceptual, methodological revolution that is sweeping all computation-related fields. Analyzing metaphors and the social and political work done by language is not a luxury that we can leave in the hands of literary theorists. It is a tool we need to add to the essential skills of computer and information scientists. The researchers cited above acquired these skills in unusual career trajectories. We now have enough examples of such successful interdisciplinary travelings to design curricula so that all technology professionals can be trained rigorously in language analysis.

2 HISTORY MATTERS

To understand how our computers came to be on fire, we must understand how we got here. The design of technological objects and the shape of the social relations that embed them are contingent; their present state was never inevitable. We must be curious about the possible emergence and death of other potential outcomes. We must engage with the lingering afterlives of ideas and enter debates that were never fully closed. History matters, but not in a simple chronological sense. Technologists cannot simply borrow from just-so stories about the glories of scientific discovery.

Historians themselves, many still committed to varieties of positivism (with roots in scientific norms of the nineteenth century), don't often acknowledge the deep constitutive link between things and words, or events and the stories we tell about them. As historian Michel-Rolph Trouillot has said about his disciplinary colleagues, "Some, influenced by positivism, have emphasized the distinction between the historical world and what we say or write about it."[7] He reminds us that the discipline of history itself bears the mark of the scientific revolution and the positivist legacy of the early twentieth century.

Trouillot notes that it is difficult, and sometimes impossible, to distinguish between events and their representations. He argues that there is a fluid, shifting, overlapping boundary between two categories we think of as easily separable: historical events as they happened, on the one hand, and historians' discovery, analyses, and representation of those processes, on the other. Most historical theorists, he notes, have treated the fuzzy overlap "between the historical process and narratives about that process" as if it could be cleared up by harder facts, better verification, or stronger empirical warrants. This fuzziness is to them a matter of embarrassment, and they believe its complete elimination is desirable, and possible. This eagerness for perfect correspondence between reality and its representation is in many ways laudable, and offers an empirical ideal toward which to strive. For many historians, this striving toward accuracy and verifiability distinguishes the profession from other humanist research, making the act of writing history closer to evidence-based scientific research than to speculative humanist interpretation. The role of this kind of historian is to "approximate the truth."[8]

Other historians, following theories of reading, writing, and power produced by literary and cultural scholars, point out that historical truth is always constructed. Cutting through the dichotomous, fractious debate that divides these competing schools of historiography, Trouillot reminds us that facts are always material and social, and that dividing them up into empirical or constructed, brute reality or airy imagination, misses the more complicated, but true-to-life, aspect of history. We can understand both facts and speculation if we track how all stories are constituted by the workings of power.

Historian of empire Raymond Betts put it pithily when he summed up: "Empire was as much stage performance as military engagement, as much the presentation of arms as the firing of them."[9] Betts does not downplay the fact that empire was won, and kept, by the sword—by the exercise of brute material power. But representational strategies (visible to us, for example, in world's fairs, expositions, museums, literature, and popular culture) were inseparable from military acts.

What kinds of facts and theories matter in the history of computation? Modern historians of computation, writing as they do in a present riven with power differences and dominating legacies, have a particular responsibility to write histories in full cognizance of the power structures and ideological legacies that shape their investigations.

3   POLITICS MATTER

When we do history with an attention to the power dynamics that shape processes and outcomes, we arrive at the difficult question of politics. Politics matters in the conversation between technology and social change—people's desires and their resistance to oppression, the power of collectives, as well as the agency of individuals in the face of structures of domination all shape the technologies we make and use.

To navigate the complexities of power, we should pin our hopes not on axiomatic, rule-based ethics or on the hope of finding value-neutral data to shape our technological practices. We should look instead to the arguments that social movements make for justice and accountability. Social movements raise political questions that cannot be mixed, stirred, brokered, or leveraged out of existence. When tech workers protest against large corporations, when gig workers form unions against the various forms of informal work enabled by the peer-to-peer paradigm, when communities form movements against predictive policing and other kinds of computationally enhanced inequality, they are asking us to radically rethink our design models for both technology and society.

Technology did not invent oppression; historical injustices are being reshaped, revivified, and redefined by technology. It is not always possible, or even effective, to find one contemporary individual who is responsible for injustices. Contemporary analyses laying responsibility at the door of big funders, national priorities, or powerful CEOs are useful organizing strategies, but they also point out the need for broader structural critique. Tech activists are often reported in the media as being separate from, or anomalous in, the history of social justice activism, but the most urgent movements are those that link both tech and justice. To formulate the terms of this debate, technologists, political theorists, and activists will have to forge new kinds of engagements, both contentious and collaborative, with each other.

AFTER THE FIRE

I opened this afterword with an allusion to the legal problem that is illustrated by a person shouting "fire!" in a crowded theater. What happens after shouting "Your Computer Is on Fire" in a crowded technological theater? The various entailments of such an act—panic, death, and destruction—make this kind of speech "dangerous," and thus disallowed. Thus the phrase is commonly invoked to suggest the limits of free speech.

It has its origins in a World War I draft resistor's 1919 Supreme Court indictment, *Schenk v. United States*. Schenk's right to free speech stopped at the point when he infringed the Sedition Act by making 15,000 anti-draft fliers to encourage anti-draft behavior. The Supreme Court found him guilty of seditious and dangerous speech. This seditious limit to free speech, however, was overturned a half-century later by the Supreme Court (*Brandenburg v. Ohio*), when it ruled that rural Ohio Ku Klux Klan activist Clarence Brandenburg's vitriolic, racialized tirade was protected speech. This counterintuitive legal vignette reminds us that rights, ethics, speech, freedoms, and other complex human situations cannot be predicted or deduced from axioms or a priori rules. We could not have guessed in advance that a pacifist war resistor would bring free speech to its limits, and that violent racists would free speech from these very limits. To understand this, we might turn to histories of state racism, the interaction between social movements and legislative history, or the larger relations between liberalism and power. Logic alone won't help us here—we need context, history, interpretation, and debate in order to understand the cultural and political context and significance of these legal debates and their changing significance over time.

Analogously, we need broad contextual debate about almost every technological outcome since the Enlightenment. A priori rules for ethics and logical axiomatic frameworks for describing technology's operation are useful for some kinds of documentation but counterproductive in most everyday messy examples of real life. We can only get the complexity we need to navigate a crowded technological theater on fire through collaboration, conversation, and rigorous exploration of all the strands of modernity's multivalent skein.

Looking back on his education, computer scientist Philip Agre wrote, "When I was a graduate student in Artificial Intelligence (AI), the humanities were not held in high regard. They were vague and woolly, they employed impenetrable jargons, and they engaged in "meta-level bickering that never decides anything." Agre recalled that many of his classmates loved reading literature (the "Appreciate the Great Books" model) but couldn't imagine how the interpretive and historical humanities might have any bearing on their task—the design and creation of machine intelligence. Agre recalled the fast-moving research field that constantly broke and rewrote systems (the "Move Fast and Break Things" model). In retrospect, he attributed this to the "'brittleness' of symbolic, rule-based AI systems," which, he observed, "derives from their tendency to fail catastrophically in situations that depart even slightly from the whole background of operating assumptions that went into the system's design." In other words, in trying to reduce the noise in their definitions of intelligence, AI researchers had created brittle, axiomatic understandings that had stripped intelligence of its historical messiness while imparting to it great mathematical elegance. Yet, recalled Agre, his colleagues rarely interpreted this brittleness as evidence that they needed to take a fundamentally different approach to the problem of understanding human thinking. Rather than embracing the messiness and historicity of human thought, they reduced human intelligence to something completely nonhuman, in order to reproduce it mechanistically. Agre understands this via the ideas of philosopher Daniel Dennett, who "spoke of the need for 'discharging the homunculus,' something he imagined to be possible by dividing the intelligent homunculus into successively less intelligent pieces, homunculi within homunculi like the layers of an onion, until one reached a homunculus sufficiently dumb to be implemented in a bit of computer code."[10]

Searching for a pure, clean signal entails wiping out all noise. It involves shaping humans in the dumb models for which computational systems have been built. It involves writing the worst kinds of imperial and gendered histories into our future forms of work. It involves mistaking the economic and infrastructural layers of history for irrelevant and old-fashioned models rather than recognizing them as shaping the structures of our present computational practices.[11]

The chapters in this book show the various ways in which wiping out "noise" from technical systems has brought unintended consequences, domination, exploitation, and waste. These were the first sparks of the fire that now threatens to burn this system to the ground. This book is a plea to embrace the noise, for it is through the acknowledgment of the incredible complexity of human thought and ingenuity of our practices that we can rethink our mistakes and create a host of other options for sociotechnical futures worth inhabiting.

NOTES

1. Steve Pyne, "Winter Isn't Coming: Prepare for the Pyrocene," *History News Network* (August 25, 2019), https://historynewsnetwork.org/article/172842; Stephen Pyne, *Fire: A Brief History* (Seattle: University of Washington Press, 2019).

2. A common objection is that it would take many lifetimes to learn the workings of every technological tool as well as their historical contexts. There are disciplines that have devised ways of knowing across disciplines that do not require lifetimes. There are shortcuts that don't dumb down or simply broker between various authoritative conclusions—see, for example, the work of science and technology studies scholars such as H. M. Collins and Trevor Pinch, *The Golem of Science: What Everyone Should Know about Science* (Cambridge: Cambridge University Press, 1993). See also studies that show how the poor have a more comprehensive experience and understanding of their condition than simple technical models can full reproduce, e.g. Virginia Eubanks, *Digital Dead End: Fighting for Social Justice in the Information Age* (Cambridge, MA: MIT Press, 2012).

3. Lilly Irani, *Chasing Innovation: Making Entrepreneurial Citizens in Modern India* (Princeton, NJ: Princeton University Press, 2019); Morgan Ames, *The Charisma Machine: The Life, Death, and Legacy of One Laptop per Child* (Cambridge, MA: MIT Press, 2019).

4. Luke Stark and Anna Hoffmann, "The Language We Use to Describe Data Can also Help Us Fix Its Problems," *Quartz*, June 18, 2019, https://qz.com/1641640/the-language-we-use-to-describe -data-can-also-help-us-fix-its-problems/.

5. Lilly Irani and M. Six Silberman, "From Critical Design to Critical Infrastructure: Lessons from Turkopticon," *Interactions* 21, no. 4 (2014): 32–35; Lilly Irani and M. Six Silberman, "Stories We Tell about Labor: Turkopticon and the Trouble with 'Design,'" *CHI '16: Proceedings of the 2016 Conference on Human Factors in Computing Systems* (New York: Association for Computing Machinery, 2016), 4573–4586; Lilly Irani and M. Six Silberman, "Turkopticon: Interrupting Worker

Invisibility in Amazon Mechanical Turk," *CHI '13: Proceedings of the SIGCHI Conference on Human Factors in Computing Systems* (New York: Association for Computing Machinery, 2013), 611–620.

6. Irani and Silberman, "Stories We Tell about Labor."

7. Michel-Rolph Trouillot, *Silencing the Past: Power and the Production of History* (Boston: Beacon Press, 1995).

8. Michel-Rolph Trouillot, *Silencing the Past*, 5.

9. Raymond F. Betts, *Decolonization* (New York: Routledge, 2008), 3.

10. Philip Agre, "The Soul Gained and Lost: Artificial Intelligence as a Philosophical Project," *Stanford Humanities Review* 4, no. 2 (July 1995): 1–19. See also Philip Agre, *Computation and Human Experience* (Cambridge: Cambridge University Press, 1997).

11. Rebecca Solnit advocated a feminist response to the COVID-19 pandemic using a similar critique of the historical limitations of liberal freedoms, in "Masculinity as Radical Selfishness: Rebecca Solnit on the Maskless Men of the Pandemic," *Literary Hub*, May 29, 2020, https://lithub .com/masculinity-as-radical-selfishness-rebecca-solnit-on-the-maskless-men-of-the-pandemic/.

HOW DO WE LIVE NOW? IN THE AFTERMATH OF OURSELVES

Benjamin Peters

Your computer—no, the world itself—is on fire. Or so says a news cycle fueled by more high-octane outrage than insight. Take a good look around. Modernity and its merciless waves of cutting-edge science and technology have brought forth many fruits—some of them sweet and many bitter: these pages survey crowds of imperial, eugenic, racialized, sexist, industrialist, dangerously distributed, and centralized cruelties of computing and new media, and the world burns not just with metaphorical rage. It is also literally burning. If the global emissions curve is not altered, climate experts predict that the wet blue ball we call home will soon be uninhabitable by human civilization as we know it. The COVID-19 pandemic has only increased the global need to collectively care for the lives of others.

Still, bleak realities do not necessarily call for more smash-the-machine thinking (or at least not all the time). Human life may soon be forfeit as we know it, and still, not all is wrong with global computing and new media: to retreat into either crude Luddite self-righteousness or burn-it-all rage would solve nothing except dispensing aspirin to the panged conscience of those of us privileged few who can choose to live with less technology in our lives. I can opt out of social networks; many others cannot. A generation ago the hip rushed online; today the self-proclaimed cool minorities are logging off because, unlike most, they can.

Still, by many standards, modern media and computing technology have ushered in a host of net positives to the world's population. Access to information and the proliferation of knowledge has never been as high and widespread in world history:

according to a recent United Nations report, far more people have access to a cell-phone (six billion) than to a flushing toilet (four and a half billion), with the majority of new mobile users hailing from the Middle East, Asia, and Africa. Since at least the sociologists Malcolm Willey and Stuart Rice's 1933 Communication Agencies and Social Life, most of the connections in a globally networked world have clearly been local. Even as real risks and new dangers attend the uneven distribution of knowledge, power, and connection worldwide, it is hard to deny that computing and new media, whatever else their costs, have kindled and rekindled real-life relations.

The puzzle remains: How then do readers singed by the fires described in these pages live in what the essayist Pankaj Mishra calls the age of anger?[1] And perhaps more vexingly, what in the world should one live for? A satisfactory response must fall, of course, well outside of the scope of an afterword, and perhaps even a single lifetime. Nevertheless, a few remarks might be offered in the spirit of the ethical reflection that drives the project of critical scholarship.

The fact is the world of human suffering has never so clearly appeared on the brink of ruin: in fact, the human race has suffered devastating problems—epidemics, famines, slave trades, world wars, and the colonization of indigenous peoples—at almost every turn in world history, but perhaps only recently has the recognition of what sociologist Luc Boltanski calls "distant suffering" become so potent and wide-spread.[2] Never before in world history, in other words, has the modern media user had such unprecedented access to witness secondhand the finitude and fragility of our human condition from afar. This mediated condition of being—both privileged to be removed from yet intimately privy to unaccountable human suffering—should shake us to our core and, with it, shed the comfort of the privilege to which so many (myself included) cling.

Media technologies and computing power tend to coincide with straightforward beliefs about progress—sometimes with an Enlightenment-era capital P. That is a mistake, or at least "progress" presents a mistakenly incomplete picture. Instead of looking for enlightenment in the distorting mirror of modern media technologies and always coming up disappointed, let us contemplate the more stoic, Zen, or, perhaps with the philosopher of technology Amanda Lagerkvist, *existential* possibility that, properly understood, new media technologies should help prepare us to die—or, better put, they should help prepare us to live now that we might die well.[3] By this I mean that new media and computing technologies help accelerate the global spread of what sociologists Paul Lazarsfeld and Robert Merton called "the narcotizing dysfunction" of mass media in 1948:[4] modern media acquaints the most privileged parts of the world with the least, while at the same time insulating the privileged. Digital connectivity bruises the sensitive souls among us while limiting our capacity, or perhaps worse, our motivation, to make a difference.

This feeling of double-bind despondency—aware but helpless, sensitive but hopeless—ensures, whether by the shock of front-page pessimism (pandemics, earthquakes, mass murders, despots) or by the exhaustion of reflecting on the enormous problems and biases that new media amplifies (racism, sexism, environmental degradation), that no one can lead a life among the menaces of modern media without receiving a kind of accidental exposure to our own and even our species's mortality. Yet instead of hiding from or leaping to embrace media as memento mori, we must learn from their call to remember in advance our own death: then we may find in computing and new media a rigorous, even healthful training for our own death and the demise of all those we care about.

A call to face and embrace one's own death may seem prima facie an over-harsh conclusion to a book about technology and media: in fact, there may be little more inimical to the modern mind that seeks prosperity, peace, and beneficent politics than such a call to reconcile ourselves with the brevity of human life and even to release that desperate desire to stave off our own annihilation.

But it is too late to wish otherwise. Encounters with death are not "for hire." It is a must. Our globe already demands it at every scale of life: climate change and pandemics, two heirs to the nuclear age, are perhaps the current species-wide issue whose profound ethical register combines the apocalyptic (the world is ending as we know it) with arithmetic (the recent trend in corporate carbon footprint scorecards and epidemiologists' models). But we cannot not act: our media compel us to it. Watch the TV news. Fight a Twitter firestorm. Organize a movement. Eat meat sparingly. Wear a mask. Measure your most recent travel in the carbon footprint and its weight in the ashes of burned dinosaur bones. On our planet, there is nowhere left to run, even online: the frontier is closed (and should have been foreclosed against long ago). The search to recombine technological solutions and efficiencies will, of course, continue apace, but it is impossible not to face the fact that, on a global scale, the needs of the many far continue to outweigh the efforts of the privileged few. The case for any individual, weighed in the balance of the globe and its problems, is a thoroughly losing proposition. The self is lost. You and I do not matter (not that we ever did).

And yet—and this is the point—far from all is lost: in fact, with the self out of the way, there is much to live for. Indeed, perhaps life begins anew with the loss of the self. Our species will no doubt insist on living, as it long has, in what we might call the aftermath of ourselves. In this view, accepting self-loss does not mandate disengagement, indifference, and ennui. In fact, it may speed the opposite: it may lead to a joining of hands, a sloughing off of the fantasies of infantile omnipotence and puffed-up power. In its place, a clarion call, with Kimberlé Crenshaw, to collective action across both the illuminating intersections of identity and the underlying unions of our species' mutual interests.[5] This is crucial, for accepting the loss of the self—and with it the concomitant bankruptcy of selfishness—also calls us to recognize and then carefully reclaim and reshape our common lot as mediated creatures. We cannot return the unasked-for gifts of new media and computing; modernity has saddled our species with gifts we neither seek nor can return (*data* means "that which is given" in Latin, Russian, and French). Nor can we escape the ever longer shadows of our individual futility in the face of the world's towering problems. But at least—and this is no small step—we can begin to recognize in the loss of a self a partial antidote to the same despair and despondency that tries to fill the void of the mediated self in the first place: from it sounds out the fire alarm, a call to arms, collective action, and even care.

So the question of life becomes how do not *I* but *we* live now? Let us learn from our media technologies how to be something other than only ourselves. For example, the couplet "people are tech, tech is people!" resonates across this book as alternately a stirring call for collective action (the people of tech, unite and throw off your chains!) and as a despairing cry (read in the voice of Sol in the final line of the 1972 B-movie *Soylent Green*, "Soylent Green is people"!). Both of these lines offer variations on that hardy perennial of principles that, despite the temptation to alternately idolize or curse tech as an independent agent of change, behind every piece of tech is, in fact, the messy interdependencies of humans. Whether or not AI ever becomes sentient, there is already a human behind every machine. Artificial intelligence, by contrast, requires, integrates with, and obscures human labor: for example, behind most self-check-out registers at the grocery store stands a hidden grocer ready to error correct the smart scanner. Behind every automated search algorithm toils a content moderator. Capital, run amok, produces social relations that are opaque to those who inhabit them, and, as every tech executive knows, *AI* is another word for investor capital.

This volume harmonizes with the chorus of critics crying out "tech is human too," but not in the sense of anthropocentric machines. Indeed, AI is often billed as an automated big "brain," when, if anything, the better organ projection for complex filtering in machine learning might be the liver—a distinctly less sexy comparison! Instead, contributors examine the stubborn humanity that persists in the private eyes that must be trained to train automated image-recognition systems on the most abhorrent images, on the dual pleasures of apprehension in facial recognition systems as well as the pleasure of workplace refusal and resistance. Other contributors also demonstrate that, under the guise of meritocracy, computing industry hiring practices have long, and likely will continue to, baked disturbing cultural traditions of sexism (on multiple ends of the British empire), racism (on both ends of Siri), Western backwardness in relationship to the Islamic and Chinese civilizations, and the faux beliefs in the generalizability of code into representations of the world.

In the debates and conferences supporting this book, authors probed whether the category of the machine helps us rethink the human: How do we protect the content moderators behind every filter algorithm? Might soldier robots help deconstruct the gender divide? Might bullets that feel press for more reasonable Second Amendment limitations among the morally mechanical who feel it when machines, not our fellow flesh and blood, suffer? Might the largely male teams of engineers designing smart tech internalize more than the obvious gender lesson that the assistant robots that serve and sense others' needs while hiding out of sight should no longer be gendered female by default (Siri, Alexa, Cortana)?

But the moral dilemmas of tech almost always eclipse the solitary self: the social consequences of how we train and educate and hire coders, content moderators, and human trainers of algorithms on trial in this volume. So too does the field of computer programming education stand accused of practicing a student bait-and-switch, in which a student programmer transforms, in the intro courses, from an aspiring universal analyst into an insider mage and specialist in more advanced topics. Behind every tech design lies vast fields of ethical, provincial, and untested baseline civilizational assumptions about what code, writing, and even language means. Embarrassingly, many of these assumptions, as the case studies on Chinese writing interfaces, Arabic keyboards, and English-inflected Siri show, are simply wrong. It seems that no matter how we try to code or paper over what we would rather keep hidden, our best representations of ourselves still betray a deeper, more disturbing reality that is also beautiful in its irreproducibility: the relations between humans are more diverse, rich, and full of problems than any palette of skins, colors, sounds, genders, or other categories could emulate.

Engineering science faces an analytic paradox: it has the tools to better know but not to solve the world's problems, even though it often acts as if the opposite were true. Its basic tools for making more efficient and more accurate representations of the world are heralded as the great problem solvers in the current age of data, and yet no representation, no matter how complete or comprehensive, can model the world in ways that must address its problems. The slippage between media subject and object, between tech "user" and "used," will continuously incentivize humans to push back against the distorted reflecting mirror of their own profiles online. Moreover, reality itself, as the philosopher Immanuel Kant recognized, cannot be limited to only that which can be represented. There is more to reality than meets even our boldest innovation. (Consider how many wires make tech "wireless" and how much coal is burned for "cloud" computing.) Yet, to these objections, the standard engineering response is to strive to represent more accurately and, especially after a public scandal, more ethically. What happens when building systems that better represent the diversity of the world only exacerbates the original problem? An emoticon that matches every iPhone user's skin color solves nothing of the struggle for the promotion and reparation of those who suffer social-structural hardships, disadvantages, and biases. Universal representation is the wrong answer to the question of unequal infrastructure. (No one gets off easy: Is, say, the critic who interprets content as representative of larger cultural problems not guilty of the same?) In a world driven by universal representation logics, tech reveals its blind spot to a key condition of human existence: there has never been justice and at once there must be less injustice now.

This volume calls to redirect attention away from the universal and global, except in the case of urgent collaboration to abate a looming environmental catastrophe, and instead to attend to the immediate and local contexts of lived experience: perhaps we can observe that the globe, as such, does not exist, just as, again and again, the uneven warp and weft of institutions and local politics do. IBM, like other multinational corporations, is transnational but also distinctive, even filial and personal, in its discourse and discontents. Sound technologies all too often mute our many bodily differences. Script technologies like keyboards miss all but the most basic on-off-hold-swipe touch of digital typing. QWERTY is anything but text-predictive, and keyboards are "mute" to all but the most limited speech strategies. Siri cannot hear your anger. Your keyboard sublimates your oral voice based on assumptions about the Latin alphabet that the majority of the world does not share. Between these two endpoints, the continuum of embodied human experience still walks the earth unsounded and scriptless.

Tech cannot be universal for other reasons too. The varieties of local labor, political economies, and the illiberalities of free markets continue to pile on to the tendency for military research to fund your favorite tech. A roboticist recently quit their field because they could not serve an innovation without a military application—from DARPA research to sex dolls for soldiers. It is also a disturbing, if understandable, fact that the largest server storing child pornography belongs to the US government and its central police force. What appears to be a stable tech infrastructure can evaporate overnight, depending on the crumble of an empire and whether the Federal Communications Commission labels the tech as "platform media." Still other contributors complicate tech universalist claims by showing that the past, present, and future of tech corporate culture belongs to local workplace resistance. While tech may stretch global, it knows even less how to be local: a world of tech is not the world of lived experience.

Tech will deliver on neither its promises nor its curses, and tech observers should avoid both utopian dreamers and dystopian catastrophists. The world truly is on fire, but that is no reason that it will either be cleansed or ravaged in the precise day and hour that self-proclaimed prophets of profit and doom predict. The flow of history will continue to surprise. As described in this volume, many attempts to diversify tech workforces, no matter how well intentioned, may end up exacerbating the very hiring biases driving the lack of diversity; as in the British case, chauvinist biases may pour in to fill the gaps of aggressive, and ultimately regressive, centralized hiring. In terms of surprising phrases, the once-untimely "series of tubes" turns out to be a suitable metaphor for the global material and mucky internet. Or, one can play countless games online while playing out no more than the lives of third-string monopolists, second-world designers, and first-person soldiers. Meanwhile, in real life, the world never fails to surprise: even Thomas Hobbes, infamous for declaring life "solitary, poor, nasty, brutish, and short," died in bed, surrounded by loved ones, at the age of ninety-one. The value of tech criticism, like all criticism, must not be mistaken for moral panic talk, whose audience often mistakes the carrier of the unfortunate news for the cause of the crisis. Uncertain times call for uncertain media.

In the ongoing search for better heresies and causal explanations, let us pause, with historian of technology Ksenia Tatarchenko, before ceding tomorrow's future to yesterday's science fiction.[6] May tech observers stop off-loading and outsourcing the imagination of better worlds without first attending to the earth. In the same spirit that Hegel noted that Minerva's owl takes flight in the dusk, so too does ethics only stretch its wings after algorithms have already been, in the military jargon, "deployed." The robots will never take over—that has never been the crisis. Rather, robotic analysis of the future took over our minds and language many decades ago.

As Lewis Mumford opened his 1934 masterpiece *Technics and Civilization*, our species became mechanical before machines changed the world. Not only is tech human, people are the original machines.

The globe is ablaze, and few have the collective language to call to put it out. This book sounds out a call for that language. The challenge of anyone who lives in our broken world is not to delay to some future date the fact that the needs of the many outweigh the privileges of the few here and now. (Pandemics may make no finer point.) The difficulty of learning to love, live with, and care for others is perhaps the problem of all those who live. Those who overdraft from the accounts of the self live unknowingly on the credit of others. The resolution, if not solution, may come to all those who learn to live together now so as to pass on a better world, and soon enough from this world.

All editors harbor conceits about their books. The four of us hope this book proves a "no crap" book in at least two tenses: first, present-day readers acknowledge its message by acting on its seriousness and urgency. Second, as a consequence of the first, readers, in the hopefully not too distant future, will also be able to look back and see its message as all too obvious. As I conclude this afterword, the internet is turning fifty years old to the day, surely a fitting moment to stage an intervention and to declare its midlife crisis in full swing. The success of this book's no-nonsense attitude will be measured by its self-evident irrelevance fifty years from now—by how we learn to live in the aftermath of ourselves. Until then, the question remains: Will its message—your computer is on fire—or the world as we know it burn up first?

NOTES

1. Pankaj Mishra, *Age of Anger: A History of the Present* (New York: Farrar, Straus and Giroux, 2017).

2. Luc Boltanski, *Distant Suffering: Morality, Media and Politics*, trans. Graham D. Burchell (Cambridge: Cambridge University Press, 1999).

3. Amanda Lagerkvist, "Existential Media: Toward a Theorization of Digital Thrownness," *New Media & Society* 19, no. 1 (June 2017): 96–110.

4. Paul F. Lazarsfeld and Robert K. Merton, "Mass Communication, Popular Taste, and Organized Social Action," in *The Communication of Ideas* (1948); reprinted in *Mass Communication and American Social Thought: Key Texts, 1919–1968*, ed. John Durham Peters and Peter Simonson (Lanham, MD: Rowman & Littlefield, 2004), 235.

5. Kimberlé Crenshaw, *On Intersectionality: Essential Writings* (New York: The New Press, 2020).

6. Ksenia Tatarchenko and Benjamin Peters, "Tomorrow Begins Yesterday: Data Imaginaries in Russian and Soviet Science Fiction," *Russian Journal of Communication* 9, no. 3 (2017): 241–251.

CONTRIBUTORS

Janet Abbate is a historian of computing with particular interests in gender and labor issues, internet history and policy, and computing as a science. She is Professor of Science, Technology, and Society at Virginia Tech.

Ben Allen is a historian of programming languages and related technologies, as well as a computer science educator. He teaches computer science at Berkeley City College.

Paul N. Edwards is a historian of information and its technologies. He writes and teaches about the history, politics, and culture of information infrastructures, especially computers and climate change data systems. He is Director of the Program on Science, Technology & Society at Stanford University and Professor of Information and History (Emeritus) at the University of Michigan.

Nathan Ensmenger is Associate Professor in the Luddy School of Informatics, Computing, and Engineering at Indiana University. His research explores the social, labor, and gender history of software workers. He is currently working on a book exploring the global environmental history of the electronic digital computer.

Mar Hicks is a historian who researches the history of computing, labor, technology, and queer science and technology studies. Their research focuses on how gender and sexuality bring hidden technological dynamics to light, and how the experiences of women and LGBTQIA people change the core narratives of the history of computing in unexpected ways. Hicks is Associate Professor of history of technology at Illinois Institute of Technology in Chicago and associate editor of the *IEEE Annals of the History of Computing*.

Halcyon M. Lawrence is an Assistant Professor of Technical Communication and Information Design at Towson University. Her research focuses on speech intelligibility and the design of speech interactions for speech technologies, particularly for underrepresented and marginalized user populations.

Thomas S. Mullaney is Professor of History at Stanford University, where he specializes in the history of modern Asia, China, transnational technology, and race and ethnicity. He is a Guggenheim Fellow and holds a PhD from Columbia University.

Safiya Umoja Noble is Associate Professor in the Departments of Information Studies and African American Studies at the University of California, Los Angeles (UCLA), and the author of *Algorithms of Oppression: How Search Engines Reinforce Racism.*

Benjamin Peters is a media historian and theorist who works on the transnational causes and consequences of the information age, especially in the Soviet century. He is the Hazel Rogers Associate Professor and Chair of Media Studies at the University of Tulsa.

Kavita Philip is a historian of science and technology who has written on nineteenth-century environmental politics in British India, information technology in postcolonial India, and the political-economic intersections of art, fiction, historiography, and technoscientific activism. She holds the President's Excellence Chair in Network Cultures at The University of British Columbia, and is Professor of History and (by courtesy) Informatics at The University of California, Irvine.

Sarah T. Roberts is Associate Professor in the Department of Information Studies at UCLA. She is known for her work on social media labor, policy, and culture, with expertise in commercial content moderation, a phenomenon she has been studying for a decade. She is cofounder, with Safiya Noble, of the UCLA Center for Critical Internet Inquiry, or C2I2.

Sreela Sarkar is Associate Professor in the Department of Communication at Santa Clara University. Based on sustained ethnographic research, her publications focus on smart cities, new forms of capitalism, labor, and identities of class, caste, and gender in India's acclaimed information economy.

Corinna Schlombs is a historian of computing technology and culture with interests in gender and labor questions who works on productivity, automation, and capitalism in transatlantic relations. She is Associate Professor at Rochester Institute of Technology.

Andrea Stanton is a cultural historian who works on the intersections of technology, postcolonial national identity, and piety in the Middle East. She is Associate Professor of Islamic Studies and Chair of Religious Studies at the University of Denver.

Mitali Thakor is an anthropologist of technology with interests in feminist and critical race studies of surveillance, policing, artificial intelligence, and robotics. She is an Assistant Professor of Science in Society at Wesleyan University.

Noah Wardrip-Fruin is a computational media researcher who works on approaches and technologies for understanding and making games, interactive narratives, and electronic literature. With Michael Mateas he co-directs the Expressive Intelligence Studio at UC Santa Cruz.

INDEX

Page numbers followed by an "f" or "t" indicate figures and tables, respectively.